Post-Agricultural Succession in the Neotropics

Randall W. Myster
Editor

Post-Agricultural Succession in the Neotropics

Springer

Randall W. Myster
Ecosystems Studies
Institute for Tropical Ecosystems Studies
University of Puerto Rico at Rio Piedras
PO Box 21910
San Juan, PR 00931–1910
rmyster@lternet.edu

ISBN: 978-0-387-33641-1 e-ISBN: 978-0-387-33642-8

Library of Congress Control Number: 2007937949

Printed on acid-free paper.

9 8 7 6 5 4 3 2 1

springer.com

To Olia, for all her love and support

"Seek simplicity, but distrust it."
Alfred North Whitehead

Preface

It was 1985, and I remember walking into the office of Peter Morin at Rutgers University. He told me of some boxes with computer punch cards that had been littering his office for years and of his desire to get rid of them. Because of my background in computer science, I said I would take a look. The cards were in bad shape: moldy, broken in places, and full of cockroaches that were living on what was left of the rubber bands that once had held them together. This was the Buell-Small successional data set. I cleaned up the cards, ran them through an old card-reader, and finally transferred the data onto magnetic tape which I could access using my office computer.

While I was at Rutgers, I also managed to add five years to the data set by going out to Hutcheson Memorial Forest with Steward Pickett and others where we sampled the old fields that comprised the Buell-Small study. Accompanying us was a kindly and cheerful old woman, the last surviving member of the team of scientists who had the foresight to set up those permanent plots in the late 1950s and early 1960s. Her name was Helen Buell. She had an infectious sense of humor—her bumper-sticker read "I brake for *Verbascum*"—which came in handy as we crawled through rose bushes and prickly raspberries. I have heard she is gone now, but I would like to remember her here.

Inspired by the work with the Buell-Small data set, I set up post-agricultural plots in the Neotropics that are reported on in several papers and in the first chapter of this book. I believe, as many others do, that such plots are critical to a complete understanding of vegetative changes after agriculture. I hope that those data and other data analyzed and interpreted by the authors of the present book will be a valuable contribution to this scientific undertaking, as has been true for the data from the plots set up so many years ago in the temperate fields of New Jersey.

R. W. M.

Contents

List of Contributors

Carlos C. Cerri
Centro de Energia Nuclear na Agricultura (CENA)
Universidade de Sao Paulo
CP.96, 13400-970
Piracicaba, (SP), Brazil

Carlos Eduardo P. Cerri
Centro de Energia Nuclear na Agricultura (CENA)
Universidade de Sao Paulo
CP.96, 13400-970
Piracicaba, (SP), Brazil

Shu-Chun Chuang
Institute of Zoology
National Taiwan University
No. 1 Roosevelt Road
Section 4, Taipei 106
Taiwan

Daniela Cusack
Department of Environmental Science, Policy & Management
University of California, Berkeley
137 Mulford Hall #3114
Berkeley, CA 94720-3114

Brigitte J. Feigl
Centro de Energia Nuclear na Agricultura (CENA)
Universidade de Sao Paulo
CP.96, 13400-970
Piracicaba, (SP), Brazil

Bruce G. Ferguson
El Colegio de la Frontera Sur Carretera Panamericana y Periférico Sur s-n
Barrio Ma. Auxiliadora San Cristóbal de Las Casas
Chiapas
México CP 29260

Gislene Ganade
Laboratório de Ecologia da Restauração Biologia
Universidade do Vale do Rio dos Sinos
CP 275, São Leopoldo
RS, Brasil 93022-000

Grizelle González
International Institute of Tropical Forestry
USDA Forest Service
1201 Ceiba St.
San Juan, PR 00926-1115

Ernestino Guarino
Embrapa Recursos Genéticos e Biotecnologia
Parque Estação Biológica - PqEB - Av. W5 Norte (final)
Caixa Postal 02372
70770-900 Brasília DF – Brazil

William A. Haber
Apartado 50-5655
Monteverde de Puntarenas
Costa Rica

Elaine R. Hooper
Department of Forestry
and Environmental Studies
Yale University
210 Prospect St
New Haven, CT 06511

Katherine Horjus
Department of Ecology
and Evolutionary Biology
University of California
Santa Cruz, CA 95062

Ching Yu Huang
Institute of Ecology
University of Georgia
Athens, GA, USA

Ingo Hübel
Laboratório de Ecologia da
Restauração, Biologia
Universidade do Vale do Rio dos
Sinos, CP 275
São Leopoldo
RS, Brasil 93022-000

Mark Kuhlmann
Department of Biology
Hartwick College
West Street, Oneonta
New York 13820

Yiqing Li
College of Agriculture,
Forestry and Natural Resources
Management
University of Hawaii
200 W. Kawili St.
Hilo, HI 96720

Tammy M. Long-Robinson
12024 E. Pleasant Lake
Manchester, MI 48158

Ariel E. Lugo
International Institute
of Tropical Forestry
USDA Forest Service
Ceiba Street 1201
Jardín Botánico Sur
Río Piedras PR 00926-1115

Erika Marín-Spiotta
Department of Environmental Science,
Policy & Management
Univeristy of California, Berkeley

Current address:
U.C. Santa Cruz, Rm. 2125,
USDA/ARS
800 Buchanan St.
Albany, CA 94710

Oscar Abelleira Martínez
International Institute
of Tropical Forestry
USDA Forest Service
Ceiba Street 1201
Jardín Botánico Sur
Río Piedras, PR 00926-1115

Clara Luz Miceli-Méndez
El Colegio de la Frontera Sur Carretera
Panamericana y Periférico Sur s-n
Barrio Ma. Auxiliadora
San Cristóbal de Las Casas
Chiapas, México
CP 29260

Florencia Montagnini
School of Forestry
and Environmental Studies
Yale University
370 Prospect St
New Haven, CT 06511

K. Greg Murray
Department of Biology
Hope College
Holland, MI 49423

Randall W. Myster
Institute for Tropical Ecosystem Studies
University of Puerto Rico at Rio Piedras
P.O. Box 21910
San Juan,
PR 00931-1910

Rebecca Ostertag
Department of Biology
University of Hawaii at Hilo
200 W. Kawili St.
Hilo, HI 96720

Marisa C. Piccolo
Centro de Energia Nuclear na Agricultura (CENA)
Universidade de Sao Paulo
CP.96, 13400-970
Piracicaba, (SP), Brazil

Neptalí Ramírez-Marcial
El Colegio de la Frontera Sur Carretera Panamericana y Periférico Sur s-n
Barrio Ma. Auxiliadora
San Cristóbal de Las Casas
Chiapas, México
CP 29260

Jason Roberts
128 Nichols St.
Lewiston, ME 04240

Alexandre B. Sampaio
IBAMA
SCEN Trecho 2 - Ed. Sede - Caixa Postal 09870
70818-9 – Brasília, DF – Brazil

Aldicir Scariot
Embrapa Recursos Genéticos e Biotecnologia
Parque Estação Biológica - PqEB - Av. W5 Norte (final)
Caixa Postal 02372
70770-900 Brasília DF – Brazil

Anderson C. Sevilha
Embrapa Recursos Genéticos e Biotecnologia
Parque Estação Biológica - PqEB - Av. W5 Norte (final)
Caixa Postal 02372
70770-900 Brasília DF – Brazil

Whendee L. Silver
Department of Environmental Science, Policy & Management
University of California Berkeley
137 Mulford Hall #3114
Berkeley,
CA 94720-3114

Daniel L. M. Vieira
Embrapa Tabuleiros Costeiros
Av. Beira Mar, 3250 - jardins
Caixa Postal 44
49025-040- Aracaju, SE - Brazil

Kathy Winnett-Murray
Department of Biology
Hope College
Holland, MI 49423

Lessandra Zanini
Laboratório de Ecologia da Restauração
Biologia, Universidade do Vale do Rio dos Sinos
CP 275, São Leopoldo
RS, Brasil 93022-000

Willow Zuchowski
Apartado 50-5655
Monteverde de Puntarenas
Costa Rica

Part I
Patterns of Living and Non-living Components of Old Fields After Abandonment

Chapter 1
Introduction

Randall W. Myster

1.1 Rationale

Ecosystems were originally defined as units of the earth's surface which includes both organisms and the physical factors forming the environment (Tansley 1935). As the science of ecosystem ecology developed, ecosystems came to be categorized by function and structure (Odum 1953), with an emphasis on integration and indirect interaction (Muller 1997). While all ecosystem functions deal with energy capture/transfer and nutrient cycling, plant-based productivity and decomposition have come to be considered among the most critical (Watt 1947).

To develop a deeper understanding of ecosystem functions, we need to know the pattern of an ecosystem's abiotic and biotic responses which define function, on the one hand, and the processes and mechanisms that regulate the responses, on the other. In particular for the biota, it has long been recognized that plants and other taxa respond not only to gradients (Whittaker 1975, Kessell 1979, Walter 1979) but also to disturbances (i.e., disruptions of ecosystem structure that occur largely on gradients: Gleason 1926, Pickett and White 1985). Consequently, the vegetation mosaic of any landscape results both from the environmental variation expressed on gradients and from historical disturbances (Risser 1987, Turner and Dale 1991).

Gradients are gradual, directional, and large-scale (from a plants' perspective) changes in environmental factors (Whittaker 1975, Hallet al. 1992). Although gradients are mainly a spatial phenomenon, they may also show seasonal variation and even be modified over long time periods, for example, by changes in climate. Individual plants of a given species exist within certain ranges of environmental conditions (Good 1953) that, when combined with others of the same species, translate into distribution patterns (usually bell-shaped, Gaussian Curves) across gradients for that species (Whittaker 1975). Even though these curves correspond with direct gradients, such as elevation, slope, and aspect, they actually reflect species-specific responses to indirect environmental factors, such as the plant resources of temperature, light, water, and soil nutrients, which correlate with the direct gradients. Such plant growth responses exist in a hierarchy of survivorship, growth, allocation, and architecture.

Unlike gradients, disturbances are local, discrete, and patchy (Pickett and White 1985). However, similar to gradients, they create combinations of resources and other influences that plants respond to. Indeed, plant distribution patterns over successional time, after a disturbance, are just as species-specific and bell-shaped as they are over gradient space (Pickett 1982, Myster and Pickett 1988).

Any mechanism involving interactions with individuals of another species taxa (e.g., dispersal, predation, herbivory) and any tolerance to a plant resource/abiotic environmental cue (e.g., light, water) that affect any part of a plant's life-cycle can form a response curve. Importantly because plants interact with each other by affecting their local resource environment, interactions such as competition and facilitation are expressed as tolerances. Taken together, response curves for each species define that species' "niche" (Whittaker 1965, Grubb 1977, Parrish and Bazzaz 1982, Bazzaz 1996) that can be quantified using yes/no questions at each life-stage and for each tolerance or mechanism that affects it. Typical questions during a tolerance investigation could be: Will this seed germinate at a given temperature? Will this seedling have a positive carbon balance at a given light level? During a mechanism investigation, typical questions include: Will this seed be dispersed at a given distance from the forest? Will this seedling survive at a given herbivore density? Each time the answer is "yes", a point that represents a particular individual plant is marked along the natural variation scale of the abiotic/biotic environmental cue being investigated, such as the level of temperature, light, distance, or predator density. When the responses of many individuals from the same life-stage and the same species are also marked on the same natural variation scale and then counted, a frequency diagram for the entire species is created.

Because these curves are frequency diagrams and because individual plants can be chosen at random, the response curves/surfaces express probabilities (after normalization: Zar 1999) of how individual plants of that species will respond to the natural variation in the working of a given mechanism or in the availability of a given resource.[1] These response curves give the individual response variation within a plant species for a given life-stage and environmental cue, and are generally wider for species that come early in successional time compared to species that appear later (Bazzaz 1979, 1996). Usually, a few of the plant mechanisms or resources dominate, making the probabilities associated with them key to how the individual plants of that species use the strategies of capture, congruence, and capacitance (Bazzaz 1996). Furthermore, the fact that the response curves have the same shape as plant distribution patterns both over space (Whittaker 1975) and over time (Pickett 1982, Myster and Pickett 1988) suggests that the distribution of a plant species is an outcome of that species' niche probabilities.

Conducted for decades over many of the most common gradients on earth, studies of succession after agriculture (also called old field succession) have helped ecologists gain insights into the processes that structure plant communities and into the role of history and initial conditions in community development (Myster and Pickett 1990b, 1994). Examination of old fields has led to the rise of important ecological theories, such as the initial floristic composition hypothesis (Egler 1954)

---

[1] As an example of a response surface which expresses probabilities, see Figure 3 in Burton and Bazzaz (1991) for emergence of five common old field trees in the American temperate zone.

and the resource ratio hypothesis (Tilman 1988). Using permanent plots established at abandonment as the backbone of old field research (e.g., the Buell-Small old field plot study in New Jersey USA: Buell et al. 1971, Myster and Pickett 1990a), scientists have been successful in finding the pattern of response over time after abandonment from crops (e.g., plant cover [Myster and Pickett 1992b] and diversity [Myster 2003b]) and in discovering many of the mechanisms that determine the pattern (Myster 1993, Bazzaz 1996). Furthermore, studies of post-agricultural succession have provided a framework for comparison and evaluation of various theories of succession and community development (Finegan 1984).

A considerable amount of old field and pasture research has taken place in the temperate regions of the United States (e.g., Oosting 1942, Bazzaz 1968, Buell et al. 1971, Pickett 1982, Tilman 1988, Myster 1993) and Europe (Castro et al. 1986, Miles 1979). The recovery of agricultural areas, however, is vital to ecosystems all over the world (Borhidi 1988). In the tropics, for example, the cutting down of wooded areas (Grainger 1988, Fearnside 1993) for agriculture and pasture (Skole and Tucker 1993) is the major cause of deforestration (Buschbacher 1986, Uhl et al. 1988, Fearnside 1993, Thomlinson et al. 1996). It should come as no surprise then that recovery of Neotropical areas after agriculture (Donfack et al. 1995, Fernandes and Sanford 1995, Quintana-Ascencio et al. 1996) is relevant to such important issues as forest regeneration (Grainger 1988, Brown and Lugo 1990, Myers 1991, Singh 1993, Skole and Tucker 1993), forest ecosystem restoration (Brown and Lugo 1990, Lugo 1992), sustainability of agriculture (Brown and Lugo 1990, Serrao and Toledo 1990), maintenance of biodiversity (Brown and Lugo 1990, Aide and Cavelier 1994), and impacts of global climate change on forest dynamics (Padoch and Vayda 1983, Hobbie 1992, Mabberley 1992, Keller et al. 1993). Neotropical areas recovering from agriculture may also serve as a buffer between "primary" forest and more intensely human-influenced areas (Brown and Lugo 1990). Last but not least, it is expected that with future increase of global warming forest disturbance in general and old field succession in particular will affect larger and larger parts of the earth (Bazzaz 1996).

The nature of the recovery of old fields depends on their disturbance "regime," which is characterized by severity, size, spatial location, and frequency (Myster 2001). The measurement of severity is best accomplished by sampling the loss of plant biomass and determining whether such loss includes loss of plant meristems (Raunkaier 1934, Pickett and White 1985, Tilman 1988, Myster 2001). Severity of old fields is moderate, compared to the more severe landslides (Myster 2001), because old fields maintain an intact soil profile, roots, and humus layer despite the removal of above-ground vegetation. In size, old fields do not usually exceed a few tens of hectares (they are larger, however, than many Neotropic treefall gaps: Brokaw 1982), and crops that are planted in them often require specific locations (e.g., Coffee grows well in the mountains while Banana and Sugarcane do best in the valleys). Frequency of old field disturbance includes the number of times a field has been cultivated, the duration of each cropping period, the time between fallow periods, and the order of crop rotation. Usually fields are fallow longer than they are in crop and can be either reused with different crops and fallow periods until abandoned or used for 5-10 years until productivity declines. As the length of time an area is in crop increases, the length of successional, or recovery, time may also lengthen. For example, a short cropping period can lead to large contributions from the seed bank and trees may regenerate quickly from stump or root sprouts, whereas an increased cropping period brings about the decay of rootstocks and tree stumps, thus providing for grass invasion and prolonged tree invasion. Among other conditions that influence old field recovery are historical effects of the past crop (a crop's "signature": Myster and Pickett 1988, 1990b, 1994), the year of abandonment, the season of abandonment, the percent of border with forest (Bazzaz 1968), and a field's plowing record (whether the field was plowed under or left fallow when abandoned: Myster and Pickett 1990a).

Studies of post-agricultural succession in the temperate zone have demonstrated the particular importance of past crops in determining old field patterns (e.g., alteration of successional pathways and species composition and abundance for up to 8 years after abandonment: Myster and Pickett 1988, 1990b, 1994). Hence, a major focus of this book will be on the examination of fields after common Neotropic crops, such as Maize (*Zea mays*), Sugarcane (*Saccharum officinarum*), Banana (*Musa* spp.), Coffee (*Coffea* spp.), Cassava (*Manihot esculenta*), Beans (*Phaseolus* spp.), and Rice (*Oryza* spp.), and after pasture grasses that usually establish themselves after cropping but can also be purposely planted (e.g., *Setaria sphacelata*). Pastures will be treated as a kind of old field succession for several reasons. First, they are usually converted from depleted agricultural fields when cows are allowed to enter. Second, they may continue to have remnants of past crops growing in them for some years. And third, they can be planted with native and exotic grasses or be colonized by local grass species due to cattle activity (e.g., trampling of vegetation, creation of hummocks, deposition of dung, and soil compaction).

Two forest-clearing techniques that are the most common in the Neotropics will also be discussed in this book. Both methods, while increasing light quantity and contact of rainfall with the soil, eventually decrease soil fertility. The first technique is mechanical clearing that aims at removing woody biomass (e.g., by using a bulldozer) and brings about changes in physical properties of the soil. The other technique is shifting, or "slash and burn" (Uhl 1987), agriculture that occurs when natural vegetation is first cut down to the ground and then burnt. Although burning produces a short pulse of nutrients like phosphorus, it generally volatilizes both organic matter and nitrogen while raising soil temperature and, in the long term, makes soil poor in both nutrients and the seed bank. Shifting agriculture also entails a rotation of fields rather than crops that is accompanied by long fallow periods (Grigg 1978). The decision as to when to rotate fields and for how long to leave them fallow depends on such factors as weed and insect loads in the fields, soil fertility, production, labor availability, and local dietary needs (Staver 1990). The implementation of either one of these forest-clearing methods also causes carbon loss, a mixing of the soil horizons, and a higher degree of soil aeration (Bazzaz 1996).

Both mechanical clearing and shifting agriculture lower the level of advanced regeneration from sapling banks and from root/stem resprouting (Uhl et al. 1988) so that seed and seedling dynamics dominate regeneration after agriculture. Consequently, an understanding of old field succession demands knowledge of invasion and establishment at the seed and seedling phases of the plant life-cycle (Grubb 1977, Myster 1993).

Old field recovery goes through several structural and conceptual stages: (1) domination by past crop and its various effects, (2) patch dynamics of grass, past crop remnants, asexual shrubs, and other plants, (3) tree invasion, and (4) development of a closed-canopy "secondary" forest. Although individual plants may or may not colonize bare ground during these stages, they always become involved in the on-going process of plant-plant replacement (Busing 1996). Plants enter this process because they grow and interact asymmetrically (Bazzaz 1996)—one individual is always gaining advantage over another. As a result, the plants that are growing best replace the plants that are growing worst, are in senescence or dead. However, plant-plant replacement does not necessarily involve two plants only, for more than one plant can be replaced by just one or only one plant can be replaced by more than one. Because individual plants are "planted" in the soil, they have a defined physical space which can be occupied and reoccupied as plant-plant replacements proceed. Every land plant in the world sooner or later assumes these roles-of the "replacer" and the "replaced."

Because plants move very slowly, we naturally do not see them and their communities changing. Yet, they are always both engaged in and are the product of the dynamic process of plant-plant replacement. Plant-plant replacements determine which plants are present at every point in space and time, for example, during a typical "static" plant growth and allocation field experiment or a "snapshot" plot sampling. Even in those cases, however, plant niches and their interaction with local environmental cues will determine how those plants grow. Fundamentally, neither species nor patches are replaced in communities. It is individual plants that are replaced by other individual plants. I suggest that plant communities do not go through continuous cycles of disturbance (regeneration) => stability (growth) => disturbance (regeneration) –like a corn field– but instead are undergoing plant-plant replacements at all times.

After a closed canopy has developed, old field plants may eventually enter a thinning state when dead plants in the canopy are replaced not with new individuals but with modified growth of existing plants. There is a physiological limit to this kind of growth, however, and at some point every plant will die and be replaced. Furthermore, although there can be a growth phase for perennial plants between replacing and being replaced, it is rare, transitory, and, consequently, over-emphasized in plant ecological studies. This focus on individual plant-plant replacements eliminates the need for many of the old dichotomies of plant ecology, such as primary/secondary forests, natural/human disturbances, successional/nonsuccessional communities, and pioneer/climax species. This conceptual model *unifies* various old theories of plant community organization because plant-plant replacements define not only the successional dynamics of old fields but the dynamics of any plant community.

Replacements are controlled by species tolerances and mechanisms that "activate" the niches of the species involved in the replacement. Tolerances include germination and growth while mechanisms encompass dispersal, seed pool, seed predation, seed pathogens, seed germination, seedling predation, seedling pathogens, seedling herbivory, and/or seedling competition (e.g., Myster and McCarthy 1989, Myster and Pickett 1993, Myster 1994, 2003a, 2003c, 2004a, 2006, 2007). The process of plant-plant replacement is at the heart of the dynamic nature of terrestrial plant communities both over successional time and over space (e.g., at the edges of plant distributions where gradients have a major effect on plants: Whittaker 1975, Myster 2001).

For those who still need to be convinced of the dynamic nature and ubiquity of plant-plant replacements, I suggest setting aside a plot of ground, marking a few hundred plants, and coming back in a few years (longer for perennial plants) to see how many individuals have been replaced. The data I have from permanent plots sampled for decades in a variety of both Neotemperate and Neotropical old fields clearly show that replacements are working everywhere and always. Replacements are also a very common phenomenon in so called "primary" Neotropical forests (e.g., in the 50 ha plot located at Barro Colorado Island, Panama http://ctfs.si.edu/datasets/bci). It seems that plants wait not to be counted (sensu Harper 1977) but to be replaced!

The working model of plant-plant replacements entails a strong possibility that a plant will be replaced by another plant (whether of the same or of a different species) when the probabilities expressed in the species' niche of that plant are lower than the probabilities expressed in the species' niche of the other plant, those probabilities being activated by the combination of abiotic and biotic conditions at a given point in space and time and for that particular life-stage in the plants' development. I suggest that it is the smallest probability—from among the many niche "slices" for each species that are activated by local environmental conditions—that will be the most limiting and, therefore, most critical in determining whether or not one plant will replace another.

Replacements create changes in plant abundances, which may also create changes in plant composition that lead to the emergence of other patterns at larger spatial and temporal scales (e.g., successional rate and direction, patch dynamics, plant distribution, exotic plant invasion, plant associations). Care must be taken, however, with the observation of these plant patterns. For example, even though individual plants continue to be replaced, preservation of species abundances and/or species composition at a larger scale of organization may still occur, giving the impression of a "coexistence" of species (when the identity or number of plant species remain the same despite changes in plant abundances) or of a plant community at "equilibrium" (when neither plant species nor plant abundances change). Existence of high-level patterns that do not change very much over time should not obscure the fact that such patterns are the product of a dynamic and ever changing process of plant-plant replacement. Any plant pattern is only a snapshot of the vegetation at a given instance in time and at a specific scale of observation, whereas plant-plant replacements occur continuously.

Because plant niches express probabilities, plant-plant replacement itself must be a probabilistic process. In fact, succession is probabilistic (Pickett 1982), for it is made up of a multitude of plant-plant replacements that take place over time (for example, when all the individuals of a given species leave a disturbed area and/or an individual of a new species arrives, creating a change in species composition: Myster and Pickett 1994, Brokaw and Busing 2000). Again, the dynamics of any plant community is probabilistic because a plant community—and any of its plant-based structures and functions—is defined by its' resident plants that have been brought forth by a probabilistic plant-plant replacement process (Pickett 1982, Myster and

Pickett 1988). Thus the ultimate reality of every plant community are the plant-plant replacements, for they produce vegetation changes at any and all scales.

Along with succession, plant associations, and species coexistence/equilibrium, distribution of plant species exemplifies how plant-plant replacements determine patterns of a plant community at higher levels of organization. Since plant distribution patterns are shaped by plant-plant replacements, there is a strong possibility that distribution of a species will *increase* if, under given conditions, the species' niche probabilities are higher than those of another species' at the edges of the two species' distribution where the species meet, or *decrease* if the first species' probabilities are lower. For example, red cedar (*Juniperus virginiana* L.) is expanding its range in the United States by replacing plants of various other species at the edges of its distribution. The success of red cedar's invasion finds its explanation in the species' regeneration niche (those parts of a plants' niche that represent seeds and seedlings: Grubb 1977) that includes higher probabilities in seed dispersal, extended seed viability in the soil, lower seed predation, lower herbivory, and higher drought resistance than many other species (reviewed in Myster 1993).

1.2 Species Associations in Permanent Plots

Studies in the temperate zone of the Americas show that the spatial and temporal distribution of old field plants is not random and that it is associated with various factors (such as the location of "nurse" trees and plant resource levels: Myster and Pickett 1992a, Myster 1993). In the course of an investigation of these temporal and spatial patterns, numerous questions present themselves: When do species appear and when do they leave? How long are they present? Are there edge effects from the forest? Do species clump together and, if they do, does this autocorrelation change with time? Chronosequences supply some evidence to answer these questions. However, inasmuch as they place different fields sampled at different times sequentially on a time line, this approach offers only, at best, a partial solution. The true answers come from regularly sampled permanent vegetation plots that are established at the abandonment of agricultural fields because such plots show the actual sequences of plant abundance and species change over time (Pickett 1982) an important aspect of the dynamic nature of plant communities (Myster 2001).

As case studies, let us consider plots in natural pasture in Puerto Rico and plots in former Banana plantations, former Sugarcane plantations, and seeded pastures in Ecuador. The first study site is a natural pasture that had been grazed for decades before it was abandoned. It borders the Luquillo Experimental Forest (LEF) of northeastern Puerto Rico, USA (18°20'N, 65°45'W: Aide et al. 1995, Liu and Zou 2002), close to the town of Sabana. The LEF, a long-term ecological research (LTER) site of the National Science Foundation (www.luq.lternet.edu), is a tropical montane wet forest with tabonuco (*Dacryodes excelsa*), ausubo (*Manilkara bidentata*) and motillo (*Sloanea berteriana*) below 600m, palo colorado (*Cyrilla racemiflora*) and palm (*Prestoea montana*) between 600m and 850m, and cloud forest above 850m (Waide and Lugo 1992). The study pasture is located at the lower elevations and receives between 2m and 5m of rainfall per year, with an average temperature of 18°C. Its soils are fertile and volcaniclastic in origin (Thomlinson et al. 1996).

Twenty-five 5m × 2m contiguous plots were laid out on the border with the forest (Myster 2003b), the long side parallel to the forest in order to maximize any edge effects and to facilitate comparison to the Buell-Small plot study in New Jersey, USA (Myster and Pickett 1990a). Since island diversity in plant species is greatly reduced compared to mainland diversity, the plots were big enough to capture diversity in these agrosystems (Myster 2004a). The plots did not have any remnant trees or sprouting tree roots at the beginning of the study, and their tree seed bank was very small (Myster 2006). Starting in May of 1997 and continuing annually for 10 years, the plots were sampled for percent cover of each plant species. Percent cover—an indication of a species' ability to capture light and, therefore, to dominate these areas in the process of becoming forested communities (Myster 2003b)—was estimated visually in relation to each plot's area.

The analysis of the data obtained from these plots showed that grass dominated throughout the first 5 years after abandonment, constituting 75% of the plant cover after 5 years, while forbs declined to 40% and ferns and woody species increased to 45 and 15% respectively. Common trees included *Syzygium jambos*, *Calophyllum calaba*, and *Tabebuia heterophylla*. Common shrubs were *Clidemia hirta* and *Miconia* spp., both of which invaded from the forest. The community parameters that peaked within the first 5 years after abandonment were productivity (at 400 $g/m^2/yr$), total basal area (at 1,000 cm^2), and richness (at 19 species: Myster 2003b).

The second study site comprises recovering Banana plantations, Sugarcane plantations, and seeded pastures at the Maquipucuna Reserve, Ecuador (0°05'N, 78°37'W; www.maqui.org; Sarmiento 1997, Rhoades et al. 1998, Rhoades and Coleman 1999, Zahawi and Augspurger 1999, Myster 2004a, 2004b). Maquipucuna lies between 1,200m and 1,800m and is classified as tropical lower montane wet forest (Edmisten 1970). It has deeply dissected drainages with steep slopes and receives between 2m and 5m of rainfall a year. The temperature ranges between 14°C and 25°C, with an average temperature of 18°C. The reserve's fertile andisol soil is developed from volcanic ash deposits (Myster and Sarmiento 1998).

In June of 1996, 6 agricultural fields were selected: two Sugarcane (*Saccharum officinarum*) plantations, two Banana (*Musa* sp.) plantations, and two pastures seeded in *Setaria sphacelata*.[2] All 6 fields were 2 ha rectangles located within a few hundred meters of each other at the lower elevations of the reserve, on the border with the primary forest. In each field, twenty-five 5m × 2m contiguous plots were laid out, with the long side on the forest border. This forest is also tropical montane wet forest.

[2] Although native to Africa, the *Setaria* grass is being planted more and more extensively in the Neotropics because it can withstand heavy grazing by spreading asexually and can form large crowns. Well accepted by cattle, the *Setaria* grass is tolerant of relatively cold temperatures and short-term waterlogged conditions, both common at the Maquipucuna Reserve.

Starting in 1997, these subplots were sampled annually for 10 years for percent cover of each plant species. Maquipucuna plant taxonomists, trained at the University of Georgia, USA, where voucher specimens are kept on file (Myster and Sarmiento 1998, Zahawi and Augspurger 1999), assisted in the identification of species by using specimens located on site. The analysis of the data from the first 5 years after abandonment showed that the resident graminoid dominated in Sugarcane (at 50% after 5 years), in Banana (at 70%), and in pasture (at 90%), that forbs and ferns were at low levels everywhere, and that woody species increased in Sugarcane (at 45% after 5 years) and in Banana (at 25%) but remained sparse in pasture (at 10%). Common trees included *Acalypha pladichephalus, Costus* sp., *Vernonia patens*, and *Piper aduncum* while common shrubs were *Pilea* sp. and *Miconia* spp. After 5 years, both total basal area and species richness were greatest after Sugarcane (at 700 cm^2 and 59 species respectively), lower after Banana (at 600 cm^2 and 49 species), and lowest after pastures (at 100 cm^2 and 25 species: Myster 2007).[3]

The plot data were used to implicate plant interactions and successional mechanisms by generating spearman coefficients of rank associations (SAS 1985, Milbau et al. 2005). The analyses showed that the actual number of associations among species was much lower than the potential number of 3,250 per field (all pairs of 26 species over the 10-year period). For example, the pasture in Puerto Rico had a total of 362 associations, 55 of which were negative (Table 1.1), while one of the two Banana fields in Ecuador had a total of 168 associations, 12 of them being negative (Table 1.2).

Because controlled field experiments have confirmed that species that interact significantly also form significant associations with each other (Tirado and Pugnaire 2005), the statistics from the case studies (Tables 1.1–1.7) point to a low level of species interactions in these fields (as was also seen in temperate old field plots in New Jersey, USA: Myster and Pickett 1992b). These results support the view of communities as loosely-organized assemblages of species that are created mainly by tolerance parts of individual species' niches (Gleason 1926, Whittaker 1975) rather than as tightly-linked entities that are structured by species interactions.

Most of the significant associations in the study plot data were positive (unlike those, mainly negative, associations that were computed from plots in temperate old fields: Myster and Pickett 1992b). This suggests that facilitation in these early-successional plots where plants are under stress (see Chapter 10) may be more important than competition among those relatively few species that are actively interacting. In the Puerto Rican pasture (Table 1.1), species that formed many positive associations included the trees *Syzgium jambos, Guarea guidonia, Ocotea leucoxylon*, and *Prestoea montana*. Among major inhibiting species were the grass *Panicum* spp., the fern *Gleichenia bifida*, and the tree *Tabebula heterophylla*. In

[3] The data from the plots in Puerto Rico (LTERDATB #97) and in Ecuador (LTERDATB#101) are housed in the archives of the LEF LTER site. Visit the Luquillo web site (www.luq.lternet.edu) for further details.

Table 1.1 Significant spearman rank correlation coefficients among all plant species in the *Puerto Rican pasture* over the first 10 years of succession. Significant positive associations are indicated by a "+" and significant negative associations are indicated by a "–". Each matrix entry contains all significant associations, with a maximum of ten, between two species in a certain field. Plant species and families are indicated as follows: Bromelia spp. (A), Guarea guidonia (B), Ocotea leucoxylon (C), Citrus frutus (D), Syzgium jambos (E), Desmodium spp. (F), Gleichenia bifida (G), Inga laurina (H), Citrus limon (I), Casearia sylvestris (J), Prestoea montana (K), Calophyllum calaba (L), Miconia prasina (M), Eugenia pseudopsidium (N), Tabebuia heterophylla (O), Eugenia malaceensis (P), Piper hisperdium (Q), Andira inermis (R), Psychotria brachiata (S), Miconia racemosa (T), Psychotria berteriana (U), Xanthosoma spp. (V), Clidemia hirta (W), Panicum spp. (X), Myrcia splendens (Y) and Ocotea sintenisii (Z).

	B	C	D	E	F	G	H	I	J	K	L	M	N	O	P	Q	R	S	T	U	V	W	X	Y	Z
A	+– –	+	–	++		+		++	–	+	+++			++			+		–					–	–++
B		++++++	++	+++++++	– ++	+++			++	+++++	++	+	–	+++				+	+	+++	–––+	++	–++	++++	++
C			+	+++		+	+	++	++++	++	+++	+	+++	++	+++							++	++++	++++	++
D				+	++	++			+	+	+		+	+–		++	+		+						
E					–+	+	++++	++		+++++	+++++++	++	+	–++	++	–––	+	–	++	++	+++–	+–	++++	+++	++
F						++		+		+	–				–		+					+	+–		
G							+	++	++	+	++++	–		––	+	+	––	+++		+	––	+	++	––	+
H										+++++		+	+++	––	++++			+		+	+	+		–	++
I										++	+	+		++	+	+++	+	+	+			+–	+	+	+++++
J											++		++	++	+				+++++				++–	++	++
K														++–	++	++	+–	+	+	++		+++–	+–	++	+++
L												++	+	+	++	+			+++			–––	––++	+++	+
M														++++	++		+	++++	+			++	+		
N														++	+					+		++	++		
O																	––	–	+		–+		–	–	+
P																			+			++	+	+++	+
Q																									
R																	++					++			
S																				+	–	+			
T																					+	++++		++	+
U																									
V																							++–	++	
W																									+
X																								++++	+
Y																									+

Table 1.2 Significant spearman rank correlation coefficients among all plant species in the *Ecuador Left Banana Plantation* over the first 10 years of succession. Significant positive associations are indicated by a "+" and significant negative associations are indicated by a "–". Plant species and families are indicated as follows: Acalypha pladichephalus (A), Begonia spp. (B), Geonoma undata (C), Cyathea spp. (D), Musa spp. (E), Pilea spp. (F), Anthurium spp. (G), Trichipterix pilosissima (H), Nectandra spp. (I), Ochroma spp. (J), Baccharis spp. (K), Anthurium spp. (L), Setaria spp. (M), Boconia frutescens (N), Piper aduncum (O), Erythrina megistophyla (P), Vernonia patens (Q), Hedyosum spp. (R), Commelina diffusa (S), Althernantera spp. (T), Siparuna piloso-lepidota (U), Solanum spp. (V), Vernonia spp. (W), Digitaria sanguinalis (X), Inga spp. (Y) and Passiflora spp. (Z).

	B	C	D	E	F	G	H	I	J	K	L	M	N	O	P	Q	R	S	T	U	V	W	X	Y	Z
A	+				+	+	++		+	+		– ++	++	–					++	+					++
B		++	++	++	++++	+	+		++	+	++	–								+	+	+			
C			+++	+	+	+++	+	+++			+–			–											
D					+	+	+	+++				–		+– –	+										
E					++	+								+			+								+
F							++	+		++	+			–					+			+			
G							+++	+++		+		–										+			
H								+++	++	+									+		+	++		+	
I									++			–								+					
J										+++	++			–				+	+	+	+	+	+	+	+
K												++	++	++									+		+
L												+					+		+	+	+		+		
M														+											+
N																									+
O															+++	+–									
P																++							+		
Q																	++	++						++	
R																		+++	+						
S																			++		+				
T																				+	+				
U																					+++++				
V																							++		
W																							++	+++	
X																								+	+
Y																									

one of the two Banana fields, key species with many positive associations included *Begonia* spp., *Trichipterix pilosissima*, and *Ochroma* spp. (Table 1.2). However, the past crop in that field (*Musa* spp.) did not play a significant role. In the other Banana field, *Begonia* spp., *Cuphea* spp., and *Brugmansia* spp. were very active (Table 1.3). In one of the two Sugarcane fields, key species included members of the families Asteraceae, Verbenaceae, and Pupillionacia. The past crop (*Saccharum officinarum*), however, was not implicated (Table 1.4). In the other Sugarcane field, *Cuphea* spp. and *Piper aduncum* were important, but the past crop was not (Table 1.5). In the Ecuadorian pastures, negative associations were more common than positive associations, especially when the planted pasture grass *Setaria sphacelata* was one of the species (Tables 1.6–1.7).

Although the results in species associations strongly suggest species interactions (see Tirado and Pugnaire 2005 and references therein) among some tropical plant species (as for some temperate plant species: Myster and Pickett 1992b), any further comment would be premature, given the lack of tropical studies involving these species (see Chapter 14). When augmented with the findings of future tropical experiments, however, these results will allow greater synthesis of research effort and more viable interpretations of post-agricultural succession in the Neotropics.

Table 1.3 Significant spearman rank correlation coefficients among all plant species in the *Ecuador Right Banana Plantation* over the first 10 years of succession. Significant positive associations are indicated by a "+" and significant negative associations are indicated by a "–". Plant species and families are indicated as follows: Acalypha pladichephalus (A), Costus spp. (B), Musa spp. (C), Solanum muricatum (D), Piperaceae (E), Setaria spp. (F), Tagetes terniflora (G), Begonia spp. (H), Cuphea cartagenensis (I), Polypodiaceae (J), Vernonia patens (K), Brugmansia spp. (L), Digitaria sanguinalis (M), Urticaceae (N), Chusquea spp. (O), Nectandra spp. (P), Piperaceae (Q), Commelina diffusa (R), Erythrina megistophyla (S), Heliotropium spp. (T), Inga spp. (U), Musa acuminate (V), Chenopodium album (W), Crataegus monogyna (X), Boconia frutescens (Y) and Cecropia monostachyma (Z).

	B	C	D	E	F	G	H	I	J	K	L	M	N	O	P	Q	R	S	T	U	V	W	X	Y	Z
A		+	–		–	+	+++	+	+		–	+			+	+						– +			
B		+	– +	+	++			– ++	+										+			++			
C						+		++			++	+	+	+						+		+			+
D									+	+			+	+		+								++	+
E						+++			– +	+	+				+						+				
F						–					–						+	+			++				
G							++++	++	+	+							+								
H								++	+	+				+	++	+									
I									+	+	+	+				+				+	++	+	+		
J										+ –		+		+										++	+
K											+++							+							
L												++	+		+	+	+	+	+	+					+
M													+++	+++					+		+				
N														+++										+	+
O																+++									
P																	+++	++							
Q																		++++	++	+				+	+
R																			+						
S																					+	+			
T																						+++	+		
U																							++	+	++
V																									
W																								+	
X																									++
Y																									

1.3 About this Book

Examined in this book are vital concerns of post-agricultural recovery in the Neotropics, such as common clearing, planting, cultivation, harvesting and abandonment practices, the post-agricultural environment and disturbance regime, and successional mechanisms. In addition, this book has advantages over other studies of tropical disturbances in that it focuses directly on post-agricultural areas and includes real data of successional pattern and process while investigating the function and structure of old fields and presenting a conceptual framework that both current and future Neotropical studies can use and add to.

All work that is reported in the book chapters has been done in the Neotropics, the American landmass and associated islands that lie between the Tropic of Cancer (23.5°N) and the Tropic of Capricorn (23.5°S). Between these latitudes, the sun remains in the sky the longest over the course of a year compared to the rest of the earth, which leads to a greatly reduced seasonality and prolonged growing period. Responding to this geographic variation, the chapter authors share their expertise

Table 1.4 Significant spearman rank correlation coefficients among all plant species in the *Ecuador Right Sugarcane Plantation* over the first 10 years of succession. Significant positive associations are indicated by a "+" and significant negative associations are indicated by a "–". Plant species and families are indicated as follows: Acalypha pladichephalus (A), Asteraceae (B), Digitaria sanguinalis (C), Polypodiaceae (D), Nectandra spp. (E), Stachys micheliana (F), Piperaceae (G), Lantana camara (H), Verbenaceae (I), Erythrina megistophyla (J), Piper aduncum (K), Rubus spp. (L), Commelina diffusa (M), Elephantopus mollis (N), Cecropia spp. (O), Costus spp. (P), Miconia spp. (Q), Passifloraceae (R), Fabaceae (S), Chusquea spp. (T), Marantaceae (U), Pilea spp. (V), Hieracium spp. (W), Sabicea spp. (X), Columnea spp. (Y) and Orchidaceae (Z).

	B	C	D	E	F	G	H	I	J	K	L	M	N	O	P	Q	R	S	T	U	V	W	X	Y	Z
A	+– –	+	+	++			++			+–	+									+	+		+		
B			+	+	+	++++	–	++														+	++	+	
C														+							++	++	– +++		
D				+			–	–+	–	+			++	++		+			+		+				
E					+	++		– ++																	+
F						+		++					+				+	+	+				+		
G								+		+							+					+			
H								+	+++									+	+	++	++				
I											+		++	+						+			+		
J										+	+		++							+	+				
K											++						+						+		+
L												–	++									+		+	
M														+	+				+++	+					
N														+									+	+	+
O																			+				+		
P																		+	+						
Q																	+++							+	
R																		++	+						+
S																			++	++	+	+	+	++	
T																				+++	+			+	
U																					+	+			
V																									
W																							+		
X																									
Y																									+

and relate studies over a wide range of the major gradients of temperature and rainfall found throughout the Neotropics.

The book is divided into three sections, each of which develops and expands upon key points of post-agricultural succession in the Neotropics:

I. Patterns of living and non-living components of old fields after abandonment.
II. Mechanisms and tolerances that cause plant-plant replacements, producing old field vegetation patterns.
III. Restoration, management, and the future of post-agricultural areas.

There are no clear-cut boundaries, however, among the three parts. The book offers a lot of "cross-fertilization," for most of its chapters discuss issues that formally belong to more than one section. This multilateral approach presents itself as the best way for a comprehensive investigation of the complex phenomenon of post-agricultural succession in the Neotropics.

We begin the book by exploring permanent plot data in areas that used to be Neotropic Banana plantations, Sugarcane plantations, and pastures. We continue by examining the above- and below-ground Carbon (C) dynamics after crop agriculture and pasture and by discussing the mechanisms of C sequestration in these areas and the methods of studying soil C dynamics. Next, we focus on species composition,

Table 1.5 Significant spearman rank correlation coefficients among all plant species in the *Ecuador Left Sugarcane Plantation* over the first 10 years of succession. Significant positive associations are indicated by a "+" and significant negative associations are indicated by a "−". Plant species and families are indicated as follows: Musa spp. (A), Costus spp. (B), Cuphea cartagenensis (C), Digitaria sanguinalis (D), Miconia spp. (E), Piper spp. (F), Rubus spp. (G), Sida rhombifolia (H), Asteraceae (I), Baccharis spp. (J), Polypodiaceae (K), Lantana camara (L), Vernonia patens (M), Acalypha pladichephalus (N), Solanum spp. (O), Saccharum officinarum (P), Piper aduncum (Q), Verbenaceae (R), Commelina diffusa (S), Erythrina megistophyla (T), Nectandra spp (U), Altus spp. (V), Orchidaceae (W), Polybotria spp. (X), Vernonia spp. (Y) and Polypodiaceae (Z).

	B	C	D	E	F	G	H	I	J	K	L	M	N	O	P	Q	R	S	T	U	V	W	X	Y	Z
A		+				+	−		+		−+			+	++	+++	+	−	+				+		+
B		++	++	++										−	+			+			+	+++	+	+	
C			+++	+	++++	++	+			+++											+			+	
D				+	+++	++	+	+			+														
E					+	+	+	−			+−											+	+		
F							+−			+	−		−+			−	+						+		
G							+		++	+							+	+	+				+	+	
H										+			+		+		+	−					+	+	
I												+			++					+		+	+	+	+
J											+	+		+	−						+				
K																+					+				
L													++++	+				+				+	+	+	
M													+	++				++		+					++
N														++		+	+	+				++	+		
O															+	++		++	+			+			
P																+	++					+			
Q																	+					++++		+	
R																		++			+			+	
S																					+	+	+		
T																					++++	++		+	++
U																						+			
V																									
W																							+++		
X																								+++	
Y																									

Table 1.6 Significant spearman rank correlation coefficients among all plant species in the *Ecuador Left Pasture* over the first 10 years of succession. Significant positive associations are indicated by a "+" and significant negative associations are indicated by a "−". Plant species and families are indicated as follows: Setaria sphacelata (A), Acalypha pladichephalus (B), Xanthosoma sagittifolia (C), Heliocarpus americanus (D), Heliconia spp. (E), Ficus spp. (F), Solanceae (G), Araceae (H), Citrus spp. (I), Vernonia patens (J), Begonia spp. (K) and Nectandra spp. (L).

	B	C	D	E	F	G	H	I	J	K	L
A	- - - -	- -	- -	-	-						
B			+		+++	+ ++	++	+	+		
C											
D					++						
E											
F									+	+	+
G											
H											
I											
J											
K											
L											

Table 1.7 Significant spearman rank correlation coefficients among all plant species in the *Ecuador Right Pasture* over the first 10 years of succession. Significant positive associations are indicated by a "+" and significant negative associations are indicated by a "–". Plant species and families are indicated as follows: Setaria sphacelata (A), Erythrina megistophyla (B) and Polpodiaceae (C).

	B	C
A	-----	---
B		
C		

soil C, and nutrients after pastures and after Sugarcane and tree plantations. The role of alien species is introduced at this point to be developed in chapters that follow. We finish the first section by concentrating on the living part of the soil: the fungi, macrofauna, and earthworms. With all three kinds of organisms, the patterns of their diversity and their abundance during succession are explored. Also examined are the effects that these organisms may have on soils and on ecosystem functioning and the possible distribution of these organisms among functional groups.

In the second section of the book, we discuss what is known about mechanisms and tolerances that lead to plant-plant replacements during succession. First, we look into the complete suite of plant mechanisms and tolerances in Neotropic wet forest. Then we move on to seed dispersal by cattle and by birds. Raised in the discussion are questions of how these types of dispersal affect viability and germination and how the animal vectors themselves affect and, in turn, are affected by the vegetation structure in old fields. Next, studies that explore variation in the working of seed predation are reviewed, and issues of facilitation and competition are investigated in field experiments. Finally, plant mechanisms and tolerances in Neotropic dry forest are examined.

In the final part of the book, we address the issues of restoration, management, and future conditions of agricultural fields. Discussed are the role of microbes in restoration and sustainability of cropping systems as well as the ways in which agricultural practices affect microbial diversity. Also explored are different kinds of pasture degradation, objectives in pasture restoration, and specific techniques, such as intercropping and not clearing trees, from Costa Rican studies. We conclude the book by assessing the *status quo* of research in the area of Neotropical succession after agriculture along with research perspectives and their relation to ecological paradigms. Finally, specific avenues for future old field studies are presented.

We hope that these studies of old fields after agricultural disturbance will also enhance an understanding of natural transition from grassland/meadow to forest and a comprehension of fluctuations in the development of plant communities. This knowledge will enable researchers and land managers not only to predict and manipulate the nature of land recovery but also to more skillfully utilize lands with various kinds of vegetation cover, from lands on which woody vegetation is desired (e.g., forests, surface mines, road-sides) to lands on which woody vegetation should be removed (e.g., rangelands, roadsides, utility corridors).

Acknowledgment I wish to thank Mr. Michael Malahy for his help in analysis of the plot data. I also wish to thank Mr. Bert Witteveen for logistic help at Maquipucuna Reserve and Dr. Nick Brokaw for logistic help at Luquillo Experimental Forest. Finally, I thank Ms. Eda Melendez and her staff for their help with the LTER data sets. I received support from grant DEB-0218039 from the National Science Foundation to the Institute of Tropical Ecosystem Studies, University of Puerto Rico, and the USDA Forest Service, International Institute of Tropical Forestry as part of the Long-Term Ecological Research Program in the Luquillo Experimental Forest. Additional support was provided by the Forest Service (U.S. Department of Agriculture) and the University of Puerto Rico.

References

Aide, T. M., and Cavelier, J. 1994. Barriers to lowland tropical forest restoration in the Sierra Nevada de Santa Marta, Colombia. Restoration Ecology 2:219–229.

Aide, T. M., Zimmerman, J. K., Herrera, L., Rosario, M., and Serrano, M. S. 1995. Forest recovery in abandoned tropical pastures in Puerto Rico. Forest Ecology and Management 77:77–86.

Bazzaz, F. A. 1968. Succession on abandoned fields in the Shawnee Hills, southern Illinois. Ecology 49(5):924–936.

Bazzaz, F. A. 1979. The physiological ecology of plant succession. Annual review of Ecology and Systematics 10:351–371.

Bazzaz, F. A. 1996. Plants in Changing Environments: Linking Physiological, Population, and Community Ecology. Cambridge: Cambridge University Press.

Borhidi, A. 1988. Vegetation dynamics of the savannization process on Cuba. Vegetatio 77:177–183.

Brokaw, N. V. L. 1982. The definition of treefall gap and its effect on measures of forest dynamics. Biotropica 14:158–160.

Brokaw, N. V. L., and Busing, R. T. 2000. Niche versus chance and tree diversity in forest gaps. Trends in Ecology and Evolution 15:183–188.

Brown, S., and Lugo, A. E. 1990. Tropical secondary forests. Journal of Tropical Ecology 6(1): 1–32.

Buell, M. F., Buell, H. F., Small, J. A., and Siccama, T. G. 1971. Invasion of trees in secondary succession on the New Jersey piedmont. Bulletin of the Torrey Botanical Club 98(2):67–74.

Burton, P. J., and Bazzaz, F. A. 1991. Tree seedling emergence on interactive temperature and moisture gradients and in patches of old-field vegetation. American Journal of Botany 78: 131–149.

Buschbacher, R. J. 1986. Tropical deforestation and pasture development. BioScience 36:2–28.

Busing, R. T. 1996. Estimation of tree replacement patterns in an Appalachian *Picea-Abies* forest. Journal of Vegetation Science 7:685–694.

Castro, I. A., Sterling, S., and Galiano, E. F. 1986. Multi-species pattern analysis of Mediterranean pastures in three stages of ecological succession. Vegetatio 68(1):37–42.

Donfack, P., Floret, C., and Pontanier, R. 1995. Secondary succession in abandoned fields of dry tropical northern Cameroon. Journal of Vegetation Science 6(4):499–508.

Edmisten, J. 1970. Some autoecological studies of *Ormosia krugii*. In A Tropical Rain Forest, ed. H. T. Odum and R. F. Pigeon, Chapter D-8. Springfield, Virginia: National Technical Information Service.

Egler, F. E. 1954. Vegetation science concepts. I. Initial floristic composition, a factor in oldfield vegetation development. Vegetatio 4:412–417.

Fearnside, P. M. S. 1993. Deforestation in Brazilian Amazonia: the effect of population and land tenure. Ambio 22(8):537–545.

Fernandes, D. N., and Sanford Jr., R. F. 1995. Effects of recent land-use practices on soil nutrients and succession under tropical wet forest in Costa Rica. Conservation Biology 9(4):915–922.

Finegan, B. 1984. Forest succession. Nature 312:109–114.

Good, R. E. 1953. The Geography of the Flowering Plants. London: Longmans.
Gleason, H. A. 1926. The individualistic concept of the plant association. Bulletin of the Torrey Botanical Club 53:1–20.
Grainger, A. 1988. Estimating areas of degraded tropical lands requiring replenishment of forest cover. International Tree Crops Journal 5(1-2):31–61.
Grigg, D. B. 1978. The Agricultural Systems of the World. Cambridge: Cambridge University Press.
Grubb, P. J. 1977. The maintenance of species richness in plant communities: the importance of the regeneration niche. Biological review of the Cambridge Philosophical Society 52:107–145.
Hall, C. A. S., Stanford, J. A., and Hauer, F. R. 1992. The distribution and abundance of organisms as a consequence of energy balances along multiple environmental gradients. Oikos 65:377–390.
Harper, J. L. 1977. Population Biology of Plants. London: Academic Press.
Hobbie, S. E. 1992. Effects of plant species on nutrient cycling. Trends in Ecology and Evolution 7:336–339.
Keller, M., Veldkamp, E. Weitz, A. M., and Reiners, W. A. 1993. Effect of pasture age on soil trace-gas emissions from a deforested area of Costa Rica. Nature 365:244–246.
Kessell, S. R. 1979. Gradient Modeling: Resource and Fire Management. New York: Springer-Verlag.
Liu, Z., and Zou, X. M. 2002. Exotic earthworms accelerate plant litter decomposition in a Puerto Rican pasture and a wet forest. Ecological Applications 12(5):1406–1417.
Lugo, A. E. 1992. Comparison of tropical tree plantations with secondary forests of similar age. Ecological Monographs 62(1):1–41.
Mabberley, D. J. 1992. Tropical Rain Forest Ecology. New York: Chapman and Hall.
Milbau, A., Nijs, I., De Raedemaecker, F., Reheul, D., and De Cauwer, B. 2005. Invasion in grassland gaps: the role of neighbourhood richness, light availability and species complementarity during two successive years. Functional Ecology 19(1):27–37.
Miles, J. 1979. Vegetation Dynamics. New York: John Wiley
Muller, F. 1997. State-of-the-art in ecosystem theory. Ecological Modeling 100:135–161.
Myers, N. 1991. Tropical deforestation: the last situation. BioScience 41:282.
Myster, R. W. 1993. Tree invasion and establishment in old fields at Hutcheson Memorial Forest. The Botanical Review 59(4):251–272.
Myster, R. W. 1994. Contrasting litter effects on old field tree germination and emergence. Vegetatio 114(2):169–174.
Myster, R. W. 2001. Mechanisms of plant response to gradients and after disturbances. The Botanical Review 67(4):441–452.
Myster, R. W. 2003a. Seed regeneration mechanisms over fine spatial scales on recovering Coffee plantation and pasture in Puerto Rico. Plant Ecology 166(2):199–205.
Myster, R. W. 2003b. Vegetation dynamics of a permanent pasture plot in Puerto Rico. Biotropica 35(3):422–428 .
Myster, R. W. 2003c. Effects of species, density, patch-type, and season on post-dispersal seed predation in a Puerto Rican pasture. Biotropica 35(4):542–546 .
Myster, R. W. 2004a. Regeneration filters in post-agricultural fields of Puerto Rico and Ecuador. Plant Ecology 172(2):199–209.
Myster, R. W. 2004b. Post-agricultural invasion, establishment and growth of Neotropical trees. The Botanical Review 70(4):381–402.
Myster, R. W. 2006. Shrub vs. grass patch effects on the seed rain and the seed bank of a five-year pasture in Puerto Rico. Ecotropica 12:12–19.
Myster, R. W. 2007. Early successional pattern and process after Banana, Sugarcane and Pasture in Ecuador. The New Zealand Journal of Botany 45:35–44.
Myster, R. W., and Pickett, S. T. A. 1988. Individualistic patterns of annuals and biennials in early successional old fields. Vegetatio 78(1-2):53–60.
Myster, R. W., and McCarthy, B. C. 1989. Effects of herbivory and competition on survival of *Carya tomentosa* (Juglandaceae) seedlings. Oikos 56(2):145–148.

Myster, R. W., and Pickett, S. T. A. 1990a. The Buell successional study at Hutcheson Memorial Forest Center. Permanent Plotter 3:2–3.
Myster, R. W., and Pickett, S. T. A. 1990b. Initial conditions, history and successional pathways in ten contrasting old fields. American Midland Naturalist 124(2):213–238.
Myster, R. W., and Pickett, S. T. A. 1992a. Effects of palatability and dispersal mode on spatial patterns of trees in old fields. Bulletin of the Torrey Botanical Club 119(2):145–151.
Myster, R. W., and Pickett, S. T. A. 1992b. Dynamics of associations between plants in ten old fields during 31 years of succession. Journal of Ecology 80(2):291–302.
Myster, R. W., and Pickett, S. T. A. 1993. Effects of litter, distance, density and vegetation patch type on postdispersal tree seed predation in old fields. Oikos 66(3):381–388.
Myster, R. W., and Pickett, S. T. A. 1994. A comparison of rate of succession over 18 yr in ten contrasting old fields. Ecology 75(2):387–392.
Myster, R. W., and Sarmiento, F. O. 1998. Seed inputs to microsite patch recovery on two Tropandean landslides in Ecuador. Restoration Ecology 6(1):35–43.
Odum, E. P. 1953. Fundamentals of Ecology. Philadelphia: Saunders.
Oosting, H. J. 1942. An ecological analysis of the plant communities of Piedmont, North Carolina. American Midland Naturalist 28:1–126.
Padoch, C., and Vayda, A. P. 1983. Patterns of resource use and human settlement in tropical forest. In Tropical Rain Forest Ecosystems. Volume A, ed. F. B. Golley, pp. 301–313. New York: Elsevier.
Parrish, J. A. D. and Bazzaz, F. A. 1982. Niche responses of early and late successional tree seedlings on three resource gradients. Bulletin of the Torrey Botanical Club 109:451–456.
Pickett, S. T. A. 1982. Population patterns through twenty years of old field succession. Vegetatio 49(1):45–59.
Pickett, S. T. A., and White, P. S. 1985. The Ecology of Natural Disturbance and Patch Dynamics. Orlando, FL: Academic Press.
Quintana-Ascencio, P. F., Gonzalez-Espinosa, M., Ramirez-Marcial, N., Dominguez-Vazquez, G., and Martinez-Ico, M. S. 1996. Soil seed banks and regeneration of Tropical rain forest from Milpa fields at the Selva Lacandona, Chiapas, Mexico. Biotropica 28(2):192–209.
Raunkaier, C. 1934. The Life forms of Plants and Statistical Plant Geography . Oxford, UK: Clarendon.
Rhoades, C. C., and Coleman, D. C. 1999. Nitrogen mineralization and nitrification following land conversion in montane Ecuador. Soil Biology and Biochemistry 31(10): 1347–1354.
Rhoades, C. C., Eckert, G. E., and Coleman, D. C. 1998. Effect of pasture trees on soil nitrogen and organic matter: implications for tropical montane forest restoration. Restoration Ecology 6(3):262–270.
Risser, P. G. 1987. Landscape ecology: state of the art. In Landscape Heterogeneity and Disturbance. Ecological studies 64, ed. M. G. Turner, pp. 123–145. New York: Springer-Verlag.
Sarmiento, F. O. 1997. Arrested succession in pastures hinders regeneration of Tropandean forests and shreds mountain landscapes. Environmental Conservation 24(1):14–23.
Serrao, E. S. A., and Toledo, J. M. 1990. The search for sustainability in Amazonian pastures. In Alternatives to Deforestation: Steps Toward Sustainable Use of the Amazon Rain Forest, ed. A. B. Anderson, pp. 195–214. New York: Columbia University Press.
SAS User's Guide: Statistics. 1985. Version 5th edition. Cary, NC: SAS Institute.
Singh, K. D. 1993. The 1990 tropical forest resources assessment. Unasylva 174:10–19.
Skole, D., and Tucker, C. 1993. Tropical deforestation and habitat fragmentation in the Amazon: Satellite data from 1978 to 1988. Science 260:1905–1910.
Staver, C. 1990. Why farmers rotate fields in maize-cassava-plantain bush fallow agriculture in the wet Peruivan Amazon. Human Ecology 17(4):401–426.
Tansley, A. G. 1935. The use and abuse of vegetational concepts and terms. Ecology 16:284–307.
Thomlinson, J. R., Serrano, M. I., del M. Lopez, T., Aide, T. M., and Zimmerman, J. K. 1996. Land-use dynamics in a post-agricultural Puerto Rican landscape (1936–1988). Biotropica 28: 525–536.

Tilman, D. 1988. Plant Strategies and the Dynamics and Structure of Plant Communities. Princeton, NJ: Princeton University Press.

Tirado, R., and Pugnaire, F. I. 2005. Community structure and positive interactions in constraining environments. Oikos 111(3):437–444.

Turner, M. G., and Dale, V. H. 1991. Modeling landscape disturbance. In Quantitative Methods in Landscape Ecology: The Analysis and Interpretation of Landscape Heterogeneity. Ecological Studies 82, ed. M. G. Turner and R. H. Gardner, pp. 134–146. New York: Springer-Verlag.

Uhl, C. 1987. Factors controlling succession following slash-and-burn agriculture in Amazonia. Journal of Ecology 75(2):377–407.

Uhl, C., Buschbacher, R., and Serrao, E. A. S. 1988. Abandoned pastures in eastern Amazonia. I. Patterns of plant succession. Journal of Ecology 76(3):663–681.

Waide, R. B., and Lugo, A. E. 1992. A research perspective on disturbance and recovery of a tropical montane forest. In Tropical Forests in Transition: Ecology of Natural and Arthropogenic Disturbance Processes, ed. J. G. Goldammer, pp. 173–190. Basel: Birkhäuser-Verlag.

Walter, H. 1979. Vegetation of the Earth and the Ecological Systems of the Geo-biosphere. New York: Springer-Verlag.

Watt, A. S. 1947. Pattern and process in the plant community. Journal of Ecology 35:1–22.

Whittaker, R. H. 1965. Dominance and diversity in land plant communities. Science 147:250–260.

Whittaker, R. H. 1975. Communities and Ecosystems. New York: Macmillan.

Zahawi, R. A., and Augspurger, C. K. 1999. Early plant succession in abandoned pastures in Ecuador. Biotropica 31(4):123–129.

Zar, J. H. 1999. Biostatistical Analysis. 4th edition. Englewood Cliffs, NJ: Prentice-Hall.

Chapter 2
Trends in Above and Belowground Carbon with Forest Regrowth After Agricultural Abandonment in the Neotropics

Erika Marín-Spiotta, Daniela F. Cusack, Rebecca Ostertag and Whendee L. Silver

2.1 Introduction

Increasing forest cover on lands which were recently forested (reforestation), as well as on lands which have not supported forest growth in recent times (afforestation), has been proposed by the Intergovernmental Panel on Climate Change (IPCC) to help mitigate anthropogenic C emissions from land-use change and fossil fuel use (Brown et al. 1995a, Watson et al. 2000, Metz et al. 2001). Recent research has suggested that these strategies would be most effective in the tropical latitudes (Gibbard et al. 2005). Tropical forests have higher potential carbon (C) uptake rates than forests in temperate or boreal biomes (Brown et al. 1995a, Watson et al. 2000). Globally, 40% of terrestrial biomass C is in tropical forests (Dixon et al. 1994), and 40% of this is in secondary forests (Brown and Lugo 1990). In addition to providing opportunities for C sequestration, reforestation can lead to the recovery of important forest ecosystem goods and services. These include, but are not limited to, watershed protection, erosion control, regional climate stabilization, wood and non-timber products, and habitat for biodiversity (Brown and Lugo 1990, Guariguata and Ostertag 2001, Naughton-Treves and Chapman 2002, De Walt et al. 2003).

Although deforestation is still a dominant trend across the tropics, secondary forests resulting from human disturbance are becoming an increasingly important forest cover type (Brown and Lugo 1990). Just as rates of deforestation are difficult to establish with certainty (see Achard et al. 2002, DeFries et al. 2002, Eva et al. 2003, Fearnside and Laurance 2003), estimates for rates of secondary forest growth also differ, although most agree that the trend is positive. In the 1980s, tropical secondary forests were estimated to cover more than 600 million hectares (ha) globally with an annual rate of formation of 9 million hectares per year (ha/yr), and growing (Brown and Lugo 1990). In 1993, the estimated area of tropical America covered by secondary forests was 165 million ha (Weaver 1995 in Kammesheidt 2002). The United Nations Food and Agriculture Organization in 1990 classified 33 million ha of previously agricultural or pasture lands in Latin America as fallow (cited in Kammesheidt 2002). These fallow lands are key components of forest regrowth

 R. W. Myster, *Post-Agricultural Succession in the Neotropics.*

and succession toward secondary forest cover. In many regions across the tropics, reforestation is in the form of agroforestry systems or plantations. Reforestation also occurs due to natural regeneration of forest species on abandoned cropfields or pastures, or by active planting and management which facilitate the process of secondary succession.

The factors behind the increase in secondary forest growth in the tropics vary by region, as do the initial drivers of deforestation, but typically include one of the following: shifts in economic bases from agriculture to industry, rural to urban migration across domestic and international borders, changes in land tenure policies, or loss of site fertility (Rudel et al. 2002, Aide and Grau 2004, Soares-Filho et al. 2004, Arroyo-Mora et al. 2005). Brown and Lugo (1990) identified regional differences in the former land use type prior to secondary forest growth. In tropical America and Africa, most secondary forests originated from agricultural fallows. In tropical Asia, logging was a more important contributor to secondary forest formation than agricultural abandonment. Tropical Africa had the smallest percentage of secondary forests from logging. The authors also documented differences in the disturbance history of the pre-conversion forest leading to secondary forests. At the time of their review, 58% of secondary forests in tropical America were first generation secondary forests, from previously undisturbed or primary forest. In contrast, 72–76% of secondary forest areas in tropical Africa and Asia had experienced multiple cycles of disturbance (Brown and Lugo 1990). However, regional trends can change as economic policies are altered, and an updated study considering current trends in globalization and its effect on reforestation would be a valuable undertaking.

In this chapter, we review the recent literature on the effects of reforestation on above and belowground C stocks in both natural and managed secondary forests and plantations growing on former agricultural and pasturelands. We revisit Silver et al.'s (2000a) analysis of the potential for C sequestration with reforestation in the tropics with new data published since their review. We limit our analysis to studies from the Neotropics, where the bulk of studies are still located, though we would like to point out that in recent years there has been an increase in publications on biomass and structural changes during secondary forest and plantation development in the African and Asian tropics (e.g., de Jong et al. 2001, Lemenih et al. 2004, Walker and Desanker 2004, Lawrence 2005a, Makana and Thomas 2006). We also highlight advances made in the study of soil C dynamics with land-use change, particularly, and provide suggestions for future research directions.

2.2 Reforestation in the Neotropics

Changes in economic policies and human migration have resulted in agricultural abandonment and reforestation across the tropics. A prime example of the increase in secondary forest cover during the last half-century is the Caribbean island of Puerto Rico, where the most reliable data on secondary forest cover exists, along with Costa Rica and the Brazilian Amazon (Kammesheidt 2002). Economic changes in Puerto Rico, beginning in the late 1940s with the transition from being a colonial dependency to a free associated state (Dietz 1986), resulted in a large-scale abandonment of agricultural and pasturelands at an approximate rate of 10,000 ha/yr from 1950 to 1990 (Grau et al. 2004). These abandoned lands generally reverted to secondary forests through natural succession (Marrero 1950, Weaver and Birdsey 1990). As a consequence, the area of Puerto Rico under forest cover, which in the mid-1940s was only 13%, increased to 30% by 1980 (Weaver and Birdsey 1990). A similar trend related to forest regeneration on degraded pastures has occurred in the moist lowlands of central and eastern Colombia (Etter et al. 2005). In the Brazilian Amazon, secondary forest area is estimated to have increased five-fold from 2,900,000 in 1978 to 16,100,000 ha in 2002 (Neeff et al. 2006). For example, one municipality (Igarapé) in Pará, Brazil reached 73% secondary forest cover by the late 1990s (Neeff et al. 2006). While often this increase in secondary forest cover represents a net gain in total forest cover with the loss of agricultural and other cleared lands, in many areas, such as the Brazilian Amazon, an increase in secondary forest cover is directly related to the loss of mature or primary forest.

An increase in secondary forest cover on formerly cleared lands for agriculture or pasture typically represents an increase in forest biomass and associated C stocks. Regional C fluxes are often estimated as gains or losses of C stocks at the landscape scale (Houghton et al. 1991, Fearnside and Guimarães 1996). Carbon gain or release at the landscape scale is estimated as the change in carbon content due to changes in biomass of the vegetative cover. Aerial and satellite-based remote sensing technologies have been used for estimating forest successional class, biomass, and geographical extent in Costa Rica (Helmer et al. 2000) and the Brazilian Amazon (e.g., Houghton et al. 2000, Neeff et al. 2006). Determination of the net carbon balance of tropical forests requires certainty in biomass estimates of the different land cover types, accurate estimates of the spatial extent at a landscape scale and on the rates of conversion, and ways to extrapolate from plot-level studies to regional scales (Powers et al. 2004). Secondary forest biomass, and hence their C stocks, are affected by a suite of environmental and ecological factors which are discussed in this chapter.

2.3 Tropical Forest Carbon Stocks and Their Measurement

The carbon budget of a tropical forest is affected by the balance of C inputs (primary production) and C outputs (respiration, decomposition, erosion, leaching losses), and can be expressed by a simple donor-controlled model:

$$\frac{\partial C}{\partial t} = \text{Inputs} - kC$$

where $\partial C/\partial t$ is the rate of change in C content over time and kC represents the output term, where C is the standing stock of carbon and k is a decay constant. In non-steady state systems, C accumulates when the rate of inputs into a particular pool exceeds outputs and/or the rate of outputs decreases, either by a reduction in the standing stock or a decrease in turnover time ($1/k$). Attempts to increase C storage in an ecosystem will often involve the transfer of C into pools with longer turnover times, which increases the amount of time a C molecule is not in the atmosphere where it can act as a greenhouse gas. The potential for tropical forests to remove C from the atmosphere and store it effectively depends on net primary productivity and the residence time of C in each reservoir. An actively growing forest functions as a C sink as trees increase in height and diameter, and the amount of C fixed is greater than C lost in respiration and in decomposition of litterfall. Carbon uptake by plants can be defined as the net annual removal of C from the atmosphere per unit of land area, which can measured by a change in biomass over time (Fearnside 1996).

A forest in C balance, or steady-state, may also store C if the residence time of C in the largest stocks is long. The largest and longest-lived C reservoirs in tropical forests are soil organic matter and standing live aboveground biomass (tree boles). Tropical soils can contain more than twice as much C as in aboveground biomass (Post et al. 1982) and have the potential to store C at great depths (Nepstad et al. 1994, Jobbágy and Jackson 2000). The residence time of C in tropical soils varies greatly with soil depth and soil type. In the highly weathered Oxisols and Ultisols, mean residence time of bulk soil C in the top 10 cm is on the order of decades (Trumbore 1993, de Camargo et al. 1999, Amundson 2001), while residence times of up to 20,000 years have been reported for mineral horizons at depth in some volcanic soils in Hawaii (Torn et al. 1997). Substantial C stocks can be stored in large trees (> 80 cm in diameter) which may live up to 1,000 years (Chambers et al. 1998), though 300–500 years may be more representative of an average lifetime for large canopy and emergent trees (Fichtler et al. 2003).

A common method for studying the successional regeneration and temporal C dynamics of secondary forest on abandoned lands is the use of chronosequences. Using this method, forest stands of different ages are selected, using age as a proxy for successional time. An important assumption of this method is that the patterns observed across the chronosequence will be comparable to the patterns occurring at one site over time. Another common method is paired-site comparisons, where secondary forest sites or plantations are compared to primary forests, typically in close proximity. In this chapter we include data from both chronosequence and paired site studies.

2.3.1 Aboveground Carbon Stocks

Carbon stocks in vegetation are generally calculated as 50% of dry biomass; or 45% for young secondary forests due to the presence of a larger proportion of early succession trees with lower wood density than in mature forests (Fearnside and Guimarães 1996, Alves et al. 1997). Plant biomass can be measured directly by harvesting and weighing all components in a specified area, or calculated from allometric equations using measured parameters such as diameter-at-breast-height, wood volume, and stem height. The advantages and uncertainties associated with different methods of measuring secondary tropical forest and plantation biomass have been evaluated in detail elsewhere (e.g., Brown and Lugo 1984, Brown et al.

1989, 1995b, Chambers et al. 2001, Keller et al. 2001, Losi et al. 2003, Brandeis and Suárez Rozo 2005). Significant stores of C can also be found in root biomass (Nepstad et al. 1994), in standing and fallen coarse woody debris (Delaney et al. 1998, Clark et al. 2002, Keller et al. 2004, Rice et al. 2004) and in lianas, particularly in disturbed forests (Gerwing and Farias 2000, Read and Lawrence 2003a).

2.3.2 Belowground Carbon Stocks

Carbon in soils exists in organic forms: soil organic matter (SOM) and dissolved organic carbon (DOC), or in inorganic forms: carbonates, with a minor proportion as CO_2 gas in soil pores or dissolved as carbonic acid in soil water. Most highly weathered tropical soils are acidic, and so the major component of soil C is organic. In the drier tropics or in young soils derived from limestone, karstic soils can contain significant amounts of inorganic C (Crowther 1980). In this chapter we discuss changes in organic C pools, which are the focus of the majority of tropical studies on highly weathered soils. Changes in C storage are affected by changes in the rate of organic C inputs (e.g., above- and belowground litter, dissolved organic C) and outputs (e.g., C mineralization, leaching losses, erosion). The residence time of C in soils is regulated by a variety of physical, biological, and chemical mechanisms, which affect microbial decomposition and losses of C by other means (reviewed in Paul 1984, Oades 1988, Sollins et al. 1996, Krull et al. 2003, Mikutta et al. 2006).

Soil organic matter (SOM) is a heterogeneous mixture of plant, animal and microbial residues in different stages of decomposition (Stevenson 1982, Kögel-Knabner 1993). In addition to acting as one of the largest terrestrial reservoirs of C (Post et al. 1982, Eswaran et al. 1993), SOM is an important contributor to soil fertility as a source of plant nutrients and through its enhancing effects on soil structure, its high water holding capacity, and buffering capacity (Tiessen et al. 1994). SOM also provides energy and nutrients for microbial metabolism (Lovley et al. 1996, Scow 1997) and plays an important role in metal contaminant and pesticide binding due to its high sorptive capacity (Balabane et al. 1999, Ahmad et al. 2001, Farenhorst 2006). SOM concentrations were traditionally measured by wet oxidation (Walkley and Black 1934). More commonly used today are the loss of ignition method for SOM (Dean 1974), where the amount of C is calculated as 50% of the SOM mass, and by gas chromatography, which is a much faster, automated method (Byers et al. 1978, Nelson and Sommers 1996). The C content of a particular soil profile is then calculated by multiplying % C by the bulk density of the soil sampled and by the depth of the soil core and summed over the total depth in question (Veldkamp 1994).

2.4 Trends in Carbon Stocks with Reforestation: Data Analysis

We surveyed the recent literature and updated secondary forest biomass and soil C data presented in Silver et al.'s (2000a) review. In that review, data of above and belowground C accumulation from secondary forests across the tropics was analyzed with respect to time by prior land use abandonment, prior land use type, and annual rainfall. We extracted their data from studies in Neotropical sites and added about 90 data points to their aboveground biomass and more than 100 data points to their soil C database from new studies published since their review. The abundance of studies on C dynamics with reforestation being published in the last 7 years testifies to the importance this topic has gained.

Data for this chapter were collected from chronosequence and paired site studies. We included forests and plantations growing on former croplands and pastures resulting from unassisted as well as assisted, unmanaged and actively managed regeneration. We tested for the effect of previous land use (agriculture, pasture or cleared), present cover type (secondary forest versus plantation), and life zone. For consistency, we followed the site classification used by Silver et al. (2000a): life zones were based on annual precipitation, and included dry (<1,000 mm/yr), moist (1,000–2,500 mm/yr), and wet (>2,500 mm/yr). While a lot more studies have been published since Silver et al.'s (2000a) review, some of the same methodological constraints persist. For example, data are still very strongly skewed towards the first two decades of regrowth and to the wet and moist life zones. In addition, aboveground biomass estimates in individual studies are likely influenced by the authors' choice of allometric equation and plot size (for discussion of uncertainties surrounding tropical forest biomass measurement and estimation, see Brown et al. 1995b, Alves et al. 1997, Chambers et al. 2001, Keller et al. 2001, Brandeis and Suárez Rozo 2005).

To identify trends in aboveground C accumulation, we ran regressions on C by life zone, past land use and cover type (planted versus secondary succession) over time since abandonment. We use site age and time since abandonment interchangeably. To avoid the effect of uneven representation of past land uses across biomes and vice versa in our dataset, we also tested the effect of land use within moist forests only, which had the highest representation of all land uses (59 agricultural sites, 21 cleared, 19 pasture), and the effect of life zone on agricultural lands only, which had the highest representation of all life zones (17 wet forests, 59 moist, 28 dry). To fulfill assumptions of normality, we used the natural log of aboveground C and site age. All secondary forest and plantation ages were used, but we did not include mature or primary forests in our regressions with time as we are uncertain of their real age.

To examine patterns in soil C with reforestation, we compiled data from studies that reported either soil C content (on a per area basis) or both soil C concentrations (on a per soil mass basis) and bulk density values. SOM values were converted to C by multiplying by 0.5. Our more recent literature search increased the range of soil depths reported (from 0.05 to 8.0 m) from the Silver et al. (2000a) review, although most studies still sampled in the top 0.50 and 0.25 m. Following Silver et al.'s (2000a, 2000b) technique, we used regressions to determine soil C relationships with depth to standardize all studies to a common depth of 0.25 m. The wet and dry life zone equations were derived from the literature dataset. Soil C in the wet life zone decreased with depth according to the linear equation $y = 1.29 + 0.84x$ ($r^2 = 0.66$, $p < 0.0001$, n = 98) where y= log C and x = log depth. For the dry forests, the relationship was $y = 2.34 + 0.59x$ ($r^2 = 0.86$, $p < 0.005$, n = 7).

The literature dataset gave a weak relationship ($r^2 = 0.28$) for the moist forests, so we used an equation derived from detailed soil C profiles sampled every 10 cm to 1.0 m depth in 18 moist forests in Puerto Rico, $y = 2.36 + 0.52x$ ($r^2 = 0.66$, $p < 0.0001$, n = 180) (Marín-Spiotta, unpublished data). These sites included primary forests and secondary forests aged 10, 20, 30, 60 and 80 years and so the equation incorporates age variability such as that found in our larger database. We used these equations to calculate multipliers for the maximum depth or the depth closest to 0.25 m, whichever was smallest reported, for each site to adjust soil C stocks to the standardized depth. We report soil C data for forests older than 3 years.

All statistical analyses were conducted on JMP IN version 5.1 software (SAS Institute). We tested for relationships between soil C and forest age (years), life zone (wet, moist, dry), and previous land use (pasture, agricultural crops, cleared and abandoned), and present cover type (plantation, secondary forest). We compared mean soil C stocks between secondary forests, plantations and mature forests used as end-members in the cited studies, and between life-zone using Wilcoxon/Kruskal-Wallis nonparametric test because C stocks were not normally distributed and variances between sample sizes were unequal.

2.4.1 Aboveground Biomass Accumulation: Results and Discussion

Aboveground C accumulation showed significant log-linear trends with time for all life zones and past land uses (Table 2.1, $p<0.0001$). Only planted cover types did not show a significant trend with time. Moist and wet forests had greater rates of C accumulation than dry forests overall. During the first 20 years of succession moist forests accumulated C fastest than wet forests, while wet forests accumulated more total C in the long-term (Fig. 2.1). When comparing forests growing on agricultural

Table 2.1 Best fit regression equations for aboveground carbon (Mg/ha) with time following tropical reforestation. See Appendix 1 for data

Parameter	Equation	r^2	p	n
All data	Aboveground C = −4.0 + 18.6*(ln AGE)	0.44	<0.01	190
Life zone				
Moist forests	Aboveground C = 3.0 + 17.7*(ln AGE)	0.53	<0.01	107
Wet forests	Aboveground C = −15.5 + 24.0*(ln AGE)	0.48	<0.01	47
Dry forests	Aboveground C = −2.9 + 5.9*(ln AGE)	0.48	<0.01	28
Past land use				
Agriculture	Aboveground C = −1.5 + 15.1*(ln AGE)	0.43	<0.01	104
Pasture	Aboveground C = −38.9 + 28.7*(ln AGE)	0.60	<0.01	41
Cleared	Aboveground C = −17.4 + 37.8*(ln AGE)	0.77	<0.01	25
Cover type				
Plantations	n.s.			10
Secondary forests	Aboveground C = −4.3 + 18.6*(ln AGE)	0.44	<0.01	182

Note: n.s., not significant

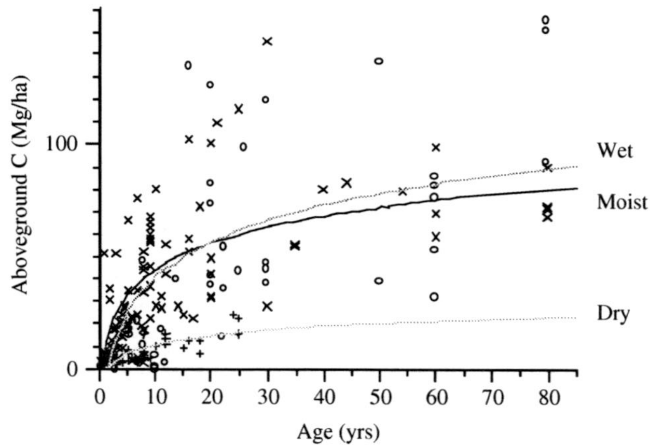

Log curve fits are shown for aboveground C accumulation in secondary tropical forests. + = dry forest, x = moist forest, o = wet forest.

Fig. 2.1 Aboveground carbon accumulation during secondary succession by life zone

land only to control for the effects of past land use, wet forests clearly had the highest rates of aboveground C accumulation, while dry forests had the lowest (Fig. 2.2).

Among past land uses, cleared lands showed the highest overall rates of aboveground C accumulation during 80 years of succession (Fig. 2.3). During the first 10 years of succession, agricultural land appeared to have had higher rates of aboveground C accumulation than pastures, but over longer succession pastures surpassed agricultural lands (Fig. 2.3). This trend was maintained when analyzing only moist forests, which had the most even representation of all past land uses (data not

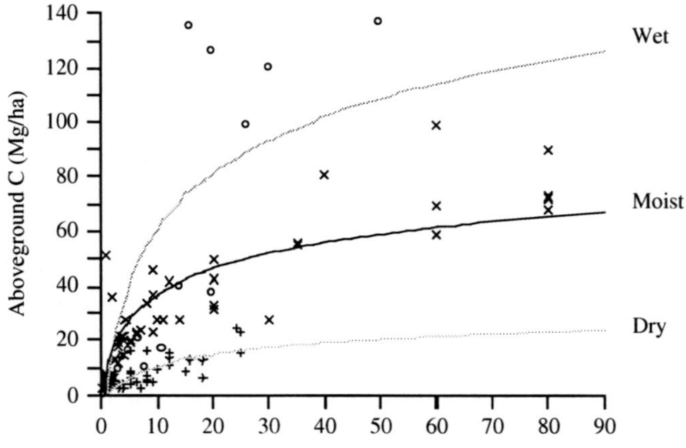

Log curve fits are shown for aboveground C accumulation in secondary tropical forests. + = dry forest, x = moist forest, o = wet forest.

Fig. 2.2 Aboveground carbon accumulation by life zone for agricultural lands only

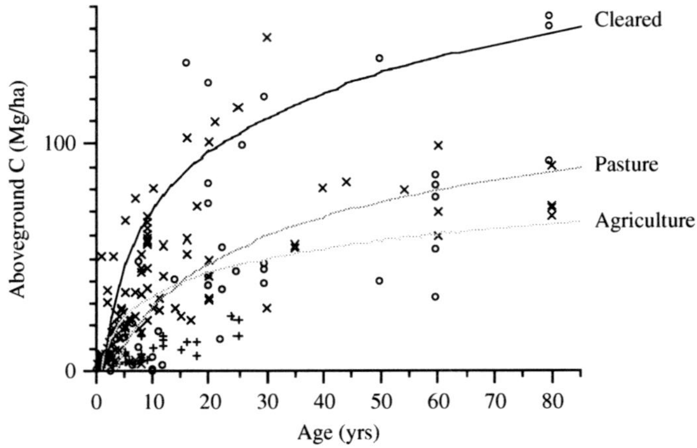

Log curve fits are shown for aboveground C accumulation in secondary tropical forests. + = dry forest, x = moist forest, o = wet forest.

Fig. 2.3 Aboveground carbon accumulation during secondary succession by past land use

shown). In an ANOVA including all factors, time, life zone, past land use and cover type were all significant ($p < 0.05$, $r^2 = 0.55$).

Greater aboveground C accumulation in wet and moist forest life zones than dry ones is consistent with the first two biomes being more productive overall (Brown and Lugo 1982). The differentiation between wet and moist forests was much clearer when examining only agricultural land than when all past land uses were included, showing the importance of past land use for regeneration in these forests. This also implies that wet and moist forests may have similar potentials for aboveground C accumulation during secondary succession depending on the local mosaic of land use history.

Our results comparing previous land uses were more surprising, and were in contrast to Silver et al. (2000a), who found that agricultural land had the highest rates of C accumulation aboveground and that cleared areas had the lowest rates. This difference could be related to a number of new data points added to the study that had secondary succession on land that had been cleared and then allowed to grow back within < 2 years in the Brazilian Amazon (Lucas et al. 2002), and a number of new data points for land regenerating on agricultural land in a dry forest in the Yucatan (Read and Lawrence 2003a). However, our test of past land use in moist forests only showed that the trend for highest regeneration potential on recently cleared land held, with agricultural and pasture land regenerating at slower rates. Many of the cleared forests were allowed to grow back within a short time period (Lucas et al. 2002), while other land uses like agriculture were maintained for over 50 years in some cases (Schroth et al. 2002). This implies that the length of land use prior to abandonment could be as important, or more so, than the type of previous land use for aboveground C accumulation.

2.4.2 Patterns in Soil Carbon with Reforestation: Results and discussion

On average, moist forests had significantly higher soil C stocks to 0.25 m (95.0 ± 6.4 Mg/ha, mean ± 1 standard error, n = 81) than wet forests (67.2 ± 2.8 Mg/ha, n = 100) and dry forests had intermediate values (67.7 ± 4.03 Mg/ha, n = 12). There were no significant differences between forests growing on former agricultural lands (95.7 ± 8.2 MgC/ha, n = 61) and those on abandoned pastures (71.9 ± 2.5 MgC/ha, n = 83), or between forests ≥ 20 years old (86.6 ± 6.6 MgC/ha, n = 55) and those > 20 years old (71.9 ± 3.1 MgC/ha, n = 109). Across all life-zones, there were no significant differences in soil C between mature forests (90.8 ± 12.7 Mg/ha, n = 29), plantations (72.1 ± 6.5 Mg/ha, n = 31) and all secondary forests grouped together (80.6 ± 4.9 Mg/ha, n = 90).

Time since abandonment had a significant (p< 0.01) but very weak (r^2= 0.05) effect on soil C stocks to 0.25 m (Table 2.2). The strongest relationship we found between age and soil C was for the dry forests, although this life-zone had the smallest number of data points. Wet forests, which comprised the larger sample size, did not show any significant trend, similar to Silver et al.'s (2000) findings. Past land-use also had a significant effect on soil C, with forests on formerly cleared-only land showing the strongest relationship (r^2= 0.41, p = 0.06) while those on former agricultural sites showed a weaker relationship (Table 2.2 and Fig. 2.4). Sites re-growing on former pastures did not show any trend over time since abandonment. With respect to current cover type, both secondary forests and plantations showed significant, but very weak relationships between soil C stocks and site time.

Silver et al. (2000a) reported a slightly more robust relationship (r^2= 0.11, p<0.05) between time and soil C for all ages with a smaller dataset (n = 57) compiled from pantropical studies. While sites with different land use history and across the

Table 2.2 Best fit regression equations for soil carbon (Mg/ha) (in the top 0.25 m) with time following tropical reforestation. See Appendix 2 for data

Parameter	Equation	r^2	p	n
All ages				
All data	SOIL C = 42.8 + 9.3*(ln AGE)	0.05	<0.01	164
Life zone				
Moist forests	SOIL C = 34.9 + 9.3*(ln AGE)	0.09	0.01	70
Wet forests	n.s.			85
Dry forests	SOIL C = 15.0 + 17.0*(ln AGE)	0.64	<0.01	9
Past land use				
Agriculture	SOIL C = 27.1 + 17.2*(ln AGE)	0.12	<0.01	57
Pasture	n.s.			83
Cleared	SOIL C = −23.4 + 35.6*(ln AGE)	0.41	=0.06	9
Cover type				
Plantations	SOIL C = 25.9 + 15.0*(ln AGE)	0.15	<0.05	31
Secondary forests	SOIL C = 38.1 + 10.2*(ln AGE)	0.05	<0.05	90

Note: n.s., not significant

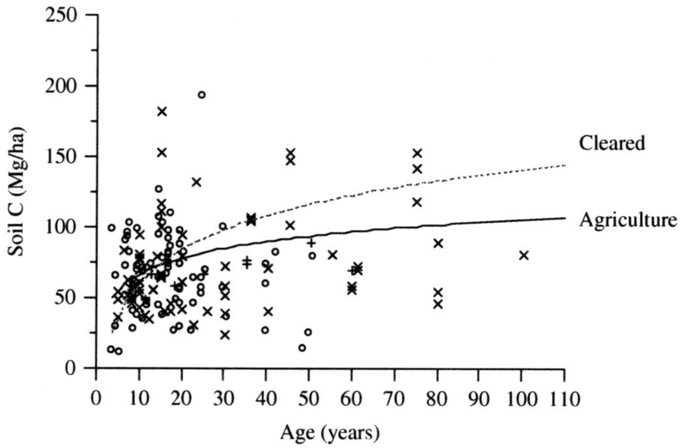

Log curve fits are shown for changes in soil C in secondary tropical forests. Former pastures showed no significant trend. + = dry forest, x = moist forest, o = wet forest.

Fig. 2.4 Soil carbon stocks during secondary succession by past land use

three climatic zones studied accumulated aboveground biomass C during secondary succession, soil C stocks changed very little, or not at all, with time since abandonment. Soils appeared to respond very differently than aboveground biomass and may be more resilient to changes in land use and plant cover. Formerly cleared sites appeared to accumulate soil C faster than agricultural sites, similar to the pattern observed in aboveground biomass. The low predictive power of time suggests that there are other factors influencing rates and direction of soil C changes with secondary succession in the tropics. In the following sections we explore other factors from individual studies that potentially determine the rate and magnitude of soil C change during reforestation.

2.5 Land-Use Change and Tropical Forest Carbon

At a global scale, aboveground biomass and soil C stocks are influenced by climate (precipitation and temperature), soil and vegetation type, and ecosystem age (Schlesinger 1977, Post et al. 1982, Eswaran et al. 1993, Torn et al. 1997, Malhi et al. 1999, Jobbágy and Jackson 2000). Human activities, such as logging and deforestation for agriculture or pasture establishment, mining, dam construction or urban development also affect C storage in forests (Parrotta and Knowles 1999, Caspersen et al. 2000, Houghton and Goodale 2004). The loss of aboveground C during deforestation has been documented extensively (e.g., McWilliam et al. 1993, Fearnside 1996, Houghton et al. 2000, Hughes et al. 2000). After the removal of a disturbance, C stocks may recover or attain a new steady-state below or above pre-disturbance levels. In areas where climate is favorable to forest growth and a seed source is available (Uhl 1987, Holl et al. 2000, Chazdon 2003), natural regeneration of intensively used lands is possible. Many areas of tropical forest which had been previously considered to be untouched primary forest, are instead located on sites of intense human agricultural activity and are in different stages of recovery from disturbance (Bush and Colinvaux 1994, Chazdon 2003).

Multiple environmental, ecological, and anthropogenic factors interact to determine the rate of natural or managed reforestation at any one place, making it difficult to predict the success of reforestation at a particular site, or to estimate regional and global C sequestration potentials of regrowing forests. At the broadest scale, climate affects primary productivity and forest regrowth after a disturbance (Johnson et al. 2000, Anderson et al. 2006). Whether the driving forces behind the initial change in land use and land cover are extrinsic (economic policy) or intrinsic (soil degradation) is likely to affect the regenerative potential of a site. Previous land-use history, the type, intensity, and duration of use, can affect the successional trajectory of a site, and its potential to recover from a disturbance (Nepstad et al. 1991, Aide et al. 2000, Silver et al. 2000). Furthermore, the same factors may have opposite effects on rates of above and belowground C accumulation (Silver et al. 2000). The frequency of a natural or anthropogenic disturbance, such as hurricanes or fire, may also slow down forest regrowth (Zarin et al. 2005).

In the following sections, we discuss recent literature on the factors affecting tree biomass, coarse woody debris, litterfall, roots, and soil C stocks during reforestation in the Neotropics. We refer the readers to Post and Kwon's (2000) review for the potential for C sequestration after agricultural abandonment in soils under natural or replanted perennial vegetation (forests and grasslands) across different global climatic zones. More recent reviews by Pregitzer and Euskirchen (2004) and Anderson et al. (2006) explore ecosystem C and biomass C changes with forest age globally by biome, although they do not distinguish between the type of disturbance before succession.

2.6 Trends in Aboveground Carbon

2.6.1 Tree Biomass

2.6.1.1 Natural Regeneration

The success and rate of secondary forest regeneration depends on a multitude of factors, including prior land use, climate, available seed source, and soil fertility. While early secondary succession can be fast, with biomass accumulations estimated up to 25–50 t/ha on 5-year-old sites, and up to 75–150 t/ha on 15-yr-old sites throughout the Amazon basin (Neeff 2005), subsequent recovery may be slower, and can depend on species composition and forest type (Gehring et al. 2005). Typically, forest regeneration on former croplands and pastures results in net C sequestration aboveground due to greater C stocks in longer-lived trees with the accumulation of woody biomass. In Silver et al. (2000a), prior land use had a significant effect on aboveground biomass accumulation, with C accumulation on abandoned agricultural fields > abandoned pastures > previously cleared sites. Compared to other land uses, grazing can lead to greater soil compaction (Spiel 1996) and slower recovery of forest height and basal area (Aide et al. 1995, Zimmerman et al. 1995, Stern et al. 2002, Chazdon 2003). In agricultural fields, different crops can have "signatures," such as root exudates, associated species, and effects of fertilizers and herbicides, all of which can influence soil properties and affect subsequent succession (Myster 2004). The distinction between prior land-uses is sometimes confused because conversion of forest first to agriculture and then to pasture once soils become nutrient depleted is common across tropical forests (Buschbacher 1986, Uhl et al. 1988, Fearnside 1993, Thomlinson et al. 1996, Carpenter et al. 2001). In some areas, like the Brazilian Amazon and in Colombia, direct conversion to cattle pasture was more common in the last 40 years (Fearnside and Guimarães 1996, Etter et al. 2005), but today, soybean cultivation is one of the main drivers of deforestation (Hecht 2005, Nepstad et al. 2006, Steward 2007). Other factors affecting regeneration include disturbance history, landscape mosaic, and physical environment (Lugo 1988, Murphy and Lugo 1995, Marcano-Vega et al. 2002).

Land use history also has an effect on C storage through its influence on successional trajectories in secondary Neotropical forests. Although successional forests may recover structural characteristics in as little as 20–30 years (Guariguata and Ostertag 2001, Kennard 2002), species composition may remain distinct in old secondary forests (Zou et al. 1995, Aide et al. 2000, Lugo and Helmer 2004, Marín-Spiotta et al. 2007). In a study of secondary dry tropical forests, Read and Lawrence (2003a) related aboveground biomass to forest age, while forest structure was most affected by water availability and anthropogenic disturbances. After a disturbance, different regeneration pathways could affect biomass accumulation (Uhl et al. 1981, Lucas et al. 2002). In a wet secondary forest chronosequence on abandoned pastures in Puerto Rico, 80-year old secondary forests had greater aboveground biomass than primary forests because of the dominance of woody species in the former, and a heavy dominance of palms in primary forests (Marín-Spiotta et al. 2007). Shifts in life-form and an increase in the abundance of palms in the seedling community with secondary succession have been described in wet tropical forests in Costa Rica (Capers et al. 2005), which may affect aboveground structure and biomass as the forests age. A comparison of species composition scenarios for a forest on Barro Colorado Island, Panama, by Bunker et al. (2005) found that future aboveground C storage could vary by several hundred percent, depending on species composition. Thus, which species predominate during secondary succession may have a large influence on the future C storage of these lands. A chronosequence study in the Brazilian Amazon showed that a history of slash and burn agricultural had a large effect on forest structural characteristics, affecting the full recovery of aboveground biomass (Gehring et al. 2005). Use of fire and mechanization during clearing have been shown to slow down recovery of biomass (Uhl et al. 1981, 1988, Zarin et al. 2005). The legacy of land-use history can persist through a natural disturbances such as hurricanes (Boucher et al. 2000, Pascarella et al. 2004).

Forest growth generally occurs at different rates during early versus later stages of succession. Early phases of succession are generally observed to be the times of greatest C accumulation in aboveground biomass, as observed in high-altitude tropical forests of Ecuador (Fehse et al. 2002a), and stem density has been seen to peak at intermediate age classes in moist tropical forests (Saldarriaga et al. 1988, Mizrahi et al. 1997, Denslow and Guzmán 2000, Kennard 2002, Ruiz et al. 2005,). Silver et al. (2000a) found that overall aboveground biomass had significantly faster biomass accumulation during the first 20 years of succession (6.17 Mg ha^{-1} yr^{-1}) than the subsequent 60 years. Nonetheless, tropical forests generally show relatively fast aboveground growth, up to 70% of mature forest height and basal area can be reached in 25 years, as observed on prior agricultural lands in moist forests of the Bolivian Amazon (Peña-Claros 2003) and dry Bolivian forests (Kennard 2002). Rapid early growth during secondary succession can lead to 50% recovery of aboveground biomass in a variety of biomes, from moist (Gehring et al. 2005) to dry (Read and Lawrence 2003a), though Read and Lawrence (2003a) estimated that mature-forest levels of biomass could take over 100 years to reach in dry forests in the Yucatan because of slower growth at later stages of succession. However, many secondary forests are much younger than 25 years, and may never reach maturity before they are cleared again, with great implications for long-term C sequestration. Neeff et al. (2006) estimated that the average age of secondary forests in the Brazilian Amazon in 2002 was 5 years.

Soil fertility can be an important factor affecting forest regeneration. Tropical soils are typically old, highly weathered Ultisols and Oxisols that are poor in mineral-derived nutrients such as phosphorous, calcium, and other base cations (Brady and Weil 2002). Prior land uses such as repeated fire and long-term pasture use may also lead to losses of nitrogen from soils. In a fertilization experiment on abandoned pastures in the Brazilian Amazon, Davidson et al. (2004) found that both grasses and early successional trees during early succession (6 years) responded positively to nutrients, especially nitrogen. Another study in the Brazilian Amazon (Gehring et al. 1999) also found a positive response of vegetation in secondary succession to nutrient additions, but in this case phosphorous was the main limiting nutrient. Gehring et al. (1999) found that different woody species had varying biomass responses to fertilization, but that grasses had the best response and gained a competitive advantage with increased nutrients. Similarly, Uhl (1987) found that nutrient additions favored only grasses on abandoned pastures in Venezuela. In all cases, it is clear that soil nutrient availability can affect both the trajectory of succession and the amount of biomass accrual.

2.6.1.2 Plantations

Planting native and exotic timber species on abandoned pastures is a popular approach for reforestation across the Neotropics. Because of C sequestered in plantations, the proven success of prior large-scale plantations, and a developed professional forestry sector, Latin America has been identified as a potentially key player in global C trading (Wright et al. 2000). Planting woody species can greatly increase the rates of aboveground C accumulation relative to natural succession, especially in areas where herbaceous species such as grasses and ferns can dominate unmanaged succession on abandoned pastures (Aide et al. 1995, Cusack and Montagnini 2004, Myster 2004). Herbaceous species have much lower aboveground C accumulation than woody species (Aide et al. 1995). Planted woody species can improve physical conditions, shading out herbaceous species (Holl 1998, 1999), reducing soil temperatures, and increasing soil nutrient and physical properties, increasing recruitment of other woody species in the understory and promoting forest succession (Oliver and Larson 1996). Prior land use and extent of pasture use can have a strong impact on subsequent C and biomass accumulation in plantations as well, similar to issues discussed above for natural succession (Silver et al. 2004).

There has been extensive study of how the species selected for planting affect forest regeneration and C accumulation. Different species have different growth patterns, making rates of aboveground C accumulation highly variable across plantation types. Aboveground C accumulation can range from as little as 0.8 to as much as 15 Mg C ha^{-1} yr^{-1} during the first two decades of plantation establishment (Lugo et al. 1988). The species planted also have a strong effect on later composition of the stand, with planted timber species an important component of forest species diversity 60 years after planting (Silver et al. 2004). Both native and exotic species have been used to plant abandoned pastures. One advantage of some exotics is that they are fast growing and rapidly accumulate C. For example, a 20-year old teak plantation (exotic species) in Panama averaged 120 Mg/ha in aboveground C (Kraenzel et al. 2003). Due to the importance of planted species in the composition of the forest, there has been an emphasis on planting native species in monoculture and mixed stands. A study of 12 and 13-yr old native species timber plantations in Costa Rican lowland forests found that most species had greater C accumulation in mixed stands than in single species stands. Across three experimental blocks with combinations of 9 native species, mixed stands (64 ± 13 Mg C/ha) had greater C accumulation after 12+ years than pure stands (50 ± 9 Mg C/ha) ($p = 0.1$) (Redondo-Brenes and Montagnini 2006). In this study, plantations with fast-growing species accumulated more C in early stages of plantation development (<10 years), while plantations with slower-growing species were projected to store more C in the long-term. Similarly, mixed species plantations in Hawai'i with Eucalyptus and Albizia (an N-fixer) had greater biomass accumulation after 17 years than monocultures of either species (Kaye et al. 2000). Since species are planted, successional patterns and rates of aboveground biomass accumulation can differ from natural secondary regeneration. Silver et al. (2004) found that timber trees grew fastest during the second 33 years than the first 22 years, indicating that C accumulation is important during later stages of succession. This contrasts with findings for natural succession, as discussed above.

2.6.2 Litterfall

Despite a considerable amount of work on changes in vegetation structure during tropical secondary forest development using chronosequences, where sites vary in time since forest regrowth, most studies of litterfall do not use this approach (but see Lawrence 2005b). More commonly, studies compare secondary forest of a single age to plantations or to primary forest (e.g., Cuevas et al. 1991, Lugo 1992, Li et al. 2005), or compile data worldwide across forests of different ages (e.g., Brown and Lugo 1990). Our understanding of litterfall nutrient inputs to the forest floor and soil during tropical secondary succession seres is therefore limited.

Secondary forests are generally categorized as establishing high rates of litterfall relatively quickly, within the first 25 years of succession (Ramakrishnan and Toky 1983, Brown and Lugo 1990, Guariguata and Ostertag 2001, Lawrence 2005b), and soon after plateau to a production rate of a primary forest (Ewel 1976). Moreover, litterfall mass is likely to reflect development of biomass during succession (Ewel 1976), with young forests generally having high rates of litterfall in the first 20 years after abandonment (Brown and Lugo 1990, Guariguata and Ostertag 2001). For some recent litterfall studies, see McDonald and Healey 2000, Sánchez-de León and Zou 2004, Silver et al. 2004, Li et al. 2005. Typical values for Neotropical secondary forests range from 6–10 Mg ha/yr, but there is wide variation depending on forest type, age, and climate. Avenues for further study might be developing synthetic models to address how these environmental variables affect litter mass and nutrients, seasonal patterns (Read and Lawrence 2003b), and more long-term studies that directly follow a given site or chronosequences over time. Litterfall has also been shown to influence forest tree seedling and sapling mortality in mature forests (Clark and Clark 1991, Guariguata 1998, Drake and Pratt 2001, Gillman et al. 2004), and its effects on secondary forest community structure would be worth investigating.

2.6.3 Coarse Woody Debris

Very little work has been done on fine and coarse woody debris inputs, stocks, or decomposition during tropical succession. Unfortunately, not all studies standardize the diameter ranges of coarse and fine woody debris, but the term usually includes both standing dead and fallen debris (Harmon and Sexton 1996). During secondary succession, coarse woody debris should be accumulating, but initially may be quite low as decomposition rates and slow inputs may initially diminish stocks after land clearing, and thus has been described as a U-shaped curve (Janisch and Harmon 2002). It has been suggested that coarse woody debris does not accumulate in an ecosystem until trees are at least 10 cm diameter at breast height (Eaton and Lawrence 2006).

Typical stocks of woody debris over succession are unknown due to a paucity of chronosequence studies. Eaton and Lawrence (2006) measured coarse woody debris in forests of different age and land uses in Mexican dry forest. Woody debris stocks in 1–16 year old secondary forests (agricultural fallows) were 5.35–8.61 Mg/ha, with approximately 75% of the total woody debris as coarse in all sites. Using different methods, DeWalt et al. (2003) report downed coarse woody debris increased with forest age, ranging from a volume of 6.5 m^3/ha in 20 year old forest to 35 m^3/ha in old-growth forest. One complication in some of these studies is woody debris is affected by that disturbance histories (e.g., Harmon et al. 1995) as well as by life zone (Delaney et al. 1998), and these often co-vary with secondary forest age.

More data are available on stocks of woody debris in old-growth forests, although studies vary in methodology and whether they report only fallen debris or include standing dead trees. For mature Neotropical forests, values range from 2.43 Mg/ha in a Venezuelan dry forest to 61.44 Mg/ha in floodplain forest (for some examples, see Harmon et al. 1995, Delaney et al. 1998, Clark et al. 2002, 2003, Keller et al. 2004, Rice et al. 2004). In addition, the allocation patterns of woody debris are not well studied. Estimates range from 13.5% to 33% of total aboveground biomass in tropical forests (Harmon et al. 1995, Clark et al. 2002, Eaton and Lawrence 2006). A significant portion of total woody debris biomass may also be in the fine woody fraction ($\leq$ 10 cm diameter); this was estimated to be 25% in Mexican dry forests (Eaton and Lawrence 2006). These numbers indicate that woody debris can be a significant C pool that is often overlooked.

2.7 Trends in Belowground Carbon

2.7.1 Roots

Root biomass during secondary succession has not received as much attention as aboveground biomass. Like other areas, most studies of root biomass compare a single aged forest to a mature forest, or a secondary forest to a plantation (generally monocultures). Thus, there are almost no studies that examine root biomass along a chronosequence (but see Jaramillo et al. 2003 for a 30 years chronosequence), and how patterns of root biomass in mixed species plantation stands compare to those in naturally regenerating stands is generally unknown.

It is hypothesized that early in succession more biomass should be allocated to fine roots and leaves for resource capture, and thus, that ratios of roots to shoots should be greater in younger secondary forests (Yan et al. 2006). Jaramillo et al. (2003) showed that in 8–30 year old secondary forests there was very little biomass allocated to roots below 40 cm in depth, suggesting that these forests may be quite susceptible to disturbances. Hertel et al. (2003) found that secondary montane forests in Costa Rica had shallower organic layers, lower root biomass, and substantially lower nutrients than mature forests. It appears that some forests such as this montane one do not return to biomass levels after several decades of succession, while other forests may (Carvalheiro and Nepstad 1996); however, as mentioned previously our understanding of the recovery process is quite limited. For some recent studies in secondary forests, see Hertel et al. 2003, Jaramillo et al. 2003, Powers 2004, and Adachi et al. 2006. Unfortunately, biomass estimates are difficult to compare among studies due to differences in the definition of fine roots, different sampling depths, and different procedures regarding the separation of live and dead roots.

2.7.2 Soil Carbon

Changes in soil carbon stocks during succession are more difficult to observe than changes in aboveground biomass, because the longer residence time of carbon in the soil may outlast the duration of land use, and because of potential storage at greater depths (beyond 1 m) than those usually sampled (Nepstad et al. 1994, Trumbore et al. 1995, Silver et al. 1996). Both net gains and no net change in soil C have been reported with reforestation in the Neotropics (e.g. Mexico, Hughes et al. 1999 and Ecuador, Rhoades et al. 2000). The wide variation in results may be attributed to differences in land use type and intensity, soil properties, and time since land-cover conversion, although the mechanisms for these observed patterns have not been well defined. Land use change can also affect the turnover time and distribution of different soil C fractions, even when changes in the bulk soil C pool are undetectable (Bashkin and Binkley 1998, Binkley and Resh 1999, Marín-Spiotta 2006, see following section on advances in soil methods). In addition to serving as a reservoir for C, soil organic matter (SOM) is a main source of plant nutrients and an important contributor to soil fertility (Tiessen et al. 1994), thus understanding how agricultural abandonment and reforestation affect soil C stocks is important not only for C sequestation but also for restoration of degraded soils.

The type and intensity of prior land use may affect the direction and rate of soil C content change during reforestation (Lugo et al. 1986). In a review of pantropical secondary forests, Silver et al. (2000a) previously reported slower soil C accumulation rates on former croplands than on pastures, opposite to trends observed for aboveground biomass C. Soil C accumulation in cultivated soils may be slower because conventional agricultural practices typically leads to a severe decline in soil C stocks (Mann 1986) relative to pastures, where belowground C inputs may equal or rival those of forests (Fisher et al. 1994, Trumbore et al. 1995, Neill et al. 1997, Schwendenmann and Pendall 2006). Some studies in Costa Rica (Guggenberger and Zech 1999) and Puerto Rico (Lugo et al. 1986, Weaver et al. 1987, Silver et al. 2004) have reported increases in soil C associated with an increase in plant biomass after pasture abandonment. Rhoades et al. (2000) reported that lower montane secondary forests in Ecuador on abandoned sugar cane and pasturelands accumulated soil C at a rate of 1.9 Mg/ha/yr, leading to the recuperation of primary forest soil C stocks in as little as 20 years. In contrast, a study in tropical wet Australia, found long-lasting effects of soil physicochemical changes after pasture use that inhibited recuperation of undisturbed forest soil C stocks (Rasiah et al. 2004). Even after 40 years under abandoned pasture with no grazing, and 30 years of abandoned pasture plus 10 years under replanted forest with native species, soil C stocks remained significantly below those of undisturbed rainforest, likely due to aluminum toxicity and low soil pH.

While aboveground biomass typically increases with secondary forest age (Hughes et al. 1999, Read and Lawrence 2003a), soil C pools often show no relationship with time since agricultural or pasture abandonment, as our data analysis showed. In a seasonally dry region of Mexico, soil C stocks in the top 10 cm did not differ between primary forests, early-successional (10–15 years), mid-successional (20–30 years), or late-successional (60 years) forests regrowing on land formerly cultivated for maize by slash-and-burn (Saynes et al. 2005). Microbial biomass C was greater in early- and mid-successional forests, and also showed seasonal trends. Other studies reported no significant differences in soil C pools between pasture, especially, and secondary or primary forests. In the Cayey mountains of Puerto Rico, gains of new secondary forest C were compensated by losses of pasture-derived C from the soil, resulting in no net change with reforestation or during 80 years of secondary succession (Marín-Spiotta 2006). Using density fractionation and radiocarbon modeling, Marín-Spiotta (2006) found that the majority of soil C was associated with mineral surfaces, which are known to have a stabilizing effect on soil C, and had long residence times, thus the bulk of the C pool was resilient to land-use change. No significant change in soil C stocks has also been reported during secondary succession after abandoned pastures in the Brazilian Amazon (Feldpausch et al. 2004), and slash-and-burn agriculture in Los Tuxtlas, Mexico (Hughes et al. 1999). Hughes et al. (1999) attributed their results to the high C storage capacity of the young volcanic soils in their study sites. These studies suggest that some tropical soils may be resilient to changes in land use and to disturbances that dramatically affect aboveground biomass.

Soil type, in particular differences in texture and mineral content, not only have an effect on bulk soil C content (Post et al. 1982, Eswaran et al. 1993, Neufeldt et al. 2002), but also on its stability and response to land-use change. Using stable C isotopes, López-Ulloa et al. (2005) found that secondary forests on Andisols in northwestern Ecuador lost pasture-derived C and gained forest-derived-C at much faster rates than similarly aged forests on Inceptisols, which appeared to be more insensitive to land-use change.

Land-use history has important effects on soil C stocks, with type and intensity potentially over-riding time since abandonment. A well-replicated study in Ecuador found that the direction of changes in soil C stocks following pasture to forest conversion was best explained by the time a site had been under pasture use (de Koning et al. 2003). Younger pasture soils (< 10 years) had on average 9.3 Mg/ha more soil C than paired secondary forest and plantation sites, while pastures > 20 years had lower soil C content than the forested sites. The difference between pasture and reforested soil C content decreased with pasture age before conversion, pastures 20–30 years old had 18.8 Mg/ha less than forests, and pastures > 30 years had 15.8 Mg/ha less than forests.

The rate and direction of soil C change may vary at different stages of succession, and with soil depth. In a cloud forest chronosequence growing on abandoned maize fields in Oaxaca, Mexico, soils down to 40 cm had the greatest C accumulation rates in the first 15 years of succession, which also coincided with the largest changes in aboveground forest structure. In the next 30 years, the top 20 cm had a net loss of soil C, followed by a smaller increase after 45 years of secondary forest cover (Bautista-Cruz and del Castillo 2005). The lower depths (20–40 cm) showed no patterns with forest age after the first 15 years.

Species composition and diversity can have effects on soil C, in addition to the effects on aboveground biomass discussed previously. Reforestation of native cerrado with pine plantations in Brazil resulted in lower soil C stocks, while reforestation with Eucalyptus increased soil C (Neufeldt et al. 2002). In Hawaii, no net soil C sequestration was documented in young (1 year, Binkley and Resh 1999) or first-rotation (8-year, Binkley et al. 2004) Eucalyptus plantations growing on former sugarcane fields, even after fertilization. Using the natural abundance ^{13}C method, the authors measured equal rates of new plantation C3 inputs and former agricultural C4 soil C losses. Mixed species plantations (Eucalyptus and Albizia, an N-fixer) did, in contrast, accumulate soil C over 17 years, with soil C content positively related to increasing percentage of Albizia stems (Kaye et al. 2000). The authors attributed this to inhibited decomposition of residual sugarcane C in the presence of an N-fixer. Paul et al. (2002) also identified species effects on soil C stocks in plantations established on former cultivated or pastoral lands. Their review suggested that soil C accumulated under hardwoods and some softwoods and nitrogen fixing species, while soil C was lost under some pine plantations.

2.8 Advances in Methods to Study Effects of Land-Use Change on Soil Carbon

Detecting changes in bulk soil C pools with land-use change can be difficult because of the large size of the pool. Differences in methodologies used across studies, including a failure to correct for changes in bulk density associated with changes in soil C concentrations (Veldkamp 1994), also contribute to the wide range of reported results. Estimating changes in C pools by measuring changes in inputs and outputs allows for a more precise assessment of soil C dynamics. Modeling and experimental studies on SOM decomposition and the effects of cultivation on soil C have provided evidence that bulk soil C is a heterogeneous mixture of pools with different chemical properties and residence times in the soil (Jenkinson and Rayner 1977, Parton et al. 1987, Trumbore 1993). As shown in these studies, at least three distinct pools have been identified in soils across biomes and climatic zones: an active or labile pool turning over in one to 5 years, a slow or intermediate pool turning over on a decadal time scale, and a passive pool with turnover time of centuries to millennia. A number of physical and chemical separation methods have been developed to explain the dynamics of the different soil C pools, or soil C fractions (e.g., Greenland and Ford 1964, Elliott and Cambardella 1991, Golchin et al. 1994, Trumbore and

Zheng 1996, Christensen 2001, Paul et al. 2006). Chemical fractionation typically separates soil C pools based on their solubility in water, strong acids, or bases, while physical fractionation methods use differences in density, particle- or aggregate-size.

When coupled with isotopic and other analytical and spectroscopic techniques, fractionation approaches have been successful at describing short- and long-term dynamics of soil C in response to land-use change and other disturbances. Enriched isotope tracers (^{13}C, ^{14}C, ^{15}N) have been applied extensively to track the movement of C or N atoms through different ecosystem components, especially leaf litterfall, microbial biomass, and soil C pools (e.g., Swanston et al. 2002, 2005, Bird et al. 2003, Hanson et al. 2005). Differences in the natural abundance of ^{13}C in C3 and C4 photosynthetic plants can be used to track soil C dynamics when a change in vegetation occurs (Balesdent et al. 1987, Farquhar et al. 1989, Bernoux et al. 1998). For example, in the tropics, most forage grasses in pastures are C4 plants, as are sugarcane and maize crops, while woody plants and forest species are C3. The inputs of different land cover types to SOM can be calculated from differences in ^{13}C /^{12}C ratios (see Vitorello et al. 1989), which can be used as a natural tracer for describing changes in soil C turnover with land-use change in the tropics (Martin et al. 1990, Veldkamp 1994, Neill et al. 1997). In Puerto Rico, Marín-Spiotta (2006) used this technique to show that even though bulk soil C stocks appeared unchanged during a reforestation chronosequence, the SOM pool was not static.

The mean residence time of both above and belowground C stocks can also be estimated from measurements of radiocarbon concentrations in SOM and modeling the decay of ^{14}C emitted from nuclear weapons testing in the 1960s and incorporated into plant biomass during photosynthesis as $^{14}C{-}CO_2$ (Trumbore 1993, 2000, Chambers et al. 1998). Using stable and radiocarbon isotope analyses of low and high density soil C fractions, de Camargo et al. (1999) found that soil C dynamics in a secondary forest resembled those of a primary forest after 16 years, even though aboveground C had still not recovered primary forest levels. In the same Puerto Rican chronosequence mentioned earlier, Marín-Spiotta (2006) used ^{14}C to estimate faster turnover rates of soil aggregates in pastures than in forests.

Spectroscopic techniques have greatly informed our understanding of the chemical makeup of SOM and of the processes leading to its formation and stabilization in soils. Advances in solid-state cross-polarization ^{13}C- nuclear magnetic resonance spectroscopy (NMR) spectroscopy allow for the study of SOM in situ without chemical modification (Kögel-Knabner 2000, Preston 2001, Helfrich et al. 2006). ^{13}C-NMR can provide a first approximation of the relative abundance of organic C functional groups, such as alkyl, O-alkyl, aromatic and carbonyl C, which can be used as indicators for the presence of different compounds, such as carbohydrates, proteins, lignin, lipids, and charred materials (see Baldock et al. 2004). This method, combined with physical fractionation, has been applied successfully to identify chemical changes in different soil C fractions with land-use change across a wide range of climates and soil types (Oades et al. 1988, Golchin et al. 1994, 1995, Guggenberger et al. 1995a, 1995b, Helfrich et al. 2006). Other methods for SOM characterization are Fourier transform infrared spectroscopy (FTIR) and pyrolysis gas chromatography/mass spectrometry (GC/MS). Nierop et al. (2001)

used pyGC/MS to determine the source and degree of decomposition of soil C profiles under organic farming, conventional tilling and pasture. Coupled with isotope ratio MS, pyrolysis-GC/MS has shed light on the fate of individual compound groups during the incorporation of plant litter into SOM (Gleixner et al. 1999).

Chromatographic techniques, such as XAD fractionation (Croué 2004), solid-phase extraction on C-18 columns (Louchouarn et al. 2000), and high-pressure size exclusion chromatography (Chin et al. 1994) are useful for separating soil extracts and dissolved organic matter from plants and in soils by polarity and molecular weight, properties which are related to chemical recalcitrance and decomposability. Generally, low-molecular weight and hydrophobic compounds are thought to represent the most labile fractions of plant litter and SOM (carbohydrates, amino acids, small proteins, organic acids), while the high-molecular weight and hydrophobic compounds represent more recalcitrant materials. These types of compounds identified by chromatography columns have been shown to accumulate under different vegetation types and land-use regimes (Sanger et al. 1997). These and other analytical techniques which can be used to enhance our understanding of soil C dynamics with land-use change have been reviewed recently by Kögel-Knabner (2000), Northcott and Jones (2000) and Poirier et al. (2005).

Results from the studies mentioned above and others using these techniques suggest that bulk soil C measurements are often too coarse to detect changes over forest successional time periods, and may lead to the erroneous conclusion that the soil C pool has not responded to land-use change. The soil C fraction most sensitive to land-use is typically the low density, particulate C fraction, also known as the labile or active pool (Cambardella and Elliott 1993, Alvarez et al. 1998, Guggenberger and Zech 1999, Compton and Boone 2000, Baisden et al. 2002, John et al. 2005, Helfrich et al. 2006). While perhaps making up a small proportion of the total C pool, the light fraction may cycle large amounts of C and thus dominate soil-atmosphere feedback interactions (McGroddy and Silver 2000). A number of studies have detected changes in the distribution and chemistry of different soil C fractions, even when the bulk soil C pool appeared unchanged in tropical, temperate and boreal soils (Binkley and Resh 1999, Neff et al. 2002, Li et al. 2005, Marín-Spiotta 2006). The implementation of physical and chemical fractionation, isotopic and spectroscopic techniques, has increased our understanding of the composition and location of C compounds in the soil organo-mineral matrix, and elucidated many of the chemical, physical and biological stabilization mechanisms controlling root biomass and soil C formation and turnover. The application of these methods to soil C studies in the tropics, however, is still limited.

2.9 Future of Secondary Forests

The C sequestration potential of secondary forests depends on their future fate. As the area of secondary forest cover increases, these will become the main source of land available for future human use. In many regions, most secondary forests are in a fallow stage, that is, they have only temporarily been taking out of the production cycle, so reconversion to non-forest land uses is highly likely (Kammesheidt 2002, Schroth et al. 2002, Walker 2003, Soares-Filho et al. 2004). In Asia, the length of time of swidden fallow secondary forest systems is decreasing, and these are increasingly being converted to cash-crop and cash-tree plantations (de Jong et al. 2001). The age of secondary forests in the Colombian lowland landscape is also increasingly becoming younger (Etter et al. 2005), implying that secondary forests are experiencing multiple cycles of disturbance. With increasing frequency of disturbance, the potential for net loss of C as CO_2 to the atmosphere increases. In regions where the economy has shifted from agriculture to an industrial or service-based economy, such as in Puerto Rico, conversion to agriculture or pasture is no longer a likely threat to secondary forests. Instead, these are being increasingly cleared for urban uses, such as residential and commercial development, or golf courses (Thomlinson and Rivera 2000, Helmer 2004).

Secondary forests have been traditionally undervalued, particularly due to the predominance of exotic species, and so they typically fall below the radar of conservationists and environmental planners. Although species composition of secondary forests may not return to that of primary forests (Lugo and Helmer 2004), there is increasingly more evidence that secondary forests can reach structural and potentially functional characteristics of primary forests in as little as 20 years (Brown and Lugo 1990, Guariguata and Ostertag 2001). The important role of secondary forests in C sequestration, in restoration of forest ecosystem goods and services, and as habitat for biodiversity in fragmented landscapes (Dunn 2004, Veddeler et al. 2005) is increasingly being recognized.

2.10 Summary and Recommendations for Future Studies

We compiled data on aboveground biomass and soil C stocks from Neotropical secondary forests and plantations growing on former agricultural or pasturelands and examined the effect of time since abandonment on C accumulation. Our analysis showed that the predictive power of time since abandonment on soil C stocks was very low. We review a number of studies that did not find changes in soil C with site age. Land-use history (type and intensity), frequency of disturbance, and the use of fire, are potentially more important than regenerative time for determining success, direction, and rates of C sequestration belowground. An identification of these factors at the landscape level coupled with the creation or enforcement of regional land use and zoning plans is important for predicting where resources for reforestation should be focused. In many areas, forest regrowth will occur unassisted in a relatively short timescale. When barriers exist to natural regeneration, conversion to plantations or managed restoration can facilitate the recovery of aboveground biomass. The conservation value of secondary forests needs to be better explored at the landscape level as well, and incorporated into environmental and agricultural planning.

Studies that combine isotopic and molecular techniques with more traditional ecological methods increase our ability to detect changes in soil C stocks and better understand soil C dynamics with land-use change. The application of these techniques to studies in the tropics is still limited. Our understanding of the factors controlling root biomass and soil C with reforestation is a lot less than those on aboveground dynamics. There are many studies across the tropics on different aspects of regeneration, species diversity, seedling survival, structural characteristics (stem density, basal area), but a lot fewer examine soil C. We encourage interdisciplinary collaborations between plant and community ecologists and biologists with soil scientists and chemists to provide a more comprehensive picture of forest dynamics with reforestation. Soil studies should report carbon content on a per area basis or carbon concentrations with bulk density values for the proper conversion to carbon content. Unfortunately, too many studies report %C (or %N) alone per depth, when potential changes in bulk density, especially during land-use change, hinder comparisons of concentrations across treatments.

Unlike for soil C, it is clear that the potential for C accumulation aboveground is high during secondary succession in Neotropical forests, and C accumulation follows a log-linear pattern, with rapid early accumulation. This C accumulation reaches an eventual plateau, which appears to occur within 50 years in dry forests, but can take over 100 years on wet and moist forests. Given similar land use histories, wet tropical forests clearly have the greatest potential for aboveground C accumulation. Our data also suggest that lands that are cleared and rapidly returned to successional processes have the potential to accumulate more C over time than lands that have been in sustained agriculture or pasture. A previous study found that agricultural lands had slightly greater potential for forest regeneration (Silver et al. 2000a), and the authors suggested that these lands were selected for agriculture because they were the richest in soil nutrients. This effect appears to be dependent on the length of time land is held in agriculture, so it may be useful to focus reforestation efforts on areas with minimal time since conversion.

Most studies on aboveground dynamics with reforestation also report basal area, and few report biomass stocks. We encourage authors to attempt to provide biomass values using published allometric equations. While using allometric equations certainly has its caveats and sources of uncertainty, these have been well identified and there is a very large number of biomass equations available for different forest types and life zones, especially for Neotropical sites. The dry tropics are a particularly understudied life zone. We also found almost twice as many studies conducted in the first 20 years after abandonment, but our data shows that older forests may behave differently, so it is important to include sites beyond the first two decades whenever available. As the age of secondary forests decreases, these older secondary forests are becoming a threatened habitat, like primary forests.

Appendices

Appendix 2.1 Aboveground C pools in secondary forests and plantations of known age following abandonment of agricultural and pasturelands and in nearby mature forests in the Neotropics. Life zones are wet (W), moist (M), and dry (D) based on mean annual precipitation descriptions given by authors. Past land uses are agriculture (A), pasture (P). Cover types are secondary forests (S) or plantations (P). Plantations are identified as mixed (M) or single species (S), and native (N) or exotic (E). Aboveground C was calculated as 50% of dry biomass

Age (y)	Life zone	Past Land use	Cover Type	Mixed vs. Single spp	Native vs. exotic	Aboveground Biomass C (Mg/ha)	Location	Reference
0.3	M	A	S			2	Peru	Szott et al. 1994
0.5	W	A	S			1	Mexico	Hughes et al. 1999
0.7	M	A	S			2	Peru	Szott et al. 1994
0.8	M	A	S			2	Guatemala	Tergas and Popenoe 1971
0.8	M	A	S			5	Guatemala	Tergas and Popenoe 1971
0.8	M	A	S			6	Guatemala	Tergas and Popenoe 1971
0.8	M	A	S			50	Guatemala	Tergas and Popenoe 1971
0.8	M	A	S			5	Guatemala	Tergas and Popenoe 1971
0.8	M	A	S			4	Guatemala	Tergas and Popenoe 1971
0.8	M	A	S			2	Guatemala	Tergas and Popenoe 1971
0.8	M	A	S			7	Guatemala	Tergas and Popenoe 1971
0.8	M	A	S			7	Guatemala	Tergas and Popenoe 1971
0.8	M	A	S			7	Guatemala	Tergas and Popenoe 1971
0.8	W	A	S			2	Mexico	Williams-Linera 1983
1	M	A	S			4	Guatemala	Snedaker 1970
1	M	A	S			3	Brazil	Uhl et al. 1988
1	W	C	S			3	Puerto Rico	Silver et al. 1996
1	W	A	S			4	Brazil	Uhl 1987
1.3	M	A	S			4	Brazil	Gehring et al. 1999
1.4	M	A	S			7	Peru	Szott et al. 1994
1.9	M	A	S			6	Brazil	Gehring et al. 1999

2	D	A	S	6	Mexico	Read and Lawrence 2003
2	M	A	S	4	Brazil	Alves et al. 1997
2	M	A	S	6	Costa Rica	Ewel 1971
2	M	S		9	Colombia	Folster et al 1976
2	M	C	S	30	Brazil	Gehring et al. 2005
2	M	A	S	36	Brazil	Lucas et al. 2002
2	M	A	S	7	Guatemala	Snedaker 1970
2	W	A	S	6	Brazil	Uhl 1987
2.3	M	A	S	6	Brazil	Gehring et al. 1999
2.4	M	A	S	12	Peru	Szott et al. 1994
2.5	M	P	S	8	Brazil	Uhl et al. 1988
2.5	M	P	S	4	Brazil	Uhl et al. 1988
2.8	M	C	S	17	French Guyana	Maury-Lechon 1982
3	D	A	S	3	Mexico	Read and Lawrence 2003
3	D	A	S	5	Mexico	Read and Lawrence 2003
3	M	A	S	19	Brazil	Alves et al. 1997
3	M	A	S	17	Brazil	Alves et al. 1997
3	M	C	S	50	Brazil	Gehring et al. 2005
3	M	A	S	11	Guatemala	Snedaker 1970
3	W	A	S	10	Brazil	Uhl 1987
3	W	C	S	4	Brazil	Uhl et al. 1982
3	W	C	S	0	Brazil	Uhl et al. 1982
3	W	C	S	6	Brazil	Uhl et al. 1982
3.5	M	P	S	8	Brazil	Uhl et al. 1988
3.5	M	P	S	15	Brazil	Uhl et al. 1988
3.5	M	P	S	4	Brazil	Uhl et al. 1988
3.6	M	A	S	20	Peru	Szott et al. 1994
4	D	A	S	4	Mexico	Read and Lawrence 2003
4	D	A	S	3	Mexico	Read and Lawrence 2003
4	M	A	S	19	Costa Rica	Ewel 1971
4	M	A	S	21	Mexico	Hughes et al. 1999
4	M	P/A	S	24	Brazil	Lucas et al. 2002

(continued)

Appendix 2.1 (continued)

Age (y)	Life zone	Past Land use	Cover Type	Mixed vs. Single spp	Native vs. exotic	Aboveground Biomass C (Mg/ha)	Location	Reference
4	M	A	S			14	Guatemala	Snedaker 1970
4	M	P	S			9	Brazil	Uhl et al. 1988
4	W		P	S	E	20	Puerto Rico	Lugo 1992
4	W	A	S			14	Brazil	Uhl 1987
4.4	M	A	S			27	Peru	Szott et al. 1994
4.5	M	P	S			27	Brazil	Uhl et al. 1988
5	D	A	S			4	Mexico	Read and Lawrence 2003
5	D	A	S			9	Mexico	Read and Lawrence 2003
5	D	A	S			16	Mexico	Read and Lawrence 2003
5	D	A	S			8	Mexico	Read and Lawrence 2003
5	D	A	S			6	Mexico	Read and Lawrence 2003
5	M	A	S			19	Brazil	Alves et al. 1997
5	M		S			34	Colombia	Folster et al 1976
5	M	C	S			65	Brazil	Gehring et al. 2005
5	M	A	S			18	Guatemala	Snedaker 1970
5	W	A	S			17	Brazil	Uhl 1987
5.5	W		S			16	Puerto Rico	Lugo 1992
6	D	A	S			5	Mexico	Read and Lawrence 2003
6	M	P	S			3	Brazil	Davidson et al. 2004
6	M	A	S			21	Costa Rica	Ewel 1971
5	W	A	S			17	Brazil	Uhl 1987
5.5	W		S			16	Puerto Rico	Lugo 1992
6	D	A	S			5	Mexico	Read and Lawrence 2003
6	M	P	S			3	Brazil	Davidson et al. 2004
6	M	A	S			21	Costa Rica	Ewel 1971
6	M	A	S			22	Guatemala	Snedaker 1970
7	D	A	S			3	Mexico	Read and Lawrence 2003
7	M	P	S			4	Brazil	Davidson et al. 2004

7	M	C	S	75	Brazil	Gehring et al. 2005
7	M	C	S	35	Brazil	Lucas et al. 2002
7	M	A	S	23	Guatemala	Snedaker 1970
7	W	P	S	3	Puerto Rico	Aide et al. 1995
7	W	A	S	22	Mexico	Williams-Linera 1983
7.5	M	P	S	19	Brazil	Uhl et al. 1988
8	D	A	S	7	Mexico	Read and Lawrence 2003
8	D	A	S	5	Mexico	Read and Lawrence 2003
8	D	A	S	6	Mexico	Read and Lawrence 2003
8	D	A	S	16	Mexico	Read and Lawrence 2003
8	M	P	S	4.8	Brazil	Davidson et al. 2004
8	M	P/A	S	52	Brazil	Lucas et al. 2002
8	M	A	S	33	Guatemala	Snedaker 1970
8	M	P	S	44	Brazil	Uhl et al. 1988
8	M	P	S	43	Brazil	Uhl et al. 1988
8	M	P	S	16	Brazil	Uhl et al. 1988
8	M	P	S	2	Brazil	Uhl et al. 1988
8	W	A	S	11	Mexico	Hughes et al. 1999
8	W	P	S	49	Mexico	Hughes et al. 1999
9	D	A	S	5	Mexico	Read and Lawrence 2003
9	M	A	S	45	Brazil	Alves et al. 1997
9	M	C	S	67	Brazil	Lucas et al. 2002
9	M	C	S	65	Brazil	Lucas et al. 2002
9	M	C	S	57	Brazil	Lucas et al. 2002
9	M	C	S	59	Brazil	Lucas et al. 2002
9	M	C	S	57	Brazil	Lucas et al. 2002
9	M	C	S	62	Brazil	Lucas et al. 2002
9	M	C	S	56	Brazil	Lucas et al. 2002
9	M	C	S	62	Brazil	Lucas et al. 2002
9	M	A	S	22	Colombia/Venezuela	Saldarriaga et al. 1988
9	M	A	S	36	Guatemala	Snedaker 1970

(continued)

Appendix 2.1 (continued)

Age (y)	Life zone	Past Land use	Cover Type	Mixed vs. Single spp	Native vs. exotic	Aboveground Biomass C (Mg/ha)	Location	Reference
10	D	A	S			10	Mexico	Read and Lawrence 2003
10	M	C	S			80	Brazil	Gehring et al. 2005
10	M	A	S			27	Guatemala	Snedaker 1970
10	W	P	S			7	Puerto Rico	Marin-Spiotta et al. in press
10	W	P	S			2	Puerto Rico	Marin-Spiotta et al. in press
10	W	P	S			1	Puerto Rico	Marin-Spiotta et al. in press
11	M	P	S			32	Brazil	Alves et al. 1997
11	M	A	S			26	Colombia/Venezuela	Saldarriaga et al. 1988
11	W	A	S			18	Puerto Rico	Cuevas et al. 1991
12	D	A	S			11	Brazil	Read and Lawrence 2003
12	D	A	S			14	Brazil	Read and Lawrence 2003
12	D	A	S			15	Brazil	Read and Lawrence 2003
12	M	C	S			55	Brazil	Lucas et al. 2002
12	M	A	S			41	Colombia/Venezuela	Saldarriaga et al. 1988
12	W	P	S			3	Puerto Rico	Aide et al. 1995
13	W	P	P	M	N	32	Costa Rica	Redondo-Brenes and Montagnini, 2006
13	W	P	P	S	N	25	Costa Rica	Redondo-Brenes and Montagnini, 2007
14	M	A	S			27	Colombia/Venezuela	Saldarriaga et al. 1988
14	W	A	S			40	Brazil	Schroth et al. 2002
15	D	A	S			9	Mexico	Read and Lawrence 2003
15	M	P	S			23	Ecuador	Fehse et al 2002
16	D	A	S			13	Brazil	Read and Lawrence 2003
16	M		S			52	Brazil	Alves et al. 1997
16	M		S			101	Colombia	Folster et al 1976
16	M	C	S			58	Brazil	Lucas et al. 2002
16	W	A	S			136	Mexico	Hughes et al. 1999

17	M	P	S			22	Brazil	de Koning et al. 1999
17	W	A	P	S	E	50	Hawai'I	Kaye et al. 2000
17	W	A	P	M	E	85	Hawai'I	Kaye et al. 2000
17	W		P	S	E	51	Puerto Rico	Lugo 1992
18	D	A	S			13	Mexico	Read and Lawrence 2003
18	D	A	S			7	Mexico	Read and Lawrence 2003
18	M		S			72	Brazil	Alves et al. 1997
18.5	W		P	S	E	85	Puerto Rico	Lugo 1992
20	M	C	S			100	Brazil	Gehring et al. 2005
20	M	A	S			31	Colombia/Venezuela	Saldarriaga et al. 1988
20	M	A	S			49	Colombia/Venezuela	Saldarriaga et al. 1988
20	M	A	S			32	Colombia/Venezuela	Saldarriaga et al. 1988
20	M	A	S			42	Colombia/Venezuela	Saldarriaga et al. 1988
20	W		P	S	E	60	Panama	Kraenzel et al. 2003
20	W	A	S			38	Mexico	Hughes et al. 1999
20	W	A	S			127	Mexico	Hughes et al. 1999
20	W	P	S			75	Puerto Rico	Marin-Spiotta et al. in press
20	W	P	S			84	Puerto Rico	Marin-Spiotta et al. in press
20	W	P	S			43	Puerto Rico	Marin-Spiotta et al. in press
21	M	C	S			109	Brazil- AMZ	Lucas et al. 2002
22	W	P	S			15	Puerto Rico	Aide et al. 1995
22.5	W		S			36	Puerto Rico	Lugo 1992
22.5	W		S			55	Puerto Rico	Lugo 1992
24	D	A	S			24	Mexico	Read and Lawrence 2003
25	D	A	S			23	Mexico	Read and Lawrence 2003
25	D	A	S			15	Mexico	Read and Lawrence 2003
25	M	C	S			115	Brazil	Gehring et al. 2005
25	W	P	S			44	Puerto Rico	Aide et al. 1995
26	W	A	S			100	Mexico	Hughes et al. 1999
30	M	C	S			146	Brazil	Lucas et al. 2002
30	M	A	S			27	Colombia/Venezuela	Saldarriaga et al. 1988
30	W	A	S			121	Mexico	Hughes et al. 1999

(continued)

Appendix 2.1 (continued)

Age (y)	Life zone	Past Land use	Cover Type	Mixed vs. Single spp	Native vs. exotic	Aboveground Biomass C (Mg/ha)	Location	Reference
30	W	P	S			39	Puerto Rico	Marin-Spiotta et al. in press
30	W	P	S			48	Puerto Rico	Marin-Spiotta et al. in press
30	W	P	S			45	Puerto Rico	Marin-Spiotta et al. in press
35	M	A	S			55	Colombia/Venezuela	Saldarriaga et al. 1988
35	M	A	S			54	Colombia/Venezuela	Saldarriaga et al. 1988
40	M	A	S			80	Colombia/Venezuela	Saldarriaga et al. 1988
44	M		S			83	Puerto Rico	Jordan and Farnsworth 1982
49	W		P	S	E	62	Puerto Rico	Lugo 1992
50	W	A	S			137	Mexico	Hughes et al. 1999
50	W		S			40	Puerto Rico	Lugo 1992
54	M	P	S			79	Puerto Rico	Silver et al. 2004
60	M	A	S			58	Colombia/Venezuela	Saldarriaga et al. 1988
60	M	A	S			99	Colombia/Venezuela	Saldarriaga et al. 1988
60	M	A	S			69	Colombia/Venezuela	Saldarriaga et al. 1988
60	W	P	P		both	40	Puerto Rico	Silver et al. 2004
60	W	P	S			33	Puerto Rico	Aide et al. 1995
60	W	P	S			54	Puerto Rico	Aide et al. 1995
60	W	P	S			87	Puerto Rico	Marin-Spiotta et al. in press
60	W	P	S			77	Puerto Rico	Marin-Spiotta et al. in press
60	W	P	S			83	Puerto Rico	Marin-Spiotta et al. in press

80	M	A	S		67	Colombia/Venezuela	Saldarriaga et al. 1988
80	M	A	S		89	Colombia/Venezuela	Saldarriaga et al. 1988
80	M	A	S		72	Colombia/Venezuela	Saldarriaga et al. 1988
80	M	A	S		71	Colombia/Venezuela	Saldarriaga et al. 1988
80	W	P	S		93	Puerto Rico	Marin-Spiotta et al. in press
80	W	P	S		152	Puerto Rico	Marin-Spiotta et al. in press
80	W	P	S		156	Puerto Rico	Marin-Spiotta et al. in press
Mature	D	A	S		30	Mexico	Read and Lawrence 2003
Mature	D	A	S		35	Mexico	Read and Lawrence 2003
Mature	D	A	S		31	Mexico	Read and Lawrence 2003
Mature	D	A	S		41	Mexico	Read and Lawrence 2003
Mature	D	A	S		33	Mexico	Read and Lawrence 2003
Mature	D	A	S		39	Mexico	Read and Lawrence 2003
Mature	D	A	S		34	Mexico	Read and Lawrence 2003
Mature	D	A	S		30	Mexico	Read and Lawrence 2003
Mature	M	P	S		132	Brazil	de Koning et al. 1999
Mature	M	C	S		229	Brazil	Gehring et al. 2005
Mature	W	P	S		84	Puerto Rico	Marin-Spiotta et al. in press
Mature	W	P	S		92	Puerto Rico	Marin-Spiotta et al. in press
Mature	W	P	S		66	Puerto Rico	Marin-Spiotta et al. in press

Appendix 2.2 Soil C pools in secondary forests and plantations of known age following abandonment of agricultural and pasturelands and in nearby mature forests in the Neotropics. Depth refers to maximum soil depth reported for the measurement of soil C content. Soil C 0-25 cm are soil C values adjusted to a common depth using a regression approach (see text). Life zones are wet (W), moist (M), and dry (D) based on mean annual precipitation descriptions given by authors. Past land uses are agriculture (A), pasture (P). Cover types are secondary forests (S) or plantations (P)

Age (y)	Depth (cm)	Life Zone	Past Land Use	Cover type	Soil C (Mg/ha)	Soil C 0–25 cm (Mg/ha)	Location	Reference
4	30	W	A	S	116	100	Mexico	Hughes et al. 1999
4	100	W		P	44	14	Puerto Rico	Lugo 1992
5	40	M	A	S	45	47	Brazil	Sommer et al. 2000
5	45	M	P	S	65	64	Brazil	Feldpausch et al. 2004
5	10	W	P	S	14	30	Costa Rica	Reiners et al. 1994
5	30	W	P/A	P	77	66	Panama	Potvin et al. 2004
5.5	100	W		S	39	12	Puerto Rico	Lugo 1992
6	10	M	P	S	51	110	Dominican Republic	Templer et al. 2005
7	25	M	P	S/P	62	62	Ecuador	de Koning et al. 2003
7	45	M	P	S	68	67	Brazil	Feldpausch et al. 2004
7	20	W	C	S	75	90	Mexico	Williams-Linera 1983
7	25	W	P	S/P	52	52	Ecuador	de Koning et al. 2003
7	30	W	P	P	85	73	Costa Rica	Powers 2004
8	25	M	P	S/P	55	55	Ecuador	de Koning et al. 2003
8	25	M	P	S/P	52	52	Ecuador	de Koning et al. 2003
8	50	M	P	S	65	61	Brazil	Buschbacher et al. 1988
8	50	M	P	S	68	64	Brazil	Buschbacher et al. 1988
8	50	M	P	S	70	66	Brazil	Buschbacher et al. 1988
8	25	W	P	S/P	60	60	Ecuador	de Koning et al. 2003
8	30	W	A	S	71	61	Mexico	Hughes et al. 1999
8	30	W	P	S	120	103	Mexico	Hughes et al. 1999
8	45	W	A	P	135	82	Hawaii	Binkley et al. 2004
9	25	M	P	S/P	59	59	Ecuador	de Koning et al. 2003

9	10	W	P	S	13	28	Costa Rica	Reiners et al. 1994
9	10	W	P	S	19	41	Costa Rica	Reiners et al. 1994
9	15	W	A	P	32	49	Costa Rica	Russell et al. 2004
9	15	W	A	P	35	54	Costa Rica	Russell et al. 2004
9	15	W	A	P	34	52	Costa Rica	Russell et al. 2004
9	15	W	A	P	37	57	Costa Rica	Russell et al. 2004
9	15	W	A	P	38	58	Costa Rica	Russell et al. 2004
9	15	W	A	P	41	63	Costa Rica	Russell et al. 2004
10	25	M	P	S/P	69	69	Ecuador	de Koning et al. 2003
10	25	M	P	S/P	55	55	Ecuador	de Koning et al. 2003
10	25	M	P	S/P	76	76	Ecuador	de Koning et al. 2003
10	25	M	P	S/P	80	80	Ecuador	de Koning et al. 2003
10	30	M	P	S	103	126	Puerto Rico	Marin-Spiotta 2006
10	30	M	P	S	67	82	Puerto Rico	Marin-Spiotta 2006
10	30	M	P	S	84	103	Puerto Rico	Marin-Spiotta 2006
10	30	M	A	S	57	70	Honduras	Paniagua et al. 1999
10	20	W	A	S	32	39	Brazil	Johnson et al. 2001
10	25	W	A	P	70	70	Puerto Rico and US Virgin Islands	Brown and Lugo 1990b
10	25	W	P	S/P	73	73	Ecuador	de Koning et al. 2003
10	25	W	P	S/P	99	99	Ecuador	de Koning et al. 2003
10	25	W	A	P	60	60	Puerto Rico and US Virgin Islands	Brown and Lugo 1990b
11	25	D	C	S	50	50	Puerto Rico and US Virgin Islands	Brown and Lugo 1990b
11	10	M	P	S	23	50	Brazil	Salimon et al. 2004
11	25	M	P	S/P	46	46	Ecuador	de Koning et al. 2003
11	25	W	P	S/P	71	71	Ecuador	de Koning et al. 2003
11	25	W	P	S/P	69	69	Ecuador	de Koning et al. 2003
11	30	W	P	P	41	35	Puerto Rico	Cuevas et al. 1991
11	30	W	C	S	42	36	Puerto Rico	Cuevas et al. 1991
12	40	M	A	S	44	46	Brazil	Sommer et al. 2000
12	10	W	P	S	21	45	Costa Rica	Reiners et al. 1994
12	10	W	C	S	22	47	Costa Rica	Reiners et al. 1994

(continued)

Appendix 2.2 (continued)

Age (y)	Depth (cm)	Life Zone	Past Land Use	Cover type	Soil C (Mg/ha)	Soil C 0–25 cm (Mg/ha)	Location	Reference
12	25	W	A	P	72	72	Hawaii	Bashkin and Binkley 1998
12.5	10	D	A	S	39	67	Mexico	Saynes et al. 2005
13	45	M	P	S	74	73	Brazil	Feldpausch et al. 2004
13	25	W	P	S/P	74	74	Ecuador	de Koning et al. 2003
14	25	M	P	S/P	78	78	Ecuador	de Koning et al. 2003
15	20	M	A	S	98	148	Mexico	Bautista-Cruz and del Castillo 2005
15	20	M	A	S	162	245	Mexico	Bautista-Cruz and del Castillo 2005
15	20	M	A	S	136	205	Mexico	Bautista-Cruz and del Castillo 2005
15	25	M	P	S/P	63	63	Ecuador	de Koning et al. 2003
15	25	M	P	S/P	100	100	Ecuador	de Koning et al. 2003
15	25	M	P	S/P	66	66	Ecuador	de Koning et al. 2003
15	25	M	P	S/P	64	64	Ecuador	de Koning et al. 2003
15	25	M	P	S/P	116	116	Ecuador	de Koning et al. 2003
15	5	W	P	S	33	128	Ecuador	Fehse et al. 2002
15	10	W	C	S	21	45	Costa Rica	Reiners et al. 1994
15	25	W	P	S/P	64	64	Ecuador	de Koning et al. 2003
15	25	W	P	S/P	108	108	Ecuador	de Koning et al. 2003
15	25	W	P	S/P	95	95	Ecuador	de Koning et al. 2003
16	25	M	P	S/P	40	40	Ecuador	de Koning et al. 2003
16	25	W	P	S/P	104	104	Ecuador	de Koning et al. 2003
16	30	W	P	S	77	66	Costa Rica	Guariguata et al. 1997
16	30	W	A	S	92	79	Mexico	Hughes et al. 1999
16	30	W	P	S	44	38	Costa Rica	Werner 1984
16.5	30	W	P	S	79	68	Costa Rica	Guariguata et al. 1997
17	10	M	P	S	25	54	Brazil	de Camargo et al. 1999
17	100	M		P	93	61	Puerto Rico	Lugo 1992
17	25	W	P	S/P	93	93	Ecuador	de Koning et al. 2003
17	25	W	P	S/P	75	75	Ecuador	de Koning et al. 2003

17	30	W	P	S	114	98	Costa Rica	Guariguata et al. 1997
17	50	W	A	P	127	71	Hawaii	Kaye et al. 2000
17	50	W	A	P	127	71	Hawaii	Kaye et al. 2000
17	50	W	A	P	135	75	Hawaii	Kaye et al. 2000
17	50	W	A	P	145	81	Hawaii	Kaye et al. 2000
17	50	W	A	P	148	83	Hawaii	Kaye et al. 2000
17	50	W	A	P	128	72	Hawaii	Kaye et al. 2000
17.5	30	W	P/A	S	129	111	Ecuador	Rhoades et al. 2000
17.5	30	W	P/A	S	101	87	Ecuador	Rhoades et al. 2000
18	25	D	P	S/P	59	59	Ecuador	de Koning et al. 2003
18.5	100	W		P	89	28	Puerto Rico	Lugo 1992
19	25	W	P	S	49	49	Puerto Rico	Lugo et al. 1986
20	25	M	P	S/P	79	79	Ecuador	de Koning et al. 2003
20	30	M	P	S	103	126	Puerto Rico	Marin-Spiotta 2006
20	30	M	P/A	S	45	55	Puerto Rico	Marin-Spiotta 2006
20	30	M	P	S	67	82	Puerto Rico	Marin-Spiotta 2006
20	20	W	A	S	25	30	Brazil	Johnson et al. 2001
20	25	W	P	S/P	98	98	Ecuador	de Koning et al. 2003
20	25	W	A	S	57	57	Puerto Rico	Li et al. 2005
20	25	W	A	P	56	56	Puerto Rico	Li et al. 2005
20	30	W	A	S	103	88	Mexico	Hughes et al. 1999
20	30	W	A	S	87	75	Mexico	Hughes et al. 1999
20	70	W		P	110	46	Panama	Kraenzel et al. 2003
21	25	W	P	S/P	83	83	Ecuador	de Koning et al. 2003
22.5	100	M		S	61	40	Puerto Rico	Lugo 1992
22.5	100	W		S	86	27	Puerto Rico	Lugo 1992
23	20	M	C	P	117	177	Brazil	Smith et al. 1998
23	25	W	A	S	65	65	Puerto Rico and US Virgin Islands	Brown and Lugo 1990b
23	25	W	P	S	47	47	Puerto Rico	Lugo et al. 1986
25	10	D	A	S	39	67	Mexico	Saynes et al. 2005

(continued)

Appendix 2.2 (continued)

Age (y)	Depth (cm)	Life Zone	Past Land Use	Cover type	Soil C (Mg/ha)	Soil C 0–25 cm (Mg/ha)	Location	Reference
25	10	W	A	S	90	194	Jamaica	McDonald and Healey 2000
25	25	W	P	S/P	57	57	Ecuador	de Koning et al. 2003
25	25	W	P	S/P	54	54	Ecuador	de Koning et al. 2003
25	25	W	P	S/P	58	58	Ecuador	de Koning et al. 2003
25	25	W	P	S/P	65	65	Ecuador	de Koning et al. 2003
26	25	M	A	S	40	40	Puerto Rico and US Virgin Islands	Brown and Lugo 1990b
26	30	W	A	S	82	70	Mexico	Hughes et al. 1999
30	25	M	P	S/P	57	57	Ecuador	de Koning et al. 2003
30	25	M	P	S	23	23	Puerto Rico	Lugo et al. 1986
30	30	M	P	S	79	97	Puerto Rico	Marin-Spiotta 2006
30	30	M	P	S	56	68	Puerto Rico	Marin-Spiotta 2006
30	30	M	P	S	42	51	Puerto Rico	Marin-Spiotta 2006
30	25	W	P	S/P	58	58	Ecuador	de Koning et al. 2003
30	30	W	A	S	117	100	Mexico	Hughes et al. 1999
31	30	W	A	S	43	37	Costa Rica	Werner 1984
35	25	D	A	S	75	75	Puerto Rico and US Virgin Islands	Brown and Lugo 1990b
35	25	D	P	S	77	77	Puerto Rico	Lugo et al. 1986
36	20	M	C	P	92	139	Brazil	Smith et al. 1998
36	20	M	C	P	93	140	Brazil	Smith et al. 1998
36	20	M	C	P	95	143	Brazil	Smith et al. 1998
40	25	M	P	S	70	70	Puerto Rico and US Virgin Islands	Brown and Lugo 1990b
40	40	M	A	S	50	53	Brazil	Sommer et al. 2000
40	20	W	A	S	22	27	Brazil	Johnson et al. 2001
40	25	W	A	S	75	75	Puerto Rico and US Virgin Islands	Brown and Lugo 1990b
40	25	W	P	S	60	60	Puerto Rico	Lugo et al. 1986
42.5	25	W	A	S	83	83	Puerto Rico and US Virgin Islands	Brown and Lugo 1990b
45	20	M	A	S	130	196	Mexico	Bautista-Cruz and del Castillo 2005
45	20	M	A	S	90	136	Mexico	Bautista-Cruz and del Castillo 2005
45	20	M	A	S	135	204	Mexico	Bautista-Cruz and del Castillo 2005
49	100	W		P	45	14	Puerto Rico	Lugo 1992

50	25	D	A	P	90	90	Puerto Rico and US Virgin Islands	Brown and Lugo 1990b
50	25	D	P	S	90	90	Puerto Rico	Lugo et al. 1986
50	100	W		S	81	25	Puerto Rico	Lugo 1992
51	25	W	A	P	80	80	Puerto Rico and US Virgin Islands	Brown and Lugo 1990b
55	25	M	A	S	80	80	Puerto Rico and US Virgin Islands	Brown and Lugo 1990b
60	10	D	A	S	41	70	Mexico	Saynes et al. 2005
60	30	M	P	S	61	75	Puerto Rico	Marin-Spiotta 2006
60	30	M	P	S	64	78	Puerto Rico	Marin-Spiotta 2006
60	30	M	P	S	60	73	Puerto Rico	Marin-Spiotta 2006
61	10	M	P	S	43	93	Puerto Rico	Silver et al. 2004
61	30	M	P	S	78	95	Puerto Rico	Silver et al. 2004
75	20	M	A	S	135	204	Mexico	Bautista-Cruz and del Castillo 2005
75	20	M	A	S	105	159	Mexico	Bautista-Cruz and del Castillo 2005
75	20	M	A	S	126	190	Mexico	Bautista-Cruz and del Castillo 2005
80	30	M	P	S	97	119	Puerto Rico	Marin-Spiotta 2006
80	30	M	P	S	49	60	Puerto Rico	Marin-Spiotta 2006
80	30	M	P	S	59	72	Puerto Rico	Marin-Spiotta 2006
100	25	M	A	S	80	80	Puerto Rico and US Virgin Islands	Brown and Lugo 1990b
Mature	10	D			36	62	Mexico	Saynes et al. 2005
Mature	25	D			60	60	Puerto Rico and US Virgin Islands	Brown and Lugo 1990b
Mature	25	D			45	45	Puerto Rico and US Virgin Islands	Brown and Lugo 1990b
Mature	10	M			26	56	Brazil	de Camargo et al. 1999
Mature	10	M			18	39	Brazil	Salimon et al. 2004
Mature	10	M			23	50	Brazil	Salimon et al. 2004
Mature	20	M			189	285	Mexico	Bautista-Cruz and del Castillo 2005
Mature	20	M			106	160	Brazil	Smith et al. 1998
Mature	20	M			210	317	Mexico	Bautista-Cruz and del Castillo 2005
Mature	20	M			126	190	Mexico	Bautista-Cruz and del Castillo 2005
Mature	30	M			70	86	Puerto Rico	Marin-Spiotta 2006
Mature	30	M			44	54	Puerto Rico	Marin-Spiotta 2006

(continued)

Appendix 2.2 (continued)

Age (y)	Depth (cm)	Life Zone	Past Land Use	Cover type	Soil C (Mg/ha)	Soil C 0–25 cm (Mg/ha)	Location	Reference
Mature	30	M			73	89	Puerto Rico	Marin-Spiotta 2006
Mature	40	M			57	60	Brazil	Sommer et al. 2000
Mature	5	W			29	112	Ecuador	Fehse et al. 2002
Mature	10	W			14	30	Costa Rica	Reiners et al. 1994
Mature	10	W			17	37	Costa Rica	Reiners et al. 1994
Mature	10	W			28	60	Costa Rica	Reiners et al. 1994
Mature	20	W			23	28	Brazil	Johnson et al. 2001
Mature	25	W			89	89	USA-Hawaii	Bashkin and Binkley 1998
Mature	25	W			110	110	Puerto Rico and US Virgin Islands	Brown and Lugo 1990b
Mature	25	W			85	85	Puerto Rico and US Virgin Islands	Brown and Lugo 1990b
Mature	25	W			62	62	Puerto Rico	Lugo et al. 1986
Mature	30	W			87	75	Costa Rica	Guariguata et al. 1997
Mature	30	W			91	78	Costa Rica	Guariguata et al. 1997
Mature	30	W			116	100	Costa Rica	Guariguata et al. 1997
Mature	30	W			90	77	Ecuador	Rhoades et al. 2000
Mature	30	W			109	94	Ecuador	Rhoades et al. 2000
Mature	30	W			49	42	Costa Rica	Werner 1984

References

Achard, F., Eva, H. D., Stibig, H.-J., Mayaux, P., Gallego, J., Richards, T., and Malingreau, J.-P. 2002. Determination of deforestation rates of the world's humid tropical forests. Science 297:999–1002.

Adachi, M., Bekku, Y. S., Rashidah, W., Okuda, T., and Koizumi, H. 2006. Differences in soil respiration between different tropical ecosystems. Applied Soil Ecology 34:258–265.

Ahmad, R., Kookana, R. S., Alston, A. M., and Skjemstad, J. O. 2001. The nature of soil organic matter affects sorption of pesticides. 1. Relationships with carbon chemistry as determined by ^{13}C CPMAS NMR spectroscopy. Environment, Science and Technology 35:878–884.

Aide, T. M., Zimmerman, J. K., Herrera, L., Rosario, M., and Serrano, M. 1995. Forest recovery in abandoned tropical pastures in Puerto Rico. Forest Ecology and Management 77:77–86.

Aide, T. M., Zimmerman, J. K., Pascarella, J. B., Rivera, L., and Marcano-Vega, H. 2000. Forest regeneration in a chronosequence of tropical abandoned pastures: implications for restoration ecology. Restoration Ecology 8:328–338.

Aide, T. M., and Grau, H. R. 2004. Globalization, migration, and Latin American ecosystems. Science 305:1915–1916.

Alvarez, C. R., Alvarez, R., Grigera, M. S., and Lavado, R. S. 1998. Associations between organic matter fractions and the active soil microbial biomass. Soil Biology and Biochemistry 30:767–773.

Alves, D. S., Vianei Soares, J., Amaral, S., Mello, E. M. K., Almeida, S. A. S., Fernandes da Silva, O., and Silveira, A. M. 1997. Biomass of primary and secondary vegetation in Rondônia, Western Brazilian Amazon. Global Change Biology 3:451–461.

Amundson, R. 2001. The carbon budget in soils. Annual Review of Earth and Planetary Science 29:535–562.

Anderson, K. J., Allen, A. P., Gillooly, J. F., and Brown, J. H. 2006. Temperature-dependence of biomass accumulation rates during secondary succession. Ecology Letters 9:673–682.

Arroyo-Mora, J. P., Sánchez-Azofeifa, G. A., Rivard, B., Calvo, J. C., and Janzen, D. H. 2005. Dynamics in landscape structure and composition for the Chorotega region, Costa Rica from 1960 to 2000. Agriculture, Ecosystems and Environment 106:27–39.

Baisden, W. T., Amundson, R., Cook, A. C., and Brenner, D. L. 2002. Turnover and storage of C and N in five density fractions from California annual grassland surface soils. Global Biogeochemical Cycles 16:117, doi:10.1029/2001GB001822,2002

Balabane, M., Faivre, D., van Oort, F., and Dahmani-Muller, H. 1999. Mutual effects of soil organic matter dynamics and heavy metals fate in a metallophyte grassland. Environmental Pollution 105:45–54.

Baldock, J. A., Masiello, C. A., Gélinas, Y., and Hedges, J. I. 2004. Cycling and composition of organic matter in terrestrial and marine ecosystems. Marine Chemistry 92:39–64.

Balesdent, J., Mariotti, A., and Guillet, B. 1987. Natural ^{13}C abundance as a tracer for studies of soil organic matter dynamics. Soil Biology and Biochemistry 19:25–30.

Bashkin, M.A., and Binkley, D. 1998. Changes in soil carbon following afforestation in Hawaii. Ecology 79:828–833.

Bautista-Cruz, A., and del Castillo, R. F. 2005. Soil changes during secondary succession in a tropical montane cloud forest area. Soil Science Society of America Journal 69:906–914.

Bernoux, M., Cerri, C. C., Neill, C., and de Moraes, J. F. L. 1998. The use of stable carbon isotopes for estimating soil organic matter turnover rates. Geoderma 82:43–58.

Binkley, D. and Resh, S. C. 1999. Rapid changes in soils following Eucalyptus afforestation in Hawaii. Soil Science Society of America Journal 63:222–225.

Binkley, D., Kaye, J., Barry, M., and Ryan, M. G. 2004. First-rotation changes in soil carbon and nitrogen in a Eucalyptus plantation in Hawaii. Soil Science Society of America Journal 68:1713–1719.

Bird, J. A., van Kessel, C., and Horwarth, W. R. 2003. Stabilization of ^{13}C-Carbon and immobilization of ^{15}N-Nitrogen from rice straw in humic fractions. Soil Science Society of America Journal 67:806–816.

Boucher, D. H., Vandermeer, J. H., Granzow de la Cerda, I., Mallona, M. A., Perfecto, I., and Zamora, N. 2000. Post-agriculture versus post-hurricane succession in southeastern Nicaraguan rain forest. Plant Ecology 156:131–137.
Brady, N. C., and Weil, R. R., Eds. 2002. The Nature and Properties of Soils, 13th edition. New Jersey: Prentice Hall.
Brandeis, T. J., and Suárez Rozo, M. de R. 2005. Effects of model choice and forest structure on inventory-based estimations of Puerto Rican forest biomass. Caribbean Journal of Science 41:250–268.
Brown, S., and Lugo, A. E. 1982. The storage and production of organic matter in tropical forests and their role in the global carbon cycle. Biotropica 14:161–187.
Brown, S., and Lugo, A. E. 1984. Biomass of tropical forests: a new estimate based on forest volumes. Science 223:1290–1293.
Brown, S., Gillespie, A. J. R., and Lugo, A. E. 1989. Biomass estimation methods for tropical forests with applications to forest inventory data. Forest Science 35:881–902.
Brown, S., and Lugo, A. E. 1990. Tropical secondary forests. Journal of Tropical Ecology 6:1–32.
Brown, S., Sathaye, J., Cannell, M., and Kauppi, P., et al. 1995a. Management of forests for mitigation of greenhouse gas emissions. In Climate Change 1995: Impacts, Adaptations and Mitigation of Climate Change: Scientific-Technical Analyses. Working Group II to Second Assessment Report of the Intergovernmental Panel on Climate Change, ed. R. T. Watson, M. C. Zinyowera, and R. H. Moss, pp. 775–797. New York: Cambridge University Press.
Brown, I. F., Martinelli, L. A., Thomas, W. W., Moreira, M. Z., Cid Ferreira, C. A., and Victoria, R. A. 1995b. Uncertainty in the biomass of Amazonian forests: an example from Rondônia, Brazil. Forest Ecology and Management 75:175–189.
Bunker, D. E., DeClerck, F., Bradford, J. C., Colwell, R. K., Perfecto, I., Phillips, O. L., Sankaran, M., and Naeem, S. 2005. Species loss and aboveground carbon storage in a tropical forest. Science 310:1029–1031.
Buschbacher, R. 1986. Tropical deforestation and pasture development. Bioscience 36:23–27.
Buschbacher, R., Uhl, C., and Serrão, E.A. S. 1988. Abandoned pastures in Eastern Amazonia. II. Nutrient stocks in the soil and vegetation. Journal of Ecology 76: 682–699.
Bush, M. B., and Colinvaux, P. A. 1994. Tropical forest disturbance: paleoecological records from Darien, Panama. Ecology 75:1761–1768.
Byers, S. C., Mills, E. L., and Stewart, P. L. 1978. A comparison of methods of determining organic carbon in marine sediments, with suggestions for a standard method. Hydrobiologia 58:43–47.
Cambardella, C., and Elliott, E. T. 1993. Carbon and nitrogen distribution in aggregates form cultivated and native grassland soils. Soil Science Society of America Journal 57:1071–1076.
Capers, R. S., Chazdon, R. L., Redondo Brenes, A., and Vilchez Alvarado, B. 2005. Successional dynamics of woody seedling communities in wet tropical secondary forests. Journal of Ecology 93:1071–1084.
Carpenter, F. L., Palacios Mayorga, S., González Quintero, E., and Schroeder, M. 2001. Land-use and erosion of a Costa Rican Ultisol affect soil chemistry, mycorrhizal fungi and early regeneration. Forest Ecology and Management 144:1–17.
Carvalheiro, K. D. O., and Nepstad, D. C. 1996. Deep soil heterogeneity and fine root distribution in forests and pastures of eastern Amazônia. Plant and Soil 182:279–285.
Caspersen, J. P., Pacala, S. W., Jenkins, J. C., Hurtt, G. C., Moorcroft, P. R., and Birdsey, R. A. 2000. Contributions of land-use history to carbon accumulation in U.S. forests. Science 290:1148–1151.
Chambers, J. Q., Higuchi, N., and Schimel, J. P. 1998. Ancient trees in Amazônia. Nature 391: 135–136.
Chambers, J. Q., dos Santos, J., Ribeiro, R. J., and Higuchi, N. 2001. Tree damage, allometric relationships, and above-ground net primary production in central Amazon forest. Forest Ecology and Management 152:73–84.
Chazdon, R. L. 2003. Tropical forest recovery: legacies of human impact and natural disturbances. Perspectives in Plant Ecology, Evolution and Systematics 6:51–71.

Chin, Y.-P., Aiken, G., and O'Loughlin, E. 1994. Molecular weight, polydispersity, and spectroscopic properties of aquatic humic substances. Environment, Science and Technology 28:1853–1858.

Christensen, B. T. 2001. Physical fractionation of soil and structural and functional complexity in organic matter turnover. European Journal of Soil Science 52:345–353.

Clark, D. B., and Clark, D. A. 1991. The impact of physical damage on canopy tree regeneration in tropical rain forest. Journal of Ecology 79:447–457.

Clark, D. B., Clark, D. A., Brown, S., Oberbauer, S. F., and Veldkamp, E. 2002. Stocks and flows of coarse woody debris across a tropical rain forest nutrient and topography gradient. Forest Ecology and Management 164:237–248.

Clark, D. A., Pipers, S. C., Keeling, C. D., and Clark, D. B. 2003. Tropical rain forest tree growth and atmospheric carbon dynamics linked to interannual temperature variation during 1984–2000. Proceedings of the National Academy of Sciences 100:5852–5857.

Compton, J. E., and Boone, R. D. 2000. Long-term impacts of agriculture on soil carbon and nitrogen in New England forests. Ecology 81:2314–2330.

Croué, J.-P. 2004. Isolation of humic and non-humic NOM fractions: structural characterization. Environmental Monitoring and Assessment 92:193–207.

Crowther, J. 1980. Soil carbon dioxide and weathering potentials in tropical karst terrain, peninsular Malaysia: a preliminary model. Earth Surface Processes and Landforms 9:397–407.

Cuevas, E., Brown, S., and Lugo, A. E. 1991. Above- and belowground organic matter storage and production in a tropical pine plantation and a paired broadleaf secondary forest. Plant and soil 135:257–268.

Cusack, D., and Montagnini, F. 2004. The role of native species plantations in recovery of understory woody diversity in degraded pasturelands of Costa Rica. Forest Ecology and Management 188:1–15.

Davidson, E. A., de Carvalho, C. J. R., Vieira, I. C. G., Figuereido, R. de O., Moutinho, P., Ishida, F. Y., dos Santos, M. T. P., Guerrero, J. B., Kalif, K., and Saba, R. T. 2004. Nitrogen and phosphorus limitation of biomass growth in a tropical secondary forest. Ecological Applications 14:S150–S163.

de Camargo, P. S., Trumbore, S. E., Martinelli, L. A., Davidson, E. A., Nepstad, D. C., and Victoria, R. L. 1999. Soil carbon dynamics in regrowing forest of eastern Amazônia. Global Change Biology 5:693–702.

de Jong, W., Chokkalingam, U., and Perrera, G. A. D. 2001. The evolution of swidden fallow secondary forests in Asia. Journal of Tropical Forest Science 13:800–814.

de Koning, G. H. J., Veldkamp, E., and López-Ulloa, M. 2003. Quantification of carbon sequestration in soils following pasture to forest conversion in northwestern Ecuador. Global Biogeochemical Cycles 17:1098, doi:10.1029/2003GB002099.

Dean, W. E. 1974. Determination of carbonate and organic matter in calcareous sediments and sedimentary rocks by loss on ignition: comparison with other methods. Journal of Sedimentary Petrology 44:242–248.

DeFries, R. S., Houghton, R. A., Hansen, M. C., Field, C. B., Skole, D., and Townsend, J. 2002. Carbon emissions from tropical deforestation and regrowth based on satellite observations for the 1980s and 1990s. Proceedings of the National Academy of Sciences 99:14256–14261.

Delaney, M., Brown, S., Lugo, A. E., Torrez-Lezama, A., and Bello Quintero, N. 1997. The distribution of organic carbon in major components of forests located in five life zones of Venezuela. Journal of Tropical Ecology 13:697–708.

Delaney, M., Brown, S., Lugo, A.E., Torrez-Lezama, A., and Bello Quintero, N., 1998. The quantity and turnover of dead wood in permanent forest plots in six life gones of Venezuela. Biotropica 30:2–11.

Denslow, J. S., and Guzmán, S. G. 2000. Variation in stand structure, light and seedling abundance across a tropical moist forest chronosequence, Panama. Journal of Vegetation Science 11:201–212.

DeWalt, S. J., Maliakal, S. K., and Denslow, J. S. 2003. Changes in vegetation structure and composition along a tropical forest chronosequence: implications for wildlife. Forest Ecology and Management 182:139–151.

Dietz, J. L. 1986. Economic History of Puerto Rico: Institutional Change and Capitalist Development. Princeton, NJ: Princeton University Press.
Dixon, R. K., Brown, S., Houghton, R. A., Solomon, A. M., Trexler, M. C., and Wisniewski, J. 1994. Carbon pools and flux of global forest ecosystems. Science 263:185–190.
Drake, D. R., and Pratt, L. W. 2001. Seedling mortality in Hawaiian rain forest: the toll of small-scale physical disturbance. Biotropica 33: 319–323.
Dunn, R. R. 2004. Recovery of faunal communities during tropical forest regeneration. Conservation Biology 18:302–309.
Eaton, J. M., and Lawrence, D. 2006. Woody debris stocks and fluxes during succession. Forest Ecology and Management 232:46–55.
Elliott, E. T., and Cambardella, C. A. 1991. Physical separation of soil organic matter. Agriculture, Ecosystems and Environment 34:407–419.
Eswaran, H., van den Berg, E., and Reich, P. 1993. Organic carbon in soils of the world. Soil Science Society of America Journal 57:192–194.
Etter, A., McAlpine, C., Pullar, D., and Possingham, H. 2005. Modeling the age of tropical moist forest fragments in heavily-cleared lowland landscapes of Colombia. Forest Ecology and Management 208:249–260.
Eva, H. D., Achard, F., Stibig, H.-J., and Mayaux, P. 2003. Response to Comment on "Determination of deforestation rates of the world's humid tropical forests". Science 299:1015b.
Ewel, J. J. 1976. Litterfall and leaf decomposition in a tropical forest succession in eastern Guatemala. The Journal of Ecology 64:293–308.
Farenhorst, A. 2006. Importance of soil organic matter fractions in soil-landscape and regional assessment of pesticide sorption and leaching in soil. Soil Science Society of America Journal 70:1005–1012.
Farquhar, G. D., Ehleringer, J.R., and Hubick, K.T. 1989. Carbon isotope discrimination and photosynthesis. Annual Review of Plant Physiology and Plant Molecular Biology 40:503–537.
Fearnside, P. M. 1993. Deforestation in Brazilian Amazon: The effect of population land tenure. Ambio 22:537–545.
Fearnside, P. M. 1996. Amazonian deforestation and global warming: carbon stocks in vegetation replacing Brazil's Amazon forest. Forest Ecology and Management 80:21–34.
Fearnside, P. M., and Guimarães, W. M. 1996. Carbon uptake by secondary forests in Brazilian Amazônia. Forest Ecology and Management 80:35–46.
Fearnside, P. M., and Laurance, W. F. 2003. Comment on "Determination of deforestation rates of the world's humid tropical forests". Science 299:1015a.
Fehse, J., Hofstede, R., Aguirre, N., Paladines, C., Kooijman, A., and Sevnik, J. 2002. High altitude tropical secondary forests: a competitive carbon sink? Forest Ecology and Management 163:9–25.
Feldpausch, T. R., Rondon, M. A., Fernandes, E. C. M., Riha, S. J., and Wandelli, E. 2004. Carbon and nutrient accumulation in secondary forests regenerating on pastures in Central Amazônia. Ecological Applications 14:164–176.
Fichtler, E., Clark, D. A., and Worbes, M. 2003. Age and long-term growth of trees in an old-growth tropical rain forest, based on analyses of tree rings and ^{14}C. Biotropica 35:306–317.
Fisher, M. J., Rao, I. M., Ayarza, M. A., Lascano, C. E., Sanz, J. I., Thomas, R. J., and Vera, R. R. 1994. Carbon storage by introduced deep-rooted grasses in South African savannas. Nature 371:236–238.
Fölster, H., De Las Salas, G., and Khanna, P. 1976. A tropical evergreen forest site with perched water table, Magdalena Valley, Columbia: biomass and bioelement inventory of primary and secondary vegetation. Oecologia Planetarium 11:297–320.
Gehring, C., Denich, M., Kanashiro, M., and Vlek, P. L. G. 1999. Response of secondary vegetation in Eastern Amazônia to relaxed nutrient availability constraints. Biogeochemistry 45:223–241.
Gehring, C., Denich, M., and Vlek, P. L. G. 2005. Resiliency of secondary forest regrowth after slash-and-burn agriculture in central Amazonia. Journal of Tropical Ecology 21:519–527.
Gerwing, J. J., and Lopes Farias, D. 2000. Integrating liana abundance and forest structure into an estimate of total aboveground biomass for an eastern Amazonian forest. Journal of Tropical Ecology 16:327–335.

Gibbard, S. G., Caldeira, K., Bala, G., Phillips, T., and Wickett, M. 2005. Climate effects of global land cover change. Geophysical Research Letters 32.

Gillman, L. N., Ogden, J., Wright, S. D., Stewart, K. L., and Walsh, D. P. 2004. The influence of macro-litterfall and forest structure on litterfall damage to seedlings. Austral Ecology 29:305–312.

Gleixner, G., Bol, R., and Balesdent, J. 1999. Molecular insight into soil carbon turnover. Rapid Communications in Mass Spectrometry 13:1278–1283.

Golchin, A., Oades, J. M., Skjemstad, J. O., and Clarke, P. 1994. Study of free and occluded particulate organic matter in soils by solid state ^{13}C CP/MAS NMR spectroscopy and scanning electron microscopy. Soil Biology and Biochemistry 32:285–309.

Golchin, A., Oades, J. M., Skjemstad, J. O., and Clarke, P. 1995. Structural and dynamic properties of soil organic matter as reflected by ^{13}C natural abundance, pyrolysis mass spectrometry and solid-state ^{13}C NMR spectroscopy in density fractions of an Oxisol under forest and pasture. Australian Journal of Soil Research 33:59–76.

Grau, H. R., Aide, T. M., Zimmerman, J. K., and Thomlinson, J. R. 2004. Trends and scenarios of the carbon budget in postagricultural Puerto Rico (1936–2060). Global Change Biology 10:1163–1179.

Greenland, D. J., and Ford, G. W. 1964. Separation of partially humified organic materials by ultrasonic dispersion. Pages 137–148 in 8th International Congress of Soil Science, Transactions.

Guariguata, M. R., Chazdon, R. L., Denslow, J. S., Dupuy, J. M., and Anderson, L. 1997. Structure and floristics of secondary and old-growth forest stands in lowland Costa Rica. Plant Ecology 132:107–120.

Guariguata, M. R. 1998. Response of forest tree saplings to experimental mechanical damage in lowland Panama. Forest Ecology and Management 102:103–111.

Guariguata, M. R., and Ostertag, R. 2001. Neotropical secondary forest succession: changes in structural and functional characteristics. Forest Ecology and Management 5292:1–22.

Guggenberger, G., Zech, W., and Thomas, R. J. 1995a. Lignin and carbohydrate alteration in particle-size separates of an Oxisol under tropical pastures following native savanna. Soil Biology and Biochemistry 27:1629–1638.

Guggenberger, G., Zech, W., Haumaier, L., and Christensen, B. T. 1995b. Land-use effects on the composition of organic matter in particle-size separates of soils: II. CPMAS and solution ^{13}C NMR analysis. European Journal of Soil Science 46:147–158.

Guggenberger, G., and Zech, W. 1999. Soil organic matter composition under primary forest, pasture, and secondary forest succession, Region Huetar Norte, Costa Rica. Forest Ecology and Management 124:93–104.

Hanson, P. J., Swanston, C. W., Garten Jr., C. T., Todd, D. E., and Trumbore, S. E. 2005. Reconciling change in Oi-horizon carbon-14 with mass loss for an oak forest. Soil Science Society of America Journal 69:1492–1502.

Harmon, M. E., Whigham, D. F., Sexton, J., and Olmsted, I. 1995. Decomposition and mass of woody detritus in the dry tropical forests of the Northeastern Yucatan Peninsula, Mexico. Biotropica 27:305–316.

Harmon, M. E., and Sexton, J. 1996. Guidelines for the measurements of woody debris in forest ecosystems. U.S. LTER Publication no. 20.

Hecht, S. B. 2005. Soybeans, development and conservation on the Amazon frontier. Development and Change 36:375–404.

Helfrich, M., Ludwig, B., Buurman, P., and Flessa, H. 2006. Effect of land use on the composition of soil organic matter in density and aggregate fractions as revealed by solid-state ^{13}C NMR spectroscopy. Geoderma 136:331–341.

Helmer, E. H., Brown, S., and Cohen, W. B. 2000. Mapping montane tropical forest successional stage and land use with multi-date Landsat imagery. International Journal of Remote Sensing 21:2163–2183.

Helmer, E. H. 2004. Forest conservation and land development in Puerto Rico. Landscape Ecology 19:29–40.

Hertel, D., Leuschner, C., and Hoelscher, D. 2003. Size and structure of fine root systems in old-growth and secondary tropical montane forests (Costa Rica). Biotropica 35:143–153.

Holl, K. D. 1998. Do bird perching structures elevate see rain and seedling establishment in abandoned tropical pasture? Restoration Ecology 6:227–313.

Holl, K. D. 1999. Tropical forest recovery and restoration. Trends in Ecology and Evolution 14:378–379.

Holl, K. D., Loik, M. E., Lin, E. H. V., and Samuels, I. A. 2000. Tropical montane forest restoration in Costa Rica: Overcoming barriers to dispersal and establishment. Restoration Ecology 8:339–349.

Houghton, R. A., Skole, D. L., and Lefkowitz, D. S. 1991. Changes in the landscape of Latin America between 1850 and 1985. II. Net release of CO_2 to the atmosphere. Forest Ecology and Management 38:173–199.

Houghton, R. A., Skole, D. L., Nobre, C. A., Hackler, J. L., Lawrence, K. T., and Chomentowski, W. H. 2000. Annual fluxes of carbon from deforestation and regrowth in the Brazilian Amazon. Nature 403:301–303.

Houghton, R. A., and Goodale, C. L. 2004. Effects of land-use change on the carbon balance of terrestrial ecosystems. Ecosystems and Land Use Change. Geophysical Monograph Series 153:85–98.

Hughes, R. F., Kauffman, J. B., and Jaramillo, V. J. 1999. Biomass, carbon and nutrient dynamics of secondary forests in a humid tropical region of Mexico. Ecology 80:1892–1907.

Hughes, R. F., Kauffman, J. B., and Cummings, D. L. 2000a. Fire in the Brazilian Amazon. 3. Dynamics of biomass, C, and nutrient pools in regenerating forests. Oecologia 124: 574–588.

Hughes, R. F., Kauffman, J. B., and Jaramillo, V. J. 2000b. Ecosystem-scale impacts of deforestation and land use in a humid tropical region of Mexico. Ecological Applications 10:515–527.

Janisch, J. E., and Harmon, M. E. 2002. Successional changes in live and dead wood carbon stores: Implications for net ecosystem productivity. Tree Physiology 22:77–89.

Jaramillo, V. J., Ahedo-Hernandez, R., and Kauffman, J. B. 2003. Root biomass and carbon in a tropical evergreen forest of Mexico: Changes with secondary succession and forest conversion to pasture. Journal of Tropical Ecology 19:457–464.

Jenkinson, D. S., and Rayner, J. H. 1977. The turnover of soil organic matter in some of the Rothamsted classical experiments. Soil Science 123:298–304.

Jobbágy, E. G., and Jackson, R. B. 2000. The vertical distribution of soil organic carbon and its relation to climate and vegetation. Ecological Applications 10:423–436.

John, B., Yamashita, T., Ludwig, B., and Flessa, H. 2005. Storage of organic carbon in aggregate and density fractions of silty soils under different land uses. Geoderma 128:63–79.

Johnson, C. M., Zarin, D. J., and Johnson, A. H. 2000. Post-disturbance aboveground biomass accumulation in global secondary forests. Ecology 81:1395–1401.

Johnson, C. M., Vieira, I. C. G., Zarin, D. J., Frizano, J., and Johnson, A. H. 2001. Carbon and nutrient storage in primary and secondary forests in eastern Amazônia. Forest Ecology and Management 147:245–252.

Jordan, C. F., and Farnworth, E. G. 1982. Natural vs. plantation forests: a case study of land reclamation strategies for the humid tropics. Environmental Management 6:485–492.

Kammesheidt, L. 2002. Perspectives on secondary forest management in tropical humid lowland America. Ambio 31:243–250.

Kaye, J. P., Resh, S. C., Kaye, M. W, and Chimner, R. A. 2000. Nutrient and carbon dynamics in a replacement series of Eucalyptus and Albizia trees. Ecology 81:3267–3273.

Keller, M., Palace, M., and Hurtt, G. 2001. Biomass estimation in the Tapajos National Forest, Brazil. Examination of sampling and allometric uncertainties. Forest Ecology and Management 154:371–382.

Keller, M., Palace, M, Asner, G. P., Pereira Jr., R., and Silva, J. N. M. 2004. Coarse woody debris in undisturbed and logged forests in the eastern Brazilian Amazon. Global Change Biology 10:784–795.

Kennard, D. K. 2002. Secondary forest succession in a tropical dry forest: patterns of development across a 50-year chronosequence in lowland Bolivia. Journal of Tropical Ecology 18:53–66.

Kögel-Knabner, I. 2000. Analytical approaches for characterizing soil organic matter. Organic Geochemistry 31:609–625.

Kraenzel, M., Castillo, A., Moore, T., and Potvin, C. 2003. Carbon storage of harvest-age teak (Tectona grandis) plantations, Panama. Forest Ecology and Management 173:213–225.

Krull, E. S., Baldock, J. A., and Skjemstad, J. O. 2003. Importance of mechanisms and processes of the stabilisation of soil organic matter for modelling carbon turnover. Functional Plant Biology 30:207–222.

Lawrence, D. 2005a. Biomass accumulation after 10–200 years of shifting cultivation in Bornean rain forest. Ecology 86:26–33.

Lawrence, D. 2005b. Regional-scale variation in litter production and seasonality in tropical dry forests of southern Mexico. Biotropica 37:561–570.

Lemenih, M., Olsson, M., and Karltun, E. 2004. Comparison of soil attributes under Cupressus lusitanica and Eucalyptus saligna established on abandoned farmlands with continuously cropped farmlands and natural forest in Ethiopia. Forest Ecology and Management 195:57–67.

Li, Y., Xu, M., Zou, X., Shi, P., and Zhang,Y. 2005. Comparing soil organic carbon dynamics in plantation and secondary forest in wet tropics in Puerto Rico. Global Change Biology 11:239–248.

Losi, C. J., Siccama, T. G., Condit, R., and Morales, J. E. 2003. Analysis of alternative methods for estimating carbon stock in young tropical plantations. Forest Ecology and Management 184:355–368.

López-Ulloa, M., Veldkamp, E., and de Koning, G. H. J. 2005. Soil carbon stabilization in converted tropical pastures and forests depends on soil type. Soil Science Society of America Journal 69:1110–1117.

Louchouarn, P., Opsahl, S., and Benner, R. 2000. Isolation and quantification of dissolved lignin from natural waters using solid-phase extraction and GC/MS. Analytical Chemistry 72:2780–2787.

Lovley, D. R., Coates, J. D., Blunt-Harris, E. L., Phillips, E. J. P., and Woodward, J. C. 1996. Humic substances as electron acceptors for microbial respiration. Nature 382:445–448.

Lucas, R. M., Honzak, M., do Amaral, I., Curran, P. J., and Foody, G. M. 2002. Forest regeneration on abandoned clearances in central Amazônia. International Journal of Remote Sensing 23:965–988.

Lugo, A. E. 1988. The future of the forest. Ecosystem rehabilitation in the tropics. Environment 30:17–20, 41–45.

Lugo, A. E. 1992. Comparison of tropical tree plantations with secondary forests of similar age. Ecological Monographs 62:1–41.

Lugo, A. E., Brown, S., and Chapman, J., 1988. An analytical review of production rates and stemwood biomass of tropical forests plantations. Forest Ecology and Management 23: 179–200.

Lugo, A. E., and Helmer, E. 2004. Emerging forests on abandoned land: Puerto Rico's new forests. Forest Ecology and Management 190:145–161.

Makana, J.-R., and Thomas, S. C. 2006. Impacts of selective logging and agricultural clearing on forest structure, floristic composition and diversity, and timber tree regeneration in the Ituri Forest, Democratic Republic of Congo. Biodiversity and Conservation 15:1375–1397.

Malhi, Y., Baldocchi, D. D., and Jarvis, P. G. 1999. The carbon balance of tropical, temperate and boreal forests. Plant, Cell and Environment 22:715–740.

Mann, L. K. 1986. Changes in soil carbon storage after cultivation. Soil Science 142:279–288.

Marcano-Vega, H., Aide, T. M., and Baez, D. 2002. Forest regeneration in abandoned coffee plantations and pastures in the Cordillera Central of Puerto Rico. Plant Ecology 161:75–87.

Marín-Spiotta, E. 2006. Controls on above and belowground carbon storage during tropical reforestation. PhD Thesis. University of California, Berkeley.

Marín-Spiotta, E., Ostertag, R., and Silver, W. L. 2007. Long-term patterns in tropical reforestation: plant community composition and aboveground biomass accumulation. Ecological Applications 17:828–839.

Marrero, J. 1950. Reforestation of degraded lands in Puerto Rico. Caribbean Forester 11:3–15.

Martin, A., Mariotti, A., Balesdent, J., Lavelle, P., and Vuattoux, R. 1990. Estimate of organic matter turnover rate in a savanna soil by ^{13}C natural abundance measurements. Soil Biology and Biochemistry 22:517–523.

Maury-Lechon, G. 1982. Regeneration forestière en Guyana française: recru sur 25 ha de coupe papetière en forêt dense humide (Arbocel.). Revue Boiset Forêts des Tropiques 197:3–21.
McDonald, M. A., and Healey, J. R. 2000. Nutrient cycling in secondary forests in the Blue Mountains of Jamaica. Forest Ecology and Management 139:257–278.
McGroddy, M., and Silver, W. L. 2000. Variations in belowground carbon storage and soil CO_2 flux rates along a wet tropical climate gradient. Biotropica 32:614–624.
McWilliam, A. L. C., Robertos, J. M., Cabral, O. M. R., Leitão, M. V. B. R., de Costa, A. C. L., Maitelli, G. T., and Zamparoni, C. A. G. P. 1993. Leaf area index and above-ground biomass of terra firme rain forest and adjacent clearings in Amazonia. Functional Ecology 7:310–317.
Metz, B., Davidson, O., Swart, R., and Pan, J. 2001. Climate Change 2001. Mitigation. Contribution of Working Group III to the Third Assessment Report of the Intergovernmental Panel on Climate Change (IPCC). Cambridge: Cambridge University Press.
Mikutta, R., Kleber, M., Torn, M. S., and Jahn, R. 2006. Stabilization of soil organic matter: association with minerals or chemical recalcitrance? Biogeochemistry 77:25–56.
Mizrahi, A., Ramos Prado, J. M., and Jimenez-Osornio, J. 1997. Composition, structure, and management potential of secondary dry tropical vegetation in two abandoned henequen plantations of Yucatan, Mexico. Forest Ecology and Management 96:273–282.
Murphy, P. G., and Lugo, A. E. 1995. Dry forests of Central America and the Caribbean. In Seasonally Dry Tropical Forests, ed. S. H. Bullock, H. A. Mooney, and E. Medina, pp. 9–34. United Kingdom: Cambridge University Press.
Myster, R. W. 2004. Post-agricultural invasion, establishment, and growth of neotropical trees. The Botanical Review 70:381–402.
Naughton-Treves, L., and Chapman, C. A. 2002. Fuelwood resources and forest regeneration on fallow land in Uganda. Journal of Sustainable Forestry 14:19–32.
Neeff, T. 2005. Spatial modeling of primary and secondary forest growth in Amazonia. Forest Ecology and Management 219:149–168.
Neeff, T., Lucas, R. M., dos Santos, J. R., Brondizio, E. S., and Freitas, C. C. 2006. Area and age of secondary forests in Brazilian Amazônia 1978–2002: an empirical estimate. Ecosystems 9:609–623.
Neff, J. C., Townsend, A. R., Gleixner, G., Lehman, S. J., Turnbull, J., and Bowman, W. D. 2002. Variable effects of nitrogen additions on the stability and turnover of soil carbon. Nature 419:915–917.
Neill, C., Melillo, J. M., Steudler, P. A., Cerri, C. C., de Moraes, J. F. L., Piccolo, M. C., and Brito, M. 1997. Soil carbon and nitrogen stocks following forest clearing for pasture in the southwestern Brazilian Amazon. Ecological Applications 7:1216–1225.
Nelson, D. W., and Sommers, L. E. 1996. Total carbon, organic carbon, and organic matter. In Methods of soil analysis Part 3, ed. D. L. Sparks, pp. 961–1010. Madison, WI: Soil Science Society of America.
Nepstad, D. C., Uhl, C., and Serrão, E. A. S. 1991. Recuperation of a degraded Amazonian landscape: forest recovery and agricultural restoration. Ambio 20:248–255.
Nepstad, D. C., de Carvalho, C. R., Davidson, E. A., Jipp, P. H., Lefebvre, P. A., Negreiros, G. H., da Silva, E. D., Stone, T. A., Trumbore, S. E., and Vieira, S. 1994. The role of deep roots in the hydrological and carbon cycles of Amazonian forests and pastures. Nature 372: 666–669.
Nepstad, D. C., Stickler, C. M., and Almeida, O. T. 2006. Globalization of the Amazon soy and beef industries: opportunities for conservation. Conservation Biology 20:1595–1603.
Neufeldt, H., Resck, D. V. S., and Ayarza, M. A. 2002. Texture and land-use effects on soil organic matter in Cerrado Oxisols, Central Brazil. Geoderma 107:151–164.
Nierop, K. G. J., Pulleman, M. M., and Marinissen, J. C. Y. 2001. Management induced organic matter differentiation in grassland and arable soil: a study using pyrolysis techniques. Soil Biology and Biochemistry 33:755–764.
Northcott, G. L. and Jones, K. C. 2000. Experimental approaches and analytical techniques for determining organic compound bound residues in soil and sediment. Environmental Pollution 108:19–43.
Oades, J. M. 1988. The retention of organic matter in soils. Biogeochemistry 5:35–70.

Oades, J. M., Waters, A. G., Vassallo, A. M., Wilson, M. A., and Jones, G. P. 1988. Influence of management on the composition of organic matter in a red-brown earth as shown by ^{13}C nuclear magnetic resonance. Australian Journal of Soil Research 26:289–299.
Oliver, C., and Larson, B. 1996. Forest Stand Dynamics. New York: John Wiley and Sons.
Paniagua, A., Kammerbauer, J., Avedillo, M., and Andrews, A. M. 1999. Relationship of soil characteristics to vegetation successions on a sequence of degraded and rehabilitated soils in Honduras. Agriculture, Ecosystems and Environment 72:215–225.
Parrotta, J. A., and Knowles, O. H. 1999. Restoration of tropical moist forests on bauxite-mined lands in the Brazilian Amazon. Restoration Ecology 7:103–116.
Parton, W. J., Schimel, D. S., Cole, C. V., and Ojima, D. S. 1987. Analysis of factors controlling soil organic matter levels in Great Plains grasslands. Soil Science Society of America Journal 51:1173–1179.
Pascarella, J. B., Aide, T. M., and Zimmerman, J. K. 2004. Short-term response of secondary forests to hurricane disturbance in Puerto Rico, USA. Forest Ecology and Management 199:379–393.
Paul, E. A. 1984. Dynamics of organic matter in soils. Plant and soil 76:275–285.
Paul, K. I., Polglase, P. J., Nyakuengama, J. G., and Khanna, P. K. 2002. Change in soil carbon following afforestation. Forest Ecology and Management 168:241–257.
Paul, E. A., Morris, S. J., Conant, R. T., and Plante, A. F. 2006. Does the acid hydrolysis-incubation method measure meaningful soil organic carbon pools? Soil Science Society of America Journal 70:1023–1035.
Peña-Claros, M. 2003. Changes in forest structure and species composition during secondary forest succession in the Bolivian Amazon. Biotropica 35:450–461.
Poirier, N., Sohi, S. P., Gaunt, J. L., Mahieu, N., Randall, E. W., Powlson, D. S., and Evershed, R. P. 2005. The chemical composition of measurable soil organic matter pools. Organic Geochemistry 36:1174–1189.
Post, W. M., Emanuel, W. R., Zinke, P. J., and Stangenberger, A. G. 1982. Soil carbon pools and world life zones. Nature 298:156–159.
Post, W. M., and Kwon, K. C. 2000. Soil carbon sequestration and land-use change: processes and potential. Global Change Biology 6:317–327.
Potvin, C., Whidden, E., and Moore, T. 2004. A case study of carbon pools under three different land-uses in Panama. Climatic Change 67:291–307.
Powers, J. S. 2004. Changes in soil carbon and nitrogen after contrasting land-use transitions in northeastern Costa Rica. Ecosystems 7:134–146.
Powers, J. S., Read, J. M., Denslow, J. S., and Guzman, S. M. 2004. Estimating soil carbon fluxes following land-cover change: a test of some critical assumptions for a region in Costa Rica. Global Change Biology 10:170–181.
Pregitzer, K. S., and Euskirchen, E. S. 2004. Carbon cycling and storage in world forests: biome patterns related to forest age. Global Change Biology 10:1–26.
Preston, C. M. 2001. Carbon-13 solid-state NMR of soil organic matter - using the technique effectively. Canadian Journal of Soil Science 81:255–270.
Ramakrishnan, P. S., and Toky, O. 1983. Secondary succession following slash and burn agriculture in Northeastern India. 2. Nutrient cycling. Journal of Ecology 71:747–758.
Rasiah, V., Florentine, S. K., Williams, B. L., and Westbrooke, M. E. 2004. The impact of deforestation and pasture abandonment on soil properties in the wet tropics of Australia. Geoderma 120:35–45.
Read, L., and Lawrence, D. 2003a. Recovery of biomass following shifting cultivation in dry tropical forests of the Yucatan. Ecological Applications 13:85–97.
Read, L., and Lawrence, D. 2003b. Litter nutrient dynamics during succession in dry tropical forests of the Yucatan: regional and seasonal effects. Ecosystems 6:747–761.
Redondo-Brenes, A., and Montagnini, F. 2006. Growth, productivity, aboveground biomass, and carbon sequestration of pure and mixed native tree plantations in the Caribbean lowlands of Costa Rica. Forest Ecology and Management 232:168–178.
Reiners, W. A., Bouwman, A. F., Parsons, W. J. F., and Keller, M. 1994. Tropical rain forest conversion to pasture: changes in vegetation and soil properties. Ecological Applications 4:363–377.

Rhoades, C. C., Eckert, G. E., and Coleman, D. C. 2000. Soil carbon differences among forest, agriculture, and secondary vegetation in lower montane Ecuador. Ecological Applications 10:497–505.
Rice, A. Y., Pyle, E. H., Saleska, S. R., Hutyra, L., Palace, M., Keller, M., de Camargo, P. B., Portilho, K., Marques, D. F., and Wofsy, S. C. 2004. Carbon balance and vegetation dynamics in an old-growth Amazonian forest. Ecological Applications 14:S55–S71.
Rudel, T. K., Bates, D., and Machinguiashi, R. 2002. A Tropical forest transition? Agricultural change, out-migration, and secondary forests in the Ecuadorian Amazon. Annals of the Association of American Geographers 92:87–102.
Ruiz, J., Fandiño, M. C., and Chazdon, R. L. 2005. Vegetation structure, composition, and species richness across a 56-year chronosequence of dry tropical forest on Providencia Island, Colombia. Biotropica 47:520–530.
Russell, A. E., Cambardella, C. A., Ewel, J. J., and Parkin, T. B. 2004. Species, rotation, and life-form diversity effects on soil carbon in experimental tropical ecosystems. Ecological Applications 14:47–60.
Saldarriaga, J. G., West, D. C., Tharp, M. L., and Uhl, C. 1988. Long-term chronosequence of forest succession in the upper Rio Negro of Colombia and Venezuela. Journal of Ecology 76:938–958.
Salimon, C. I., Davidson, E. A., Victoria, R. L., and Mello, A. W. F. 2004. CO_2 flux from soil in pastures and forests in southwestern Amazonia. Global Change Biology 10:833–843.
Sánchez-de León, Y., and Zou, X. 2004. Plant influences on native and exotic earthworms during secondary succession in old tropical pastures. Pedobiologia 48:215–226.
Sanger, L. J., Anderson, J. M., Little, D., and Bolger, T. 1997. Phenolic and carbohydrate signatures of organic matter in soils developed under grass and forest plantations following changes in land use. European J of Soil Science 48:311–317.
Saynes, V., Hidalgo, C., Etchevers, J. D., and Campo, J. E. 2005. Soil C and N dynamics in primary and secondary seasonally dry tropical forests in Mexico. Applied Soil Ecology 29:282–289.
Schlesinger, W. H. 1977. Carbon balance in terrestrial detritus. Annual Review of Ecology and Systematics 8:51–81.
Schroth, G., D'Angelo, S. A., Teixeira, W. G., Haag, D., and Lieberei, R. 2002. Conversion of secondary forest into agroforestry and monoculture plantations in Amazonia: consequences for biomass, litter and soil carbon stocks after 7 years. Forest Ecology and Management 163.
Schwendenmann, L., and Pendall, E. 2006. Effects of forest conversion into grassland on soil aggregate structure and carbon storage in Panama: evidence from soil carbon fractionation and stable isotopes. Plant and Soil 288:217–232.
Scow, K. M. 1997. Soil Microbial Communities and Carbon Flow in Agroecosystems. Ecology in Agriculture. Sab Diego, CA: Academic Press.
Silver, W. L., Scatena, F. N., Johnson, A. H., Siccama, T. G., and Watt, F. 1996. At what temporal scales does disturbance affect belowground nutrient pools? Biotropica 28:441–457.
Silver, W. L., Ostertag, R., and Lugo, A. E. 2000a. The potential for carbon sequestration through reforestation of abandoned tropical agricultural and pasture lands. Restoration Ecology 8:1–14.
Silver, W. L., Neff, J., McGroddy, M., Veldkamp, E., and Keller, M. 2000b. Patterns in soil chemical properties and root biomass along a soil texture gradient in a lowland Amazonian tropical forest. Ecosystems 3:193–201.
Silver, W. L., Kueppers, L. M., Lugo, A. E., Ostertag, R., and Matzek, V. 2004. Carbon sequestration and plant community dynamics following reforestation of tropical pasture. Ecological Applications 14:1115–1127.
Smith, K., Gholz, H. L., and de Assis Oliveira, F. 1998. Litterfall and nitrogen-use efficiency of plantations and primary forest in the eastern Brazilian Amazon. Forest Ecology and Management 109:209–220.
Snedaker, S. C. 1970. Ecological studies on tropical moist forest succession in eastern lowland Guatemala. Ph.D. Thesis. University of Florida, Gainesville.
Soares-Filho, B., Alencar, A., Nepstad, D., Cerqueia, G., Diaz, M. del C. V., Rivero, S., Solórzano, L., and Voll, E. 2004. Simulating the response of land-cover changes to road paving and governance along a major Amazon highway: the Santarém-Cuiabá corridor. Global Change Biology 10:745–764.

Sollins, P., Homann, P., and Caldwell, B. A. 1996. Stabilization and destabilization of soil organic matter: mechanisms and controls. Geoderma 74:65–105.
Sommer, R., Denich, M., and Vlek, P. L. G. 2000. Carbon storage and root penetration in deep soils under small-farmer land-use systems in the Eastern Amazon region, Brazil. Plant and soil 219:231–241.
Spiel, H. 1996. The importance of unpredictable and short-term environmental extremes for biodiversity of Oribatid mites. Biodiversity Letters 3:26–34.
Stern, M., Quesada, M., and Stoner, K. E. 2002. Changes in composition and structure of a tropical dry forest following intermittent cattle grazing. Revista de Biología Tropical 50:1021–1034.
Stevenson, F. J. 1982. Organic matter transformations, distribution, and function in soil. Pages 1–25 in Humus Chemistry: Genesis, Composition, Reactions. Toronto: John Wiley and Sons.
Steward, C. 2007. From colonization to "environmental soy": a case study of environmental and socio-economic valuation in the Amazon soy frontier. Agriculture and Human Values 24:107–122.
Swanston, C. W., Caldwell, B. A., Homann, P. S., Ganio, L., and Sollins, P. 2002. Carbon dynamics during a long-term incubation of separate and recombined density fractions from seven forest soils. Soil Biology and Biochemistry 34:1121–1130.
Swanston, C. W., Torn, M. S., Hanson, P. J., Southon, J. R., Garten, C. T., Hanlon, E. M., and Ganio, L. 2005. Initial characterization of processes of soil carbon stabilization using forest stand-level radiocarbon enrichment. Geoderma 128:63–79.
Szott, L. T., Palm, C. A., and Davey, C. B. 1994. Biomass and litter accumulation under managed and natural tropical fallows. Forest Ecology and Management 67:177–190.
Templer, P. H., Groffman, P., Flecker, A., and Power, A. 2005. Land use change and soil nutrient transformations in the Los Haitises region of the Dominican Republic. Soil Biology and Biochemistry 37:215–225.
Tergas, L. E., and Popenoe, H. C. 1971. Young secondary vegetation and soil interactions in Izabal, Guatemala. Plant and Soil 34:675–690.
Thomlinson, J. R., Serrano, M. I, López, T.del M., Aide, T. M., and Zimmerman, J. K. 1996. Land-use dynamics in a post-agricultural Puerto Rican landscape (1936–1988). Biotropica 28:525–536.
Thomlinson, J. R., and Rivera, L. Y. 2000. Suburban grown in Luquillo, Puerto Rico: some consequences of development on natural and semi-natural systems. Landscape and Urban Planning 40:15–23.
Tiessen, H., Cuevas, E., and Chacon, P. 1994. The role of soil organic matter in sustaining soil fertility. Nature 371:783–785.
Torn, M. S., Trumbore, S. E., Chadwick, O. A., Vitousek, P. M., and Hendricks, D. M. 1997. Mineral control of soil organic carbon storage and turnover. Nature 389:170–173.
Trumbore, S. E. 1993. Comparison of carbon dynamics in tropical and temperate soils using radiocarbon measurements. Global Biogeochemical Cycles 7:275–290.
Trumbore, S. E., Davidson, E. A., de Camargo, P. B., Nepstad, D. C., and Martinelli, L. A. 1995. Belowground cycling of carbon in forests and pastures of Eastern Amazonia. Global Biogeochemical Cycles 9:515–528.
Trumbore, S. E., and Zheng, S. 1996. Comparison of fractionation methods for soil organic matter ^{14}C analysis. Radiocarbon 38:219–229.
Trumbore, S. 2000. Age of soil organic matter and soil respiration: radiocarbon constraints on belowground C dynamics. Ecological Applications 10:399–411.
Uhl, C., Clark, K., Clark, H., and Murphy, P. 1981. Early plant succession after cutting and burning in the upper Rio Negro region of the Amazon basin. Journal of Ecology 69:631–649.
Uhl, C., Jordan, C., Clark, K., Clark, H., and Herrera, R. 1982. Ecosystem recovery in Amazon caatinga forest after cutting, cutting and burning and bulldozer treatments. Oikos 38: 313–320.
Uhl, C. 1987. Factors controlling succession following slash-and-burn agriculture in Amazonia. Journal of Ecology 75:377–407.
Uhl, C., Buschbacher, R., and Serrão, E. A. S. 1988. Abandoned pastures in Eastern Amazonia. I. Patterns of plant succession. Journal of Ecology 76:663–681.

Veddeler, D., Schulze, C. H., Steffan-Dewenter, I., Buchori, D., and Tscharntke, T. 2005. The contribution of tropical secondary forest fragments to the conservation of fruit-feeding butterflies: effects of isolation and age. Biodiversity and Conservation 14:3577–3592.

Veldkamp, E. 1994. Organic carbon turnover in three tropical soils under pasture after deforestation. Soil Science Society of America Journal 58:175–180.

Vitorello, V. A., Cerri, C. C., Andreaux, F., Feller, C., and Victoria, R. L. 1989. Organic matter and natural carbon-13 distribution in forested and cultivated Oxisols. Soil Science Society of America Journal 53:773–778.

Walker, R. 2003. Mapping process to pattern in the landscape change of the Amazonian frontier. Annals of the Association of American Geographers 93:376–398.

Walker, S. M., and Desanker, P. V. 2004. The impact of land use on soil carbon in Miombo woodlands of Malawi. Forest Ecology and Management 203:345–360.

Walkley, A., and Black, A. 1934. An examination of the Degtjareff method for determining soil organic matter, and a proposed modification of the chromic acid titration method. Soil Science Society of America Journal 37:29–38.

Watson, R. T., Noble, I. R., Bolin, B., Ravindranath, N. H., Verando, D. J., and Dokken, D. J, Eds. 2000. Land Use, Land-Use Change, and Forestry. Special Report of the Intergovernmental Panel on Climate Change. Cambridge: Cambridge University Press.

Weaver, P. L., Birdsey, R. A. and Lugo, A. E. 1987. Soil organic matter in secondary forests of Puerto Rico. Biotropica 19:17–23.

Weaver, P. L., and Birdsey, R. A. 1990. Growth of secondary forest in Puerto Rico between 1980 and 1985. Turrialba 40:12–22.

Weaver, P. L. 1995. Secondary forest management. In Management and Rehabilitation of Degraded Lands and Secondary Forests in Amazonia. Proc. International Symposium/Workshop, ed. J. A. Parrotta and M. Kanashiro, pp. 117–128. Santarém, Pará, Brazil.

Werner, P. 1984. Changes in soil properties during tropical wet forest succession in Costa Rica. Biotropica 16:43–50.

Williams-Linera, G. 1983. Biomass and nutrient content in two successional stages of tropical wet forest in Uxpanapa, Mexico. Biotropica 15:275–284.

Wright, J. A., DiNicola, A., and Gaitan, E. 2000. Latin American forest plantations: opportunities for carbon sequestration, economic development, and financial returns. Journal of Forestry 98:20–23.

Yan, E.-R., Wang, X.-H., and Huang, J.-J. 2006. Shifts in plant nutrient use strategies under secondary forest succession. Plant and Soil 289:187–197.

Zarin, D. J., Davidson, E. A., Brondizio, E., Vieira, I. C. G, Sá, T., Feldpausch, T., Schuur, E. A. G, Mesquita, R., Moran, E., Delamonica, P., Ducey, M. J., Hurtt, G. C., Salimon, C., and Denich, M. 2005. Legacy of fire slows carbon accumulation in Amazonian forest regrowth. Frontiers in Ecology and the Environment:365–369.

Zimmerman, J. K., Aide, T. M., Rosario, M., Serrano, M., and Herrera, L. 1995. Effects of land management and a recent hurricane on forest structure and composition in the Luquillo Experimental Forest, Puerto Rico. Forest Ecology and Management 77:65–76.

Zou, X., Zucca, C. P., Waide, R. B., and McDoweel, W. H. 1995. Long-term influences of deforestation on tree species composition and litter dynamics of a tropical rain forest in Puerto Rico. Forest Ecology and Management 78:147–157.

Chapter 3
Post Sugar Cane Succession in Moist Alluvial Sites in Puerto Rico

Oscar Abelleira Martínez and Ariel E. Lugo

Abstract Sugar cane cultivation spanned over two centuries in Puerto Rico, at one time covering 14% of the island's total area. The most productive plantations were located in alluvial valleys that were deforested before there was an opportunity to record the species composition and structural development of the original native forests. This has led to speculation about the nature of these forests. With the abandonment of sugar cane cultivation in the twentieth century, sugar cane fields have been converted to pastures and a variety of secondary forests. One of these is a new forest type dominated by *Spathodea campanulata*, an invasive, wind-dispersed, and evergreen or nearly deciduous introduced species. We studied three stands of these forests at ages 25–40 years old on moist alluvial sites and found high basal area, intermediate tree density, high Complexity Index, and relatively low species richness. We recorded 19 tree species with dbh $\geq$ 2.5 cm and 25 understory species equivalent to a total of 31 species. Of these, 11 species were introduced and 20 were native, including two endemic tree species. The large tree stratum (dbh $\geq$ 10 cm) had high dominance of *S. campanulata*, 6 tree species, and a mean height of 12 m. Compared to large trees, small trees ($\geq$ 2.5 <10 cm dbh) had higher species richness, and less dominance, while the understory had the highest species richness and lowest dominance. In spite of centuries of cultivation, forest soils were nutrient-rich, which allowed trees to attain large diameters, and produce nutrient-rich litter. The structural parameters of alluvial *S. campanulata* forests are comparable with those of remnant alluvial moist forests elsewhere on the north coast of Puerto Rico but the species composition is different. Periodic inundation of these forests appears to favor flood-tolerant species such as *S. campanulata*, and extends the period of domination of this species relative to its behavior in other geologic substrates in Puerto Rico. This process is exacerbated by long-term and extensive history of cultivation that reduced the species pool of seed sources for alluvial forests in the region. It will be difficult to restore the species complement of the original native forests because the disturbance regime of alluvial sites in Puerto Rico will continue to favor the presence, albeit reduced, of introduced species in the mature phases of these forests.

R. W. Myster, *Post-Agricultural Succession in the Neotropics.*

3.1 Introduction

Sugar cane was first planted in Puerto Rico in 1953 (Álvarez Nazario 1982, Domínguez Cristóbal 2000). Considerable deforestation accompanied its development as a crop throughout the seventeenth and eighteenth centuries (Domínguez Cristóbal 1989a, 1989b). Eventually, sugar cane became the major agricultural crop in Puerto Rico for most of the nineteenth and twentieth century. At a peak production of over 1 million tons of sugar in the 1950s, the area covered by sugar cane plantations was over 120,000 ha or about 14% of the island's land area (Picó 1969). Most of this cane was planted on alluvial soils on the coastal plains of Puerto Rico. In many instances, drainage canals were dug to protect sugar cane from flooding. These drainage works allowed cultivation to extend to wetland soils and caused the displacement of coastal wetlands.

Alluvial valleys were the first lands to be significantly deforested in Puerto Rico because they represented the best agricultural conditions available (flat lands, deep fertile soils). As a result of this early activity, there is no documentation of the structure and species composition of native vegetation in these valleys before they were converted for agricultural use. The vegetation of alluvial valleys was probably arrayed along a hydroperiod gradient that included forested and non-forested wetlands in flooded areas and moist tropical forests in locations with short or no hydroperiod (Lugo 2005). Today, alluvial soils attract urban development, and forest cover on alluvial soils is minimal as is the area protected (Table 3.1).

Sugar cane cultivation declined after the 1950s and was terminated in the 1980s. The abandonment of sugar cane fields results in different succession pathways depending on the location of the field and the timing of the abandonment. On higher ground, invasions of guava trees (*Psidium guajava*) hindered initial efforts by Spaniards to establish pastures on alluvial soils (Domínguez Cristóbal 2000). Today, this tree still forms thickets on abandoned pasturelands. With the abandonment of the drainage canals, herbaceous wetlands as well as mangrove and *Pterocarpus* forests expanded in area in the coastal zone (Lugo 2006). The expansion of mangroves is confined to saline soils and the expansion of *Pterocarpus* forests has been limited to stands near rivers or other wetlands. Many sites not invaded by guava trees allowed pastures to become a major land cover type by the 1970s (Ramos and Lugo 1994). In moist transition dry alluvial valleys, mature vegetation is dominated by *Bucida buceras* and *Guazuma ulmifolia* and secondary forests with

Table 3.1 Area of lowland alluvial habitat and forest cover and protected status of forests in lowland alluvial habitats in Puerto Rico (Helmer et al. 2002). Data are for 1991–1992. There are smaller areas of alluvial habitat in montane conditions, but these areas were not included here

Statistic	Dry alluvial	Moist alluvial	All Puerto Rico
Zone area (ha)	45,179	144,767	869,156
Percent protected	5.5	4.5	
Forest area (ha)	5,368	16,224	365,160
Forest area protected (ha)	211	729	34,026
Percent of zone area	0.5	0.5	
Percent of island-wide land development	10.5	26.0	100

mixtures of native and introduced species are developing on abandoned sugar cane fields in those dry alluvial valleys (Lugo 2005). However, in moist alluvial valleys of the north and east coast, abandoned sugar cane fields have been replaced in many locations by forests dominated by *Spathodea campanulata* (Chinea 2002).

Spathodea campanulata belongs to the family Bignoniaceae and its native to tropical Africa in a band around the equator between 12°N and 12°S from sea level to 1,200 m elevation in a wide variety of soils (Little and Wadsworth 1964, Francis 2000). *Spathodea campanulata* is shade intolerant, and produces lateral roots, light seeds (125,000 seeds/kg), and buttresses (Little and Wadsworth 1964, Francis 2000). It is also a pioneer species (Whitmore 1990) that was introduced to Puerto Rico for its ornamental value and is now considered a fast-growing invasive species (Francis 2000, Chinea 2002). The earliest specimen in island herbaria is dated 1912 (UPR Botanical Garden Herbarium). There is evidence that *S. campanula*ta shows histological and biochemical adaptation to flooded soils (Medri and Correa 1985).

The invasion of deforested alluvial soils by *S. campanulata* gave us the opportunity to study post sugar cane succession on moist alluvial sites. This is important because these sites have not been forested for centuries and there is much speculation about the potentially impressive structural development of native alluvial forest stands (Lugo 2005). Gleason and Cook (1926) considered the forests on alluvial soils, termed playa forests by them, as the climax community for the northern coastal plain. They listed the following species as probable components of this forest type: *Calophyllum antillanum* (synonymous with *C. calaba*) and *Tabebuia pallida* (synonymous with *T. heterophylla*). Wadsworth (1950) added *Manilkara nitida* (synonymous with *M. bidentata*), and *Mastichodendron foetidissimum* (synonymous with *Sideroxylon foetidissimum*) as likely timber species in these forests. Wadsworth also estimated that alluvial forests attained 26 m in height.

Would *S. campanulata* forests mimic the native stands not seen in the Island since the seventeenth century? The answer could be yes, at least structurally, if sites were not degraded. On the other hand, Chinea (2002) noted that stands dominated by *S. campanulata* are floristically far removed from the original vegetation on alluvial sites and it is not clear how much time will be required for the re-establishment of the original vegetation, or if that is even possible (see Lugo and Helmer 2004). Nevertheless, in this study we describe the structure and species composition of alluvial forest stands dominated by *S. campanulata*, infer some aspects of their functioning, and compare this succession with other post agricultural successions in Puerto Rico (Aide et al. 2000, Chinea 2002) and elsewhere (Muthuramkumar et al. 2006).

3.2 The Alluvial Environment

Our study sites are located in the subtropical moist forest life zone (*sensu* Holdridge 1967) with a mean annual temperature of about 25°C (NOAA 1998) and a mean annual rainfall between 1971 and 2000 of 1,443 mm (http://www.srh.noaa.gov/sju/pr_mean_annual_pcp.jpg). The alluvial soils of the Río Cibuco valley reach maximum depths of 85.3 m. According to Abruña et al. (1977) these alluvial soils on the north coast constitute some of the finest soils of

Table 3.2 Site characteristics, stand age, and land use history. Soil information is from Acevido (1982). For the land use information we used air photos from 1963, 1971, 1977, and 1985. We also interviewed people and used Ramos and Lugo (1994) for the 1977 information. Elevation data were obtained in the field with a Trimble GPS unit at ± 2 m precision

Site	Soil order	Soil series and slope (%)	Past agricultural use and stand age (yr)	Elevation (m)
Cibuco I	Inceptisol	Coloso, clay cienosa (0)	Sugar cane (27)	17.6
Cibuco II	Mollisol	Toa Alta, clay marga (0)	Sugar cane (38)	7.0
Paso del Indio	Mollisol	Toa Alta, clay marga (0)	Sugar cane and pasture (25)	15.7

Puerto Rico and are classified as providing prime agricultural lands, which means they reach the highest standards for food production in terms of growth conditions (water, nutrients, structure, etc.) and resistance to erosion. The Río Cibuco Valley has been populated and farmed since the sixteenth century (Álvarez Nazario 1982).

The US Army Corps of Engineers (1973) documented the history of flooding at Río Cibuco and Río Indio valleys and reported the following flood levels in the vicinity of our sites for the standard flood expected under extreme storm events (m above mean sea level): Río Indio- 16.8, Cibuco I- 22, and Cibuco II- 13. During these events the river rises up to 5 m above its normal level at rates of up to 1 m per hour, for time periods of up to 8 hours, and the critical stage lasting for up to 27 hours. Given the elevations of the floodplain at our sites (Table 3.2), the sites flood between 1.1 and 6.0 m during peak floods. Historical floods are known since the 1770s when the town of El Naranjal, now Vega Baja, was established in the region. These data show a recurrent flooding pattern that trees must adapt to because floods can occur at any month of the year (Army Corps of Engineers 1973).

3.3 Site Selection and Methods

We selected three sites in the alluvial floodplains of Puerto Rico's north coast between Vega Baja and Vega Alta (Fig. 3.1). We focused on closed canopy secondary forests dominated by *S. campanulata* with 25–40 years of age and used high resolution aerial photos to locate sites, and field visits to interview land owners about past land uses of their lands. We strived for sites as similar as possible in terms of soils, life zone conditions, topography, and previous land use (Table 3.2).

At each site, we selected 10 random points in each of two parallel 100-m transects. The transects were at least 15 m apart and were also randomly located. At each point we identified to species the four trees with diameter at breast height (dbh) ≥ 10 cm nearest to the point (one in each of four imaginary quarters surrounding the point) and measured their dbh at 1.37 m and distance to the point (Cottam and Curtis 1956). We repeated this procedure for trees with dbh ≥2.5 <10 cm. The total sampling was 160 trees per site (80 ≥ 10cm dbh and 80 ≥ 2.5 cm < 10 cm dbh). For trees with multiple stems we measured each stem's diameter at dbh for the estimate of basal area but counted the tree once for density estimates. We refer to trees with

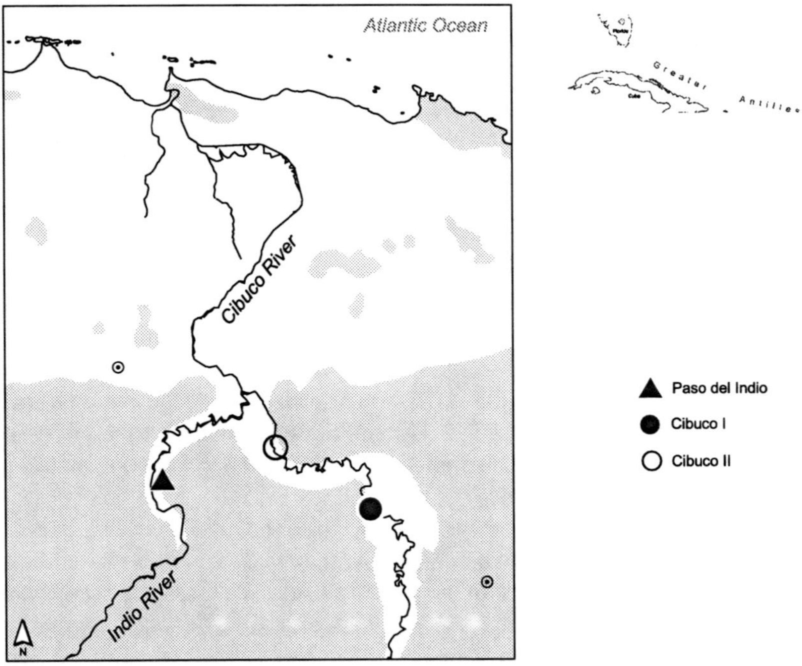

Fig. 3.1 Location of study sites

dbh ≥ 10 cm as large trees and the ones with lower dbh as small trees. We measured the height of 10 large trees in each stand with a clinometer.

To characterize understory vegetation, we randomly selected four additional points per site and established 5 × 5 m plots at each point. All woody plants with basal diameter < 2.5 cm inside these plots were identified to species and counted. These plants were mostly shrubs, small trees, saplings, and seedlings. We constructed species/area curves for each site using data from the four plots per site. Each tree and understory species was classified as native or introduced. If native, we further checked those that were endemic. We used Little and Wadsworth (1964) and Little et al. (1974) for the nomenclature of species.

With the data collected from transects, we followed Cottam and Curtis (1956) to estimate tree density, basal area, frequency and Importance Value (IV), expressed in percent. For understory data, we divided the number of plants of each species by the area of the plot to estimate stem density and estimated an IV in terms of frequency and density, expressed in percent.

At each site we collected loose litter using five 25 × 25 cm randomly located plots per site. The litter was sorted in the field into leaves, wood, and miscellaneous. Where possible, we separated *S. campanulata* leaves from other leaves, and classified leaves into recently fallen, fragmented, and old. In the laboratory, samples were oven dried at 60°C, weighed, ground, and analyzed for a suite of elements (Al, Ca, Fe, Mg, Mn, Na, P, C, S, and N) as follows. A sub sample was oven dried at 105°C for 24 hrs and a moisture factor calculated and applied to each analysis

(Wilde 1979). Ground material was digested with concentrated HNO_3 and 30% H_2O_2 and analyzed with a Spectro plasma emission spectrometer (Spectro Ciros ICP) for P, K, Ca, Mg, Al, Mn, Fe, and Na using the digestion method recommended by Luh-Huang and Schulte (1985). Total Nitrogen, total Carbon and total Sulfur were analyzed using the dry combustion method by means of a LECO CNS-2000 Analyzer. The procedure used is a modified version of the method in LECO Corporation (1995). In the dry combustion method a small weighed sample is combusted by heating it to a high temperature (1,300°C) inside a resistance furnace and in a stream of purified oxygen.

Precision for most analyses was assured by running samples of known chemical composition every forty determinations (for total C, total N and total S every 20 determinations). These control samples: citrus leaves (NBS-1572), peach leaves (NIST-1547), and pine needles (NIST-1575) were obtained from the National Institute of Standards and Technology, USA. The calibration standards: tobacco leaves, orchard leaves and alfalfa, used in the total C, N, and S analysis were obtained from LECO Corporation (St. Joseph, MI).

We also collected soil samples at two depths: 0–15 and 15–30 cm. We dug five pits per site at randomly located points. Sampling included known soil volumes for bulk density determinations and soil for chemical analyses. Samples were transported to the laboratory in pre-weighted cloth bags and dried at 100°C to constant weight. Chemical analyses included the same elements as above plus Loss on Ignition. Samples analyzed for total P, K, Ca, Mg, Al, Mn, and Fe were treated the same as litter samples above. For exchangeable ions such as Ca, Mg, Na, and Al we used the 1N KCl extraction method and for Fe, Mn, K, and P we used the Olsen-EDTA(NH_4-EDTA-$NaHCO_3$) procedure recommended by Anderson and Ingram (1993). For total N, C, and S we used the same methods as in with litter above. Soil pH was determined in a solution 1:1 (soil:water/soil:1N KCl) using an Orion Ionalyzer Model 901 with a combination pH electrode (McLean 1982).

We only report the stocks of nutrients and organic matter in loose litter and soil. To estimate element stocks in loose litter, we multiplied the mean dry weight per unit area of litter component by the corresponding element concentration. For soil element stocks, we multiplied the bulk density of each sample by depth of sampling and by the corresponding element concentration. Concentration and soil bulk density data are available from the authors.

3.4 Forest Development After Abandonment

3.4.1 Species Composition

We recorded 18 tree species above the understory (Table 3.3) and 26 species in the understory (Table 3.4) in the three alluvial stands. Of these, 13 species occurred both as trees and in the understory, five occurred only as trees but not in the understory, and 13 were found only in the understory. All species in the understory were tree or woody shrub species. Two species were endemic to Puerto Rico: *Thespesia*

Table 3.3 Importance Value of tree species in alluvial stands of *Spathodea campanulata* in the north coast of Puerto Rico. The sizes of trees correspond to the diameter at breast height (1.37 m)

	Cibuco I		Cibuco II		Paso del Indio	
	≥ 10	≥ 2.5 < 10	≥ 10	≥ 2.5 < 10	≥ 10	≥ 2.5 < 10
Species	(cm)					
*Spathodea campanulata**	90.0	49.9	91.4	83.3	91.8	89.9
*Cassia siamea**	4.4	6.9				
*Albizzia procera**	3.7				5.7	
*Terminala catappa**	1.8	6.1	6.7	5.5		
Casearia guianensis		12.8		4.0		1.8
Guarea guidonia		11.0		5.5		
Casearia decandra		2.6				
Zanthoxylum martinicense		1.8				
Spondias mombin		1.4			2.5	
*Erythrina poeppigiana**		1.3				
Casearia sylvestris		1.2				1.8
Cupania americana		1.2				
Eugenia biflora		1.2				
Ocotea coriacea		1.2				
*Thespesia grandiflora***		1.2	1.8			
Callophylum calaba				2.2		3.6
Cordia laevigata						2.1
Bucida buceras						1.8

*introduced
**endemic

grandiflora, which occurred both as large trees and in the understory, and *Roystonea borinquena*, which was found only in the understory. Thus, a total of 31 species were recorded in these alluvial stands. Five introduced species occurred as trees, 8 occurred in the understory, and two of these species were shared between strata. There were a total of 11 introduced species in the list or 35% of the species tallied. The species/area curve (Fig. 3.2) saturated with the plots sampled, suggesting that the sampling was sufficient to characterize the floristic diversity of the understory. The species composition varied among stands (Tables 3.3 and 3.4; Fig. 3.2). Cibuco II had more tree species in the understory (Table 3.4), than either Paso del Indio or Cibuco I, but Cibuco I had more tree species in the tree stratum than the other two sites (Table 3.3).

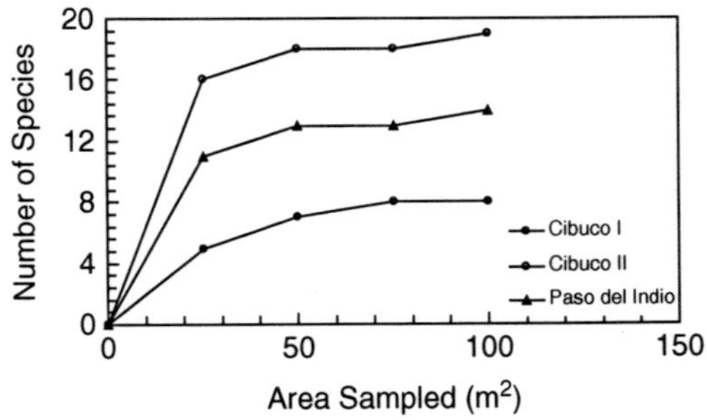

Fig. 3.2 Species/area curves for understory vegetation in alluvial stands of *Spathodea campanulata* in northern Puerto Rico. All woody plants with a dbh < 2 cm basal diameter were included

Table 3.4 Importance Value for species in the understory of alluvial stands of *Spathodea campanulata* in northern Puerto Rico

Species	Cibuco I	Cibuco II	Paso del Indio
Andira inermis	40.7	13.0	26.2
*Spathodea campanulata**	15.8	4.7	13.1
Guarea guidonia	11.9	26.1	4.5
Calophyllum calaba	10.3	6.5	14.6
Casearia guianensis	10.3	1.9	3.3
Ocotea coriacea	4.2	7.3	
*Roystonea borinquena***	3.4	1.6	4.5
Inga laurina	3.4		
*Melicoccus bijugatus**		7.0	6.5
Casearia sylvestris		6.9	
*Terminala catappa**		6.8	
Casearia decandra		3.3	
*Delonix regia**		3.1	
Cinnamomum elongatum		1.9	
Eugenia monticola		1.9	
*Cochlospermum vitifolium**		1.9	
Cupania americana		1.6	3.3
Guettarda scabra		1.6	
*Syzygium jambos**		1.6	
*Leucaena leucocephala**		1.6	
Erythroxylon brevipes			7.7
Eugenia biflora			3.3
*Thespesia grandiflora***			3.3
Bucida buceras			3.3
*Psidum guajava**			3.3
Spondias mombin			3.3

*introduced
**endemic

Table 3.5 Structure of alluvial stands of *Spathodea campanulata* in the north coast of Puerto Rico. Tree dimensions refer to diameter at breast height (1.37 m). The area sampled for understory was 100 m^2. Empty cells mean the information is not available

Site	Component	Stem density (no/ha)	Basal area (m^2/ha)	Number of species
Cibuco I	Trees ≥10 cm	1,071	77.0	4
	Trees ≥ 2.5 <10 cm	1,015	2.6	15
	All trees	2,087	79.5	15
	Understory	6,300		8
Cibuco II	Trees ≥10 cm	1,090	104.5	3
	Trees ≥ 2.5 <10 cm	1,096	2.5	5
	All trees	2,187	106.9	6
	Understory	17,200		19
Paso del Indio	Trees ≥10 cm	1,825	82.0	3
	Trees ≥ 2.5 <10 cm	1,482	4.8	6
	All trees	3,307	86.8	8
	Understory	4,200		14

3.4.2 Forest and Community Structure

Tree density and basal area were very similar across sites (Table 3.5). Most of the basal area of the stands was in the large trees, and the ranking of stands according to basal area was the same as their ranking for species richness: highest in Cibuco II followed by Paso del Indio and Cibuco I. The density of understory plants was higher in Cibuco II than in the other two stands. We found no significant difference in tree height between stands. The height of the measured 30 trees averaged 12.1 ± 0.5 m.

The *S. campanulata* forests had gradients of species dominance and species richness from the canopy to the understory (Fig. 3.3). Large trees had lower species

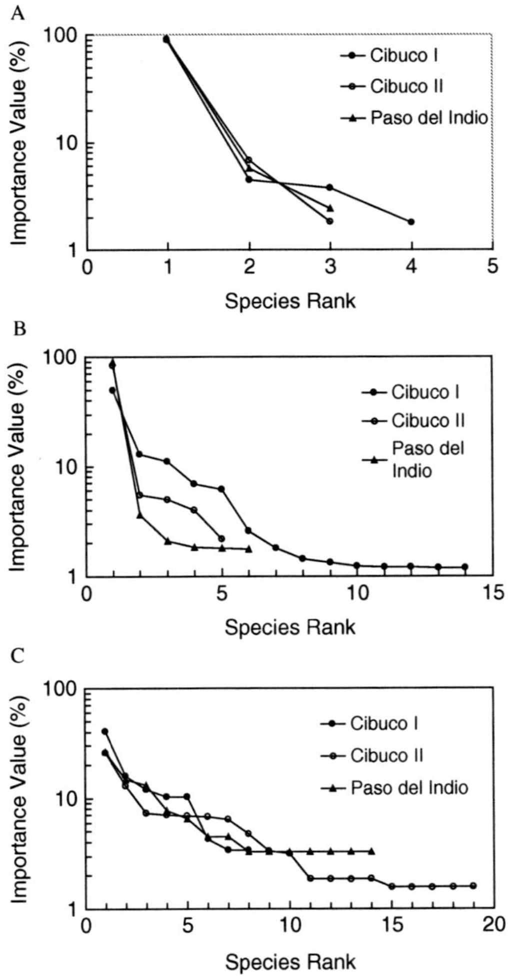

Fig. 3.3 Importance value curves for **(a)** tree species with ≥ 10 cm diameter at breast height (dbh), **(b)** tree species with dbh ≥ 2 < 10 cm dbh, and **(c)** understory vegetation with basal diameter < 2 cm in *Spathodea campanulata* forests in northern Puerto Rico

Table 3.6 Biomass of litter components of *Spathodea campanulata* alluvial forest stands in northern Puerto Rico. Data corresponds to July, 2004. Mean and standard error are included, $n = 10$. All data are in g/m^2 and values are rounded off to the nearest gram and therefore might not add to the total

Litter component	Cibuco I	Cibuco II	Paso del Indio
Recently fallen leaves	2 (0.8)	3 (1)	2 (0.5)
Old leaves	5 (2)	9 (2)	11 (3)
Fragmented leaves	38 (8)	159 (27)	114 (66)
Total leaves	45	171	127
Wood	321 (94)	364 (61)	334 (97)
Miscellaneous	36 (25)	20 (10)	9 (5)
Total	403 (90)	556 (77)	469 (125)

richness and the dominant species had a higher importance value among large trees (Fig. 3.3a), than among small trees (Fig. 3.3b), or understory plants (Fig. 3.3c). In all stands, the IV of the highest ranked large trees was uniformly higher than 95%. For smaller trees, the number of species and IV of species increased and were reduced, respectively, relative to large trees. The understory had the highest species richness and lower IV for similarly ranked species when compared to small or large trees.

3.4.3 Loose Litter and Soil Compartments

Most of the loose litter in the three alluvial stands was woody litter (Table 3.6). Leaves, mostly fragmented, were a smaller component of litter in all stands. The stock of elements in loose litter varied among stands in proportion with the total mass of loose litter (Table 3.7). Some of the differences in element stocks were due to differences in element concentrations and proportions of litter components in total loose litter. For example, leaf litter had greater element concentrations than wood and even though leaf litter was a small fraction of the total loose litter, it contained disproportionately high stocks of elements. Soil had much higher quantities of elements than loose litter and there was an insignificant difference in element stocks between the two depths studied (Table 3.7). We did not find a developed humus layer atop these soils perhaps because of insufficient time for its development, or because periodic flooding prevents the accumulation of humus.

3.5 Ecological Comparisons and Implications

3.5.1 Structure and Species Dynamics

Our data shows that even by age 25–40 years, alluvial forests dominated by *S. campanulata* develop structural complexity similar to remnant mature stands of natural alluvial forests (Table 3.8). For example, *S. campanulata* forests had a higher basal

Table 3.7 Element content of litter (kg/ha) and soil (Mg/ha) compartments of *Spathodea campanulata* alluvial forests in northern Puerto Rico. Data are rounded to the nearest kg or tenth of kg and thus total might not add. Empty cells mean no data available

	Al	Ca	Fe	K	Mg	Mn	Na	P	C	S	N	Ash
Loose litter						Paso del Indio						
Fragmented leaves	13.9	41.3	20.6	1.2	4.6	0.5	0.2	1.3	329	1.8	14.8	563
Recently fallen leaves	0.1	0.1	0.1	0.1	0.1	0.0	0.0	0.0	9	0.1	0.3	
Old leaves	0.8	2.9	1.1	0.2	0.3	0.0	0.0	0.1	44	0.2	1.6	
Total leaves	14.7	44.3	21.7	1.5	5.0	0.5	0.2	1.5	382	2.1	16.6	
Wood	9.7	39.8	14.0	2.6	4.1	0.4	0.3	2.0	1614	2.9	19.0	371
Miscellaneous	0.6	1.5	0.8	0.1	0.2	0.0	0.0	0.0	35	0.1	0.7	24
Total	25.0	85.6	36.6	4.2	9.3	1.0	0.4	3.5	2031	5.1	36.4	
						Cibuco 1						
Fragmented leaves	4.0	10.4	5.4	0.6	1.5	0.1	0.1	0.4	128	0.8	6.8	92
Recently fallen leaves	0.0	0.6	0.0	0.2	0.1	0.0	0.0	0.0	9	0.1	0.3	
Old leaves	0.2	1.6	0.3	0.1	0.2	0.0	0.0	0.0	21	0.2	0.9	
Total leaves	4.2	12.6	5.7	0.9	1.9	0.2	0.1	0.5	159	1.1	8.0	
Wood	15.5	38.6	22.3	2.5	6.1	0.6	0.2	1.7	1,448	3.5	20.9	480
Miscellaneous	0.1	1.3	0.1	2.8	0.3	0.0	0.0	0.2	184	0.4	4.1	12
Total	19.8	52.5	28.1	6.2	8.3	0.7	0.3	2.4	1,791	5.0	33.1	
						Cibuco II						
Fragmented leaves	13.3	44.6	19.8	2.2	5.4	0.5	0.2	2.2	650	4.4	35.0	432
Recently fallen leaves	0.0	1.0	0.0	0.2	0.1	0.0	0.0	0.0	14	0.2	0.4	5
Old leaves	0.1	2.3	0.2	0.3	0.3	0.0	0.0	0.1	46	0.3	1.9	11
Total leaves	13.5	48.0	20.0	2.7	5.8	0.5	0.2	2.4	710	4.9	37.3	446
Wood	10.8	55.0	15.6	3.6	5.1	0.4	0.2	3.2	1,771	5.3	37.5	409
Miscellaneous	0.7	5.1	0.9	0.3	0.4	0.0	0.0	0.3	99	0.6	5.9	27
Total	24.9	108.1	36.5	6.5	11.3	0.9	0.5	5.9	2,580	10.8	80.7	883

(continued)

Table 3.7 (continued)

Soil depth (cm)	Al	Ca	Fe	K	Mg	Mn	Na	P	C	S	N	Ash
						Paso del Indio						
0–15	20.9	15.5	31.3	0.6	5.5	0.7	0.1	0.6	126.8	28.8	0.2	2.3
15–30	20.3	12.2	32.3	0.5	5.6	0.7	0.1	0.5	110.5	18.7	0.2	1.6
Total (0–30)	41.2	27.7	63.6	1.1	11.1	1.5	0.2	1.1	237.4	47.5	0.4	3.9
						Cibuco I						
0–15	30.0	7.5	44.8	0.8	8.0	1.2	0.2	0.7	147.4	23.0	0.3	2.1
15–30	30.6	6.7	44.7	0.7	7.8	1.2	0.2	0.6	141.1	20.0	0.2	1.9
Total (0–30)	60.6	14.3	89.5	1.5	15.8	2.3	0.3	1.3	288.5	43.0	0.5	4.1
						Cibuco II						
0–15	23.2	4.8	34.3	0.7	5.9	0.8	0.1	0.6	113.9	19.2	0.2	2.0
15–30	34.5	6.2	50.2	0.9	8.5	1.3	0.2	0.7	142.1	18.7	0.2	2.2
Total (0–30)	57.7	11.0	84.5	1.5	14.4	2.1	0.3	1.2	256.0	37.9	0.5	4.2

Table 3.8 Species richness and structural parameters of lowland alluvial moist forests in Puerto Rico (Lugo 2005). Minimum diameter was 2.5 cm. The Complexity Index (CI) is the product of number of species, stem density, basal area, and height divided by 1,000. The number of species, density and basal area are per 0.1 ha. Standard error is in parenthesis. For *Pterocarpus*, *n* was 10. *Manilkara* data ($n = 9$) are from Gould et al. (2006). *Spathodea campanulata* data are from this study ($n = 3$)

Forest type	Area sampled (ha)	Number of tree species	Stem density (stems/ha)	Basal area (m^2/ha)	Height (m)	CI
*Pterocarpus**	0.1	8	1,705	42.6	18.6	118
		(2.2)	(208)	(3.7)	(1.8)	(37)
Syzygium	0.1	21	1,880	41.6	19.7	423
Manilkara	0.09	38	4,389	50.0	26	2,168
			(165)	(11.1)	(2.8)	
S.campanulata		10	2,527	91.1	17.5**	402
		(2.7)	(391)	(8.2)		

*Two *Pterocarpus* stands on alluvial soils several km east of our sites had basal areas of 73.5 and 68.9 m^2/ha (Eusse and Aide 1999)
**Mean of tallest three trees

area than any other forest on alluvial soil, but they had lower species richness and intermediate tree densities. As a result, the Complexity Index is higher than those of *Pterocarpus* wetlands, comparable to introduced *Syzygium* forest, and lower than native *Manilkara* forest. The *Manilkara* forest, a mature native stand, had taller trees, higher tree density, and higher species richness than the *S. campanulata* forest. These differences between the *S. campanulata* and *Manilkara* forest demonstrates the floristic distance between *S. campanulata* and native vegetation on alluvial soils as pointed out by Chinea (2002). However, the age differential between these two forest types is considerable as the *Manilkara* forest is probably over 200 years old compared to a maximum age of 40 years for *S. campanulata*. In spite of this difference in age, the high productivity of *S. campanulata* is evident. If we use the mean basal area in Table 3.8, a wood density of 0.26 g/cm^3 (Francis 2000), a mean stand age of 30 years, and a bole height of 10 m, we obtain a net rate of wood biomass accumulation of 8 Mg/ha.yr, a wood biomass productivity value higher than all but one of the values reported by Brown and Lugo (1982) in a review of tropical forest productivity.

Age plays a role in the development of community structure and composition among *S. campanulata* stands. The older Cibuco II stand had higher understory species richness and plant density, and larger basal area, than the other and younger stands (Table 3.5). Growth is so rapid in these forests that a decade difference in age represents significant opportunity for development. Aide et al. (2000) reported that *S. campanulata* stands on karst and volcanic parent materials are replaced by native species after 40 years. Our 40-year old stand at Cibuco II showed signs of *S. campanulata* being replaced in the near-term. *Spathodea campanulata* had the lowest IV in the understory of Cibuco II (Table 3.4). The reason why *S. campanulata* might still prevail after 40 year on alluvial soils might be the result of different conditions in the Aide et al. (2000) study. We have observed that on karstic and volcanic soils *S. campanulata* stands accumulate more species than stands under alluvial conditions. This is consistent with the mature stands studied by Aide et al. (2000). They found more species than we did (20 species of large trees per 0.1 ha), the basal areas were lower (<50 m^2/ha), and the density of *S. campanulata* was lower (<400 individual/ha) than in our alluvial stands (Table 3.5). It appears that on alluvial soils following sugar cane cultivation, flooding and absence of propagules slow forest succession. We suggest that the causal factors for this slowing of succession are flooding, which stresses trees and selects for a few species that can tolerate floods, and the history of land use, including the length of time and extensiveness of cultivation. The initial deforestation and conversion of the alluvial valleys to agricultural use affected large land areas and kept them free of trees for centuries. Such land use must have reduced the tree species pool available in the alluvial valleys as well as limiting the availability of seed sources for re-colonization of abandoned fields.

Clearly, the length and intensity of sugar cane cultivation do not appear to curtail the primary productivity of the study sites (discussed below) but cultivation does appear to have an effect on the species composition of stands. Our results reflect a high proportion of introduced species (35%) in these forests (Tables 3.3 and 3.4). Many of these species are indicators of human use such as for fruit consumption, ornamental use, or use as shade trees (*Psidium guajava*, *Delonix regia,* and *Erythrina poeppigiana*, respectively). The presence of these species in the understory suggests that they are persistent in these altered landscapes and are not likely to disappear as succession proceeds to mature states (Lugo and Helmer 2004, Lugo 2004). However, in spite of the persistence of introduced species and low species richness in *S. campanulata* alluvial forests, the species list we found suggests steady enrichment with native species, particularly the 13 species found in the understory but not yet in the small tree category.

In a study of rain forest fragments in India, Muthuramkumar et al. (2006) found results similar to ours in the sense that past land use was reflected in the species composition of the fragments. They found that *S. campanulata* was most abundant (43% of trees present) in smaller and previously disturbed fragments. Other introduced and pioneer species were associated with *S. campanulata* in these disturbed fragments. However, the presence of the introduced species did not hinder the development of stands with high species richness of native species. In abandoned wet pastures on volcanic-derived soils at 420 m elevation in Puerto Rico, *S. campanulata* did not appear as a component of the early succession (Myster 2003). Instead, the introduced *Syzygium jambos* dominates the tree layer in these successions but they lag behind herbaceous grasses, forbs, and ferns. Myster (2004) showed through experiments that seed predation and disease limited tree establishment in these pastures and not seed rain.

In the moist alluvial old fields, the early stages of succession appear to be dominated by grasses including the grass *Panicum laxum*, which Britton and Wilson (1923) identified as a grass typical of moist and wet places. Other herbaceous plants growing on those areas were *Commelina diffusa* (native and pantropical) at Cibuco I, and *Cyperus odoratus* (native and pantropical) at Cibuco II, but both were present at much lower densities than *P. laxum*. An understory herbaceous plant in the alluvial forest was *Petiveria alliacea* (native to the Americas including Puerto Rico) found at Cibuco I at low densities. *Syngonium podophyllum* (introduced from continental

America) was present and abundant as herbaceous ground cover mostly in gaps or near-gaps and grew as a climber in nearly all *S. campanulata* stems within all sites. *Dalbergia monetaria* (native to the Americas including Puerto Rico) was another climbing plant that was found at all sites although at much lower densities than *S. podophyllum*.

Spathodea campanulata is able to grow through the grass cover just described and eventually shade them out. Other fast-growing species such as *S. jambos* and the native *C. schreberiana* germinate in the understory and within gaps of *S. campanulata* forests, respectively. Apparently the filtering mechanisms (*sensu* Myster 2004) that operate in wet pasture succession at higher elevation are not as effective on moist alluvial pastures.

We anticipate a continuing floristic enrichment of *S. campanulata* stands and eventual enrichment of the canopy as other species increase in importance and *S. campanulata* declines. But the increase should not be to the levels measured in other locations without floods (Aide et al. 2000). Although *S. campanulata* and other introduced species will not grow under deep shade (*S. jambos* excepted), it is possible that these species will not disappear from these stands for at least two reasons. First, these alluvial sites flood periodically and the flooding is likely to limit the richness of species that can grown and persist under flooded conditions. At Paso del Indio, *S. campanulata* floods to depths of almost 2 meters, and these periodic floods are likely to select for low tree density, low species richness, but high basal area stands of flood-adapted species such as *S. campanulata*, *Pterocarpus* (not found in our study) or *Calophyllum calaba*. The endemic palm *Roystonea borinquena* also tolerates some flooding, as does the introduced *S. jambos* . If the species richness remains low, competition with *S. campanulata* also decreases relative to that in other soil types and it is more likely it will persist.

A second likely factor that might favor the persistence of introduced species in these stands is wind. Periodic windstorms and hurricanes will continue to open the canopy and favor the rapid growth of *S. campanulata* in canopy clearings. In a closed canopy mature forest at the Luquillo Experimental Forest, Thompson et al. (in press) recorded an increase in the density of *S. campanulata* and *S. jambos* after hurricane Georges. This mechanism might be in effect at Cibuco I, which is the only site exposed eastward to oncoming winds and which appears to have more gaps in relation to the other sites. This can be the reason for the presence of shade intolerant colonizers *Cassia siamea* and *Zanthoxylum martinicense* amongst small trees at Cibuco I and their absence at the other sites. Normally the colonizer role is played in moist and wet forests in Puerto Rico by the native *Cecropia schreberiana*, but *Cecropia* cannot invade abandoned old fields or pastures (Silander 1979). The absence of this native species in old field succession is probably one reason why *S. campanulata* has been as successful as a colonizer of abandoned sugar cane fields (its wind-dispersed seeds also help). But we have seen *Cecropia* trees growing under the shade of *S. campanulata* in alluvial stands near highway PR 2, an event that raises interesting future possibilities as continuous presence of forest conditions on alluvial soils allow this native species to again establish large populations in places where it was extirpated as a result of past cultivation of sugar cane.

3.5.2 Nutrient Dynamics

The nutrient stocks of alluvial soils of the study sites compared favorably with soils from native forests in areas with different climates and geology (Table 3.9). Taking into consideration differences in sampling depth, alluvial soils were comparatively lower in soil organic matter and N and higher in P, K, Ca, and Mg than soils of natural stands on different parent materials. Perhaps due to the annual fertilization of sugar cane fields or the continuous input of silt from the uplands, the heavily used sugar cane fields maintained soil fertility in contrast to the reductions of P shown in Table 3.9 in dry karst forests used for various other purposes. Deep alluvial soils, when allowed to be flooded periodically are thus resilient to long-term cultivation, as it appears to be the case in our study sites. Thus, a high nutrient demanding tree species like *S. campanulata* can sustain its rapid growth and high primary productivity in these fertile abandoned sugar cane fields. This is reflected in a comparison of nutrient stocks in *S. campanulata* loose litter with litter from other forest types in Puerto Rico (Table 3.10). In spite of the lower mass of accumulated loose litter in alluvial stands, the stock of P, Ca, and Mg is disproportionately high in *S. campanulata*.

Table 3.11 shows that *S. campanulata* leaf litter is high in N and P concentration relative to leaf litter of mahogany plantations and mature dry karst forest. However, alluvial soil was not as high in N concentration as the soils of those forest types, although alluvial soil was higher in P, K, and Mg. The ecological implications of these data are that the organic material produced by *S. campanulata* can circulate rapidly in the forest ecosystem given its high nutritional quality for decomposers. The presence of high densities of earthworm populations in these forests (Lugo et al.

Table 3.9 Comparison of mass and total nutrient content in soil of various forest types with contrasting use history in Puerto Rico. Soil organic matter is SOM in Mg/ha and nutrients are in kg/ha to the specified depth. The last two rows correspond to this study of *Spathodea campanulata* alluvial forests. Natural forest data are from a summary in Lugo (2005) and the rest is from Molina Colón (1998). Standard error of the mean and n are in parenthesis for SOM. Empty cells mean there are no data

Forest type/depth (cm)	Use History	SOM	N	P	K	Ca	Mg
Moist alluvial (30)	Natural	319 (63, 4)					
Moist to wet ultramaphic (30)	Natural	879 (197,6)					
Moist volcanic (30)	Natural	594					
Dry karst (15)	Natural	238	12,550	655	540	6,500	1,220
Dry karst (10)	Natural	80	5,000	569	115	4,800	263
Dry karst (10)	Baseball Park	71	4,700	6	274	8,400	659
Dry karst (10)	Charcoal pits	85	5,300	9	203	5,600	293
Dry karst (10)	Farmlands	73	5,000	12	361	6,300	460
Dry karst (10)	Houses	98	6,000	21	410	6,700	333
Moist alluvial (15)	Sugar cane	129	2,146	603	681	9,265	6,469
Moist alluvial (30)	Sugar cane	261	4,062	1,196	1,393	17,639	13,788

Table 3.10 Comparison of mass (Mg/ha) and nutrient content (kg/ha) in fine litter of various forest types with contrasting use history in Puerto Rico. The rows labeled alluvial correspond to this study of *Spathodea campanulata* alluvial forests on abandoned sugar cane fields. All other forest stands were on land previously deforested and either abandoned (dry karst and wet volcanic) or planted (mahogany plantations). Data are from Lugo and Fu (2003) and Lugo and Murphy (1986). Empty cells mean there are no data. Standard error of the mean and n are given in parenthesis followed by *n*, when sample size is > 3. For alluvial data, *n* = 3

Forest type	Mass	N	P	K	Ca	Mg
			Leaf litter			
Dry karst	12.3	165	3.7	17		
Wet volcanic		70	1.9	9.1	64	14
Mahogany	9.4 (1.8, 5)	132 (39, 4)	4.1 (1.1, 4)	27 (15, 4)	524	32
Mahogany		101	3.4	20	193	22
Alluvial	1.1 (0.4)	20 (9)	1.5 (0.5)	1.7 (0.5)	35 (11)	4 (1.2)
			Total litter			
Wet volcanic	7.4	87	2.7	8		
Wet volcanic	7.3	83	2.4	11.3	78	11
Mahogany	12.6 (1.7, 5)	163 (39, 4)	5.1 (1.0, 4)	33 (16, 4)	611	40
Mahogany	11.7	130	4.7	32	235	30
Alluvial	4.8 (0.4)	50 (15)	3.9 (1.0)	5.6 (0.7)	82 (16)	10 (0.9)

in press) also suggests that organic matter and nutrient turnover in these alluvial stands are high. This strategy of rapid nutrient flux contrasts with that of plantation forests such as mahogany and *Pinus caribaea* (Table 3.10, Lugo 1992), which tend to accumulate nutrients on sites.

In summary, our study shows that abandoned sugar cane plantations on alluvial soils favor the rapid establishment and growth of forests of *S. campanulata*, an introduced wind-dispersed species. In spite of centuries of cultivation, these sites maintain high soil fertility (Tables 3.9 and 3.11) and support a rapid rate of growth of *S. campanulata*. In 25–40 years, closed canopy stands develop a complex forest structure characterized by high basal area (Table 3.5), but low species richness (Tables 3.2 and 3.3 and Fig. 3.2). Initially *S. campanulata* forms almost pure stands (Fig. 3.3a), but these stands do not remain monodominant because other species (both native and introduced) establish populations and contribute to the diversification of the stand (Fig. 3.3b), particularly the understory (Fig. 3.3c). Because alluvial

Table 3.11 Weighted nutrient concentration (mg/g) of leaf litter and soil of various forest types in Puerto Rico. Data are derived from Tables 3.9 and 3.10. Empty cells mean there are no data

Forest type	N	P	K	Ca	Mg
			Leaf litter		
Mature dry karst	13.4	0.3	1.4		
Mahogany plantation	14.0	0.4	2.9	56	3.4
Alluvial *Spathodea campanulata*	18.2	1.4	1.5	32	3.6
			Soil		
Mature dry karst	52.7	2.8	2.3	27	5.1
Dry karst farmlands	68.5	0.2	4.9	86	6.3
Alluvial *Spathodea campanulata*	15.6	4.6	5.3	68	52.8

sites flood, it is possible that the species list in these locations will be shorter than observed in other types of abandoned agricultural fields. For example, abandoned coffee plantations, pastures on volcanic soils, or abandoned agriculture on dry or moist karst, all result in new forest with higher species counts than *S. campanulata* forest on alluvial soils (Aide et al. 2000, Lugo and Brandeis 2005). However, forests following these other types of abandoned agriculture do not develop the structural complexity that *S. campanulata* achieves on alluvial soils. High primary productivity and apparent high turnover of nutrient-rich litter suggest that *S. campanulata* forests on alluvial soils function at rapid rates commensurate with the fertility of the soil. Soil fertility and abundant water supplies implies that abandoned sugar cane fields on alluvial soils can be quickly reforested regardless of the length of time of cultivation. However, because of the long history of cultivation, the restoration of a full complement of native species on these sites appears problematic. We saw indications of a return of *Cecropia schreberiana* and we found two endemic and 20 native species in these forests, both positive developments, but we believe that restoring the original forest will be retarded by disturbances that favor the presence of introduced species such as *S. campanulata*.

Acknowledgment This study was conducted in collaboration with the University of Puerto Rico. The study is part of a larger study of *S. campanulata* forests in Puerto Rico. We thank the following people for making the study possible: Lucy Gaspar, Lydia Sáez, and Josefina Arce officials of the ALACIMA program of the National Science Foundation; Iván Vicéns, Cariel Arocho, Palma Ripole, Mark McGarrigal, Brett Wolfe, Helena Román, Arnaldo Torres, Dixon Irizarry, and Angelo Rodríguez for their field assistance; Edwin López, Mary Jeane Sánchez and Olga Ramos for laboratory assistance; Frank Axelrod for taxonomic assistance; Carlos Rodríguez for GPS support; and Lía Sánchez and Mildred Alayón for their help with the manuscript. The following colleagues reviewed the manuscript: Julie Hernández, D.J. Lodge, E. Medina, R. Myster, and F.H. Wadsworth.

References

Abruña, F., Lugo López, M. A., and Pérez Escobar, R. R. 1977. Los suelos de Puerto Rico. In Geovisión de Puerto Rico, ed. M. T. B. de Galiñales, pp. 120–157. Río Piedras: Editorial Universitaria.

Acevido, G. 1982. Soil survey of Arecibo area of northern Puerto Rico. USDA Soil Conservation Service, Washington, DC. 169 p. + maps.

Aide, T. M., Zimmerman, J. K., Pascarella, J. B., Rivera, L., and Marcano-Vega, H. 2000. Forest regeneration in a chronosequence of tropical abandoned pastures: implications for restoration. Restoration Ecology 8:328–338.

Álvarez Nazario, M. 1982. Orígen y desarrollo del español en Puerto Rico (siglos XVI y XVII). Río Piedras, PR: Editorial Universidad, Universidad de Puerto Rico.

Anderson, J. M., and Ingram, J. S. I. 1993. Tropical Soil Biology and Fertility: A Handbook of Methods, 2nd edition. Wallingford: C.A.B.

Army Corps of Engineers. 1973. Flood plain information rio Cibuco and rio Indio Vega Baja and Vega Alta, P.R. Jacksonville, FL: Unites States Army.

Britton, N. L., and Wilson, P. 1923. Botany of Porto Rico and the Virgin Islands. New York: New York Academy of Sciences.

Brown, S., and Lugo, A. E. 1982. The storage and production of organic matter in tropical forests and their role in the global carbon cycle. Biotropica 14(3):161–187.

Chinea, J. D. 2002. Tropical forest succession on abandoned farms in the Humacao municipality of eastern Puerto Rico. Forest Ecology and Management 167:195–207.
Cottam, G., and Curtis, J. T. 1956. The use of distance measures in phytosociological sampling. Ecology 37:451–460.
Domínguez Cristóbal, C. 1989a. La situación forestal de Puerto Rico durante el siglo XVII. Acta Científica 3:71–72.
Domínguez Cristóbal, C. 1989b. La situación forestal de Puerto Rico durante el siglo XVIII. Acta Científica 3:73–75.
Domínguez Cristóbal, C. M. 2000. Panorama histórico forestal de Puerto Rico. Río Piedras, PR: Editorial de la Universidad de Puerto Rico.
Eusse, A. M., and Aide, T. M. 1999. Patterns of litter production across a salinity gradient in a *Pterocarpus officinalis* tropical wetland. Plant Ecology 145:307–315.
Francis, J. K. 2000. *Spathodea campanulata* Beauv. In Bioecología de árboles nativos y exóticos de Puerto Rico y las Indias Occidentales, eds J. K. Francis and C. A. Lowe, pp. 484–487. Río Piedras, PR: Reporte Técnico General IITF-15.
Gleason, H. A., and Cook, M. T. 1926. Plant ecology of Porto Rico. In Scientific Survey of Porto Rico and the Virgin Islands. New York: New York Academy of Sciences.
Gould, W. A., González, G., and Carrero-Rivera, G. 2006. Structure and composition of vegetation along an elevational gradient in Puerto Rico. Journal of Vegetation Science 17:563–574.
Helmer, E. H., Ramos, O., López, T. d. M., Quiñones, M., and Díaz, W. 2002. Mapping forest type and land cover of Puerto Rico, a component of the Caribbean biodiversity hotspot. Caribbean Journal of Science 38:165–183.
Holdridge, L. R. 1967. Life Zone Ecology. San José, Costa Rica: Tropical Science Center.
LECO-Corporation. 1995. Carbon, nitrogen, and sulphur testing in soil/plant tissue. Organic Application Note CNS-2000, No. 203-821-002, St. Joseph, MI.
Little, E. L., and Wadsworth, F. H. 1964. Common trees of Puerto Rico and the Virgin Islands. Agriculture Handbook 249. Washington, DC: USDA Forest Service.
Little, E. L., Woodbury, R. O., and Wadsworth, F. H. 1974. Trees of Puerto Rico and the Virgin Islands, Vol. 2. Washington, DC: USDA Forest Service Agriculture Handbook 449.
Lugo, A. E. 1992. Comparison of tropical tree plantations with secondary forests of similar age. Ecological Monographs 62:1–41.
Lugo, A. E. 2004. The outcome of alien tree invasions in Puerto Rico. Frontiers in Ecology and the Environment 2:265–273.
Lugo, A. E. 2005. Los bosques. In Biodiversidad de Puerto Rico. Vertebrados terrestres y ecosistemas, ed. R. L. Joglar, pp. 395–548. San Juan, PR: Editorial del Instituto de Cultura Puertorriqueña.
Lugo, A. E. 2006. Lecciones ecológicas de una isla que lo ha visto todo. Ecotrópicos 19(2):57–71.
Lugo, A. E., and Brandeis, T. J. 2005. A new mix of alien and native species coexist in Puerto Rico's landscapes. In Biotic Interactions in the Tropics: Their Role in the Maintenance of Species Diversity, ed. D. F. R. P. Burslem, M. A. Pinard, and S. E. Hartley, pp. 484–509. Cambridge, England: Cambridge University Press.
Lugo, A. E., and Fu, S. 2003. Structure and dynamics of mahogany plantations in Puerto Rico. In Big-Leaf Mahogany: Genetics, Ecology, and Management, ed. A. E. Lugo, J. C. F. Colón, and M. Alayón, pp. 288–328. New York: Springer Verlag.
Lugo, A. E., and Helmer, E. 2004. Emerging forests on abandoned land: Puerto Rico's new forests. Forest Ecology and Management 190:145–161.
Lugo, A. E., and Murphy, P. G. 1986. Nutrient dynamics of a Puerto Rican subtropical dry forest. Journal of Tropical Ecology 2:55–76.
Lugo, A. E., Abelleira, O. J., Borges, S., Colón, L. L., Meléndez, S., and Rodríguez, M. A. In Press. Preliminary estimate of earthworm abundance and species richness in *Spathodea campanulata* Beauv. forests in northern Puerto Rico. Caribbean Journal of Science.42(3):325–333.
Luh-Huang, C. Y., and Schulte, E. E. 1985. Digestion of plant tissue for analysis by ICP emission spectroscopy. Communications in Soil Science and Plant Analysis 16:943–958.
McLean, E. O. 1982. Soil pH and lime requirement. In Methods of Soil Analysis, Part 2, pp. 199–209. Madison, WI: American Society of Agronomy.

Medri, M. E., and Correa, M. A. 1985. Aspectos histológicos e bioquímicos de *Joannesia princips* e *Spathodea campanulata*, crescendo em solos na capacidade de campo, encharcado e alagado. Semina 6:147–154.

Molina Colón, S. 1998. Long-term recovery of a Caribbean dry forest after abandonment of different land uses in Guánica, Puerto Rico. Dissertation. Río Piedras, PR: University of Puerto Rico.

Muthuramkumar, S., Ayyappan, N., Parthasarathy, N., Mudappa, D., Shankar Raman, T. R., Selwyn, M. A., and Pragasan, L. A. 2006. Plant community structure in tropical rain forest fragments of the Western Ghats, India. Biotropica 38:143–160.

Myster, R. W. 2003. Vegetation dynamics of a permanent pasture plot in Puerto Rico. Biotropica 35:422–428.

Myster, R. W. 2004. Regeneration filters in post-agricultural fields of Puerto Rico and Ecuador. Plant Ecology 172:199–209.

NOAA. 1998. Puerto Rico and the Virgin Islands. Climatological Data Annual Summary 44(13): 1–13

Picó, R. 1969. La nueva geografía de Puerto Rico. Río Piedras, PR: Editorial Universitaria, Universidad de Puerto Rico.

Ramos, O. M., and Lugo, A. E. 1994. Mapa de la vegetación de Puerto Rico. Acta Científica 8:63–66.

Silander, S. R. 1979. A study of the ecological life history of *Cecropia peltata* L., an early secondary successional species in the rain forest of Puerto Rico. Masters. University of Tennessee, Knoxville, TN.

Wadsworth, F. H. 1950. Notes on the climax forests of Puerto Rico and their destruction and conservation prior to 1900. Caribbean Forester January:38–47.

Whitmore, T. C. 1990. An Introduction to Tropical Rain Forests. Oxford: Clarendon Press.

Wilde, S. A., Corey, R. B., Iyer, J. G., and Voigt, G. K. 1979. Soil and Plant Analysis for Tree Culture. Oxford: Oxford & IBH Publishing Co.

Chapter 4
Soil Fungi and Macrofauna in the Neotropics

Yiqing Li and Grizelle González

4.1 Introduction

Decomposition is a critical ecosystem function that decomposes dead organic materials, removes wastes, recycles nutrients and renews soil fertility. In natural ecosystems most nitrogen (N) and phosphorous (P) required for plant growth are supplied through the decomposition of detritus, relying therefore on the activities of soil microbes and macrofauna. Decomposition is a consequence of interacting physical and chemical processes occurring inside and outside of living organisms in an ecosystem (Chapin III et al. 2002). The two major processes include fragmentation, breaking larger pieces of organic mater into smaller ones, and chemical alteration, the change of chemical elements of dead organic mater into others. Fragmentation is largely a consequence of feeding activity of soil macrofauna and chemical alteration is primarily associated with the activity of bacteria and fungi.

Land conversion from forests to pasture or agricultural land alters the belowground ecosystem and decomposition process, and depletion of soil organic C (Doran and Zeiss 2000). Degradation processes such as losses in soil C, nutrient depletion and reduced water holding capacity often lead to considerable change in soil microbial community and macrofuana (Bever et al. 1996, Picone 2000, Thomas et al. 2004). In the Neotropics, land conversion from tropical forest to agricultural land and then to pasture is the most common kind of land use practices (Fearnside 1993, Thomlinson 1996, Cleveland et al. 2003). Studies of the dynamics of microbes and macrofauna following land conversion are thus particularly of importance for establishing effective strategies of post-agriculture recovery and forest ecosystem restoration because soil microbial community and soil macrofauna play a key role in successional pathways.

4.2 Fungi

Together with bacteria, fungi are the main initial decomposers of dead plant and animal materials and account for 80–95% of the total decomposer biomass and heterotrophic soil respiration (Macfadyen 1963, Coleman 1985, Chapin III et al. 2002).

R. W. Myster, *Post-Agricultural Succession in the Neotropics.*

Fungi alone constitute a major portion of the soil biomass, with fungal biomass carbon (C) equal to or exceeding root biomass C (Paul and Clark 1996). Their biomass in soils ranges from 1,000 to 15,000 kg/ha in the top 15 cm (Brady and Weil 2002).

Typically, fungi produce slender filaments or hyphae with a dense concentration of cytoplasm, either branched or un-branched, and commonly multinucleate. The hyphae constitute the soma, mycelium and thallus and contain more vacuoles when resources are scare, which enables fungi to grow into new areas to explore for substrate. Fungi secrete enzymes capable of breaking down complex cell wall components, such as lignin, enabling them to penetrate the cuticle of dead leaves or the exterior of roots to gain access to more labile compounds that occur within the cell. Reproduction in fungi is sexual, asexual or both. Asexual spores are produced at hyphal tips in structures and sexual spores are produced following hyphal conjugations. In asexual reproduction, the offspring are genetically identical to the "parent" organism (clones). During sexual reproduction, a mixing of genetic material occurs so that the offspring exhibit traits of both parents. Many species can use both strategies at different times or different conditions, while others are apparently strictly sexual or strictly asexual. As heterotrophs, fungi depend on leaving or dead organic materials for their energy and C sources.

As major decomposers of organic materials in soil, fungi comprise an extremely diverse group of microorganisms. Thousands of species have been identified in soils, representing some 170 genera, and at least 1 million of fungal species in soils are waiting for discovery (Brady and Weil 2002). Two groups of fungi common to soils are the single-celled organisms called yeasts and the multicellular filamentous organisms, such as molds, mildews, smuts, and rusts. Generally, filamentous fungi are abundant in well-aerated environments, whereas yeasts inhabit in poor-aerated soils. Most filamentous fungi lack a capacity for anaerobic metabolism and therefore often absent from anaerobic soils. Mushrooms are the true reproductive structures of filamentous fungi. The aboveground fruiting structure of most mushrooms is only a small part of the total body. An extensive hyphae network of mushrooms permeates below ground. Mushroom fungi are important wood rots that attack standing and felled timber, stumps, slash and wooden structures, given appropriate moisture contents. Molds are microscopic and semi-macroscopic fungi that play a much more important role in soil organic matter decomposition than the mushroom fungi. The four most common genera of molds in soils are Penicillium, Mucor, Fusarium and Aspergillus.

In many terrestrial ecosystems, fungi are the most abundant decomposers which are more effective than bacteria. Fungi are better equipped for carrying out the decaying of the dead organic materials since their hyphae networks can penetrate new substrates and proliferate both within and between dead plant cells. Fungi also have an ability to transport nutrients through their hyphal networks to zones of exploitation. Fungi have enzyme systems capable of breaking down all types of plant compounds. All these advantages of fungi over bacteria made fungi dominant in most ecosystems. Up to 50% of the substances decomposed by fungi are used for fungal tissue, compared to around 20% for bacteria (Brady and Weil, 2002). Fungi continue to decompose various complex organic materials which bacteria and actimycetes cannot decompose.

Fungi play a critical role in the mineralization of nutrients. Fungi, together with bacteria and other microbes, break down soluble and insoluble organic matter and covert it into inorganic forms which can used by plants. The majority of N is contained in dead organic matter which depends upon the decomposition by microbes for the N availability of plants. Nitrogen fixation by microbes is another route through which plants obtain N. These microbes form nodules with the plant root where they gain carbohydrate from the plants and in turn they provide the plants with amino acids. Although bacteria play a key role in N fixation through symbiotic association with plants (legumes), the primary route by which N enters this soil is through the break down of legume leaf litter inputs that are enriched with N (Read and Perez-Moreno 2003, Bardgett 2005). The breaking down of these leaf litters are mainly completed by fungi. Phosphorous availability for plants is determined by geochemical process rather than biological process, soil microbes however are actively involved in the cycling of P (Sanders and Tinker 1971, Dodd et al. 1987, Jayachandran et al. 1992, Salas et al. 2003, Joner and Jakobsen 2005).

4.3 Soil Macrofauna

Soil macrofauna play a very important role in improving structure, organic matter content and distribution patterns of nutrient elements. Soil macrofauna may affect soil function and processes in a variety of ways, and could be used as indicators of nutrient status of soils (Vanlauwe et al. 1997, Rao et al. 1998, Martius et al. 2001). Members of the macro-fauna group of soil animals include: arthropods (isopods, amphipods, millipedes, termites), annelidans (earthworms) and molluscs (slugs and snails). These animals are mostly responsible for the initial shredding of plant remains and its redistribution within decomposer habitats (Swift et al. 1979) as their body sizes are large enough to disrupt the structure of mineral and organic soil horizons through their feeding and burrowing activities (Anderson 1988). The macro-fauna have an important role in the maintenance of soil structural stability and fertility in many natural and human-modified habitats (Gillison et al. 2003). Not surprisingly then, the soil macro-fauna are considered important ecosystem "engineers"; particularly in tropical ecosystems where their bioturbation actions are major determinant of soil processes (Lavelle 1997), climatic conditions rarely limit their activities, and their feeding strategies produce biogenic structures that can determine the activities of microorganisms and other smaller invertebrates (Lavelle et al. 2001).

The function of soil decomposer organisms may be categorized by body size, or by various physiological aspects of trophic function (Swift et al. 1979). In terms of body size, soil fauna were traditionally subdivided on the basis of body length (Wallwork 1970). However, the use of meshed litterbags to separate the effects of particular groups of the fauna on litter decomposition led to the classification of soil fauna based on their body diameter classes (Swift et al. 1979). Thus, soil invertebrates can be classified by body width into micro- (< 100 μm), meso- (< 2 mm) and macro-fauna (> 2 mm). This latter classification being more of a functional classification with respect to litter breakdown and decomposition processes (Fig. 4.1, Swift et al. 1979). These size categories are somewhat arbitrary because juveniles of some species may be classified differently as adults, and also ignores the range of trophic diversity of animals in the soil (Anderson 1988). Yet, body width rather than body length broadly defines the extent to which the feeding and burrowing activities of soil animals are constrained by, or modify, the structure of soil and litter habitats (Anderson 1988).

The size of an organism is also an important determinant of its distribution (Swift et al. 1979). It has been argued that micro-fauna species tend to have large geographic distribution patterns as a result of the size and ubiquity of the microsites they occupy, whereas the macro-fauna show a greater discontinuity in distribution (Fig. 4.2, Anderson 1977). The microorganisms and the micro-fauna may be widely dispersed by abiotic and biotic agents, whereas most of the larger sapotrophic animals are sedentary and rarely dispersed except by their locomotive efforts (Swift et al. 1979). The relative contribution of different functional groups of the soil fauna to soil / ecosystem processes is dependant on the confounding effects of their size, abundance, diversity and functionality (e.g., Hansen 1999, Heneghan et al. 1999, Irmler 2000, González and Seastedt 2001) but also by the landscape in which they

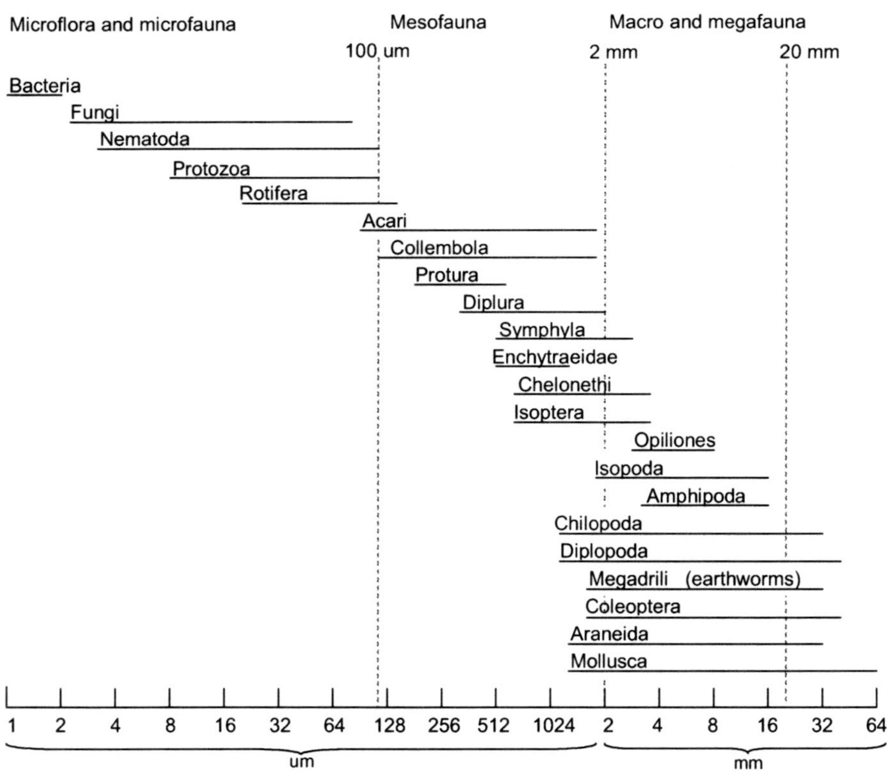

Fig. 4.1 Size classification of organisms in decomposer food webs by body width (from Swift et al. 1979)

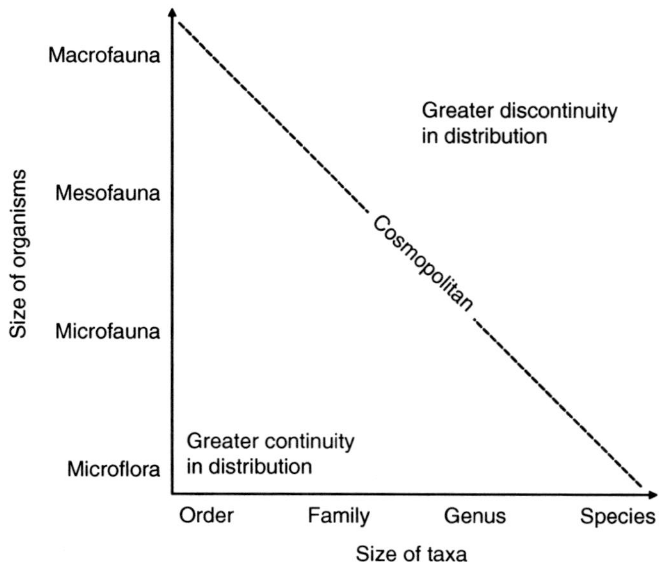

Fig. 4.2 Biogeographical distribution of soil animals in relation to body size and taxonomic level (modified from Anderson 1977)

occur (i.e., the degree of fragmentation, state of plant community succession and / or habitat alteration) as it can affect their distribution patterns.

Conversely, soil organisms can have substantial ecosystem effects on natural and human-altered habitats. This is important to consider given that (1) more than 95% of the earth's land surface is affected by human activities, including agriculture and urbanization (Moguel and Toledo 1999) and (2) the majority of animals in terrestrial habitats are invertebrate members of the decomposer community (Hansen 2000). Thus, soil macro-fauna might be more prone than smaller invertebrates to show an even greater discontinuity in distribution and more patchy effect on ecosystem processes in the context of habitat alteration and as compared to their expected distribution given size. In addition, land use intensification and fragmentation in the Neotropics can significantly modify the relative contribution of native and exotic elements within the macro-fauna; further challenging the incorporation of soil organismal effects on ecosystem processes into sound management plans and practices.

Ecosystem engineers—like the macro-fauna—are less able to withstand high levels of natural and anthropogenic environmental constraints and lower order groups may become predominant with significant differences in the function of the ecosystem (Lavelle 1997). Consequently, modifications of soil fauna communities may lead to loss of diversity and soil functions (Lavelle 1997). In the following section of this chapter, we will discuss the effects that soil macro-fauna have on soils and ecosystem functioning under different post agricultural management practices; and highlight gaps in knowledge that might stimulate further studying of soil macro-fauna effects in the Neotropics.

4.4 Discussion

4.4.1 Fungi

4.4.1.1 General Patterns in Tropical Forests and Pastures

Since the Industrial Revolution, first temperate and then tropical land-use change has significantly affected biogeochemical and biophysical processes of various ecosystems at multiple scales. Typically, when tropical forests are cut down the land is converted for cattle ranching or agricultural use. Human alteration of natural forest ecosystems for pastures is an important component of tropical land use change, and has the potential to alter soil nutrient and C cycles (Buschbacher et al. 1988, Neill et al. 1997, Holl 1999, Nepstad et al. 1999, Post and Kwon 2000, Amundson 2001, Townsend et al. 2002, Myster 2004). The nature and extent of such changes of nutrient and C cycles vary widely across gradients in climate, soil type and management strategies (Spaans et al. 1989, Groffman et al. 2001). Although the changes of biogeochemical patterns following land conversion are becoming well documented, the major mechanisms driving the variation of such changes are still poorly understood. Most previous studies have focused on environmental or plant-driven factors over biogeochemical changes following deforestation. However, soil microbial community, one of the possible mechanical controls over the biogeochemical changes, has been less documented. Land conversion disrupts the structure and activity of soil microbial community (Borneman and Triplett 1997, Nüsslein and Tiedje 1999) and microbes in turn affect the biogeochemical variables. Fungi play a significant role in regulating populations of other soil organisms and ecosystem processes (Fitter and Garbaye 1994). Thus, it is essential to understanding the relationship between patterns and activities of microbial community and biogeochemical changes following land conversion in the tropical ecosystems.

A main component of the tropical soil microbial community is fungi. Fungi are primarily responsible for mineral nutrient recycling through decomposition of organic matter and the transfer of nutrients into plants by mycorrhizal fungi (Lodge 1993, Myster 2006). The conversion of forest to pasture causes significant changes to soil environment, leading to changes in fungi community (Nüsslein and Tiedje 1999, Cleveland et al. 2003). Studies suggested the shift in overall microbial composition could be up to 50%. The change of fungal diversity following land conversion is closely related to initial fungal species in different regions even with similar climates. For example, the Neotropics have a very high fungal diversity comparing to African and Asia in the tropics (Lodge and Cantrell 1995). In addition, microbial biomass C also declines when from forest to pasture. For example, in an oxisol pasture in Costa Rica, microbial biomass (main component being fungi) is 40–60% lower than values in the forest (Cleveland et al. 2003). The decrease in microbial biomass may not entirely driven by soil organic matter because the decrease of soil organic matter in oxisol pasture was around 20% lower in pasture than in forest. Generally, the impacts of land use change on soil fungi community is lager in nutrient-poor sites and relatively modest in more fertile sites (Luizao et al. 1992, Henrot and Robertson 1994). Groffman (2001) found that land conversion did not decrease microbial biomass and activity in a relatively nutrient-rich tropical ecosystem. Fungal biomass and total biomass of all microorganisms contain significant fractions of labile nutrients in forest floor of tropical ecosystems. During the conversion from forest to pasture most of these microbial related labile nutrients are lost. Based on litter and basidiomycete nutrient concentration, 22% of litter P and 4% of K could have been immobilized in fungal biomass (Lodge 1993). Phosphorus immobilization by fungi in the Luquillo Experimental Forest is considerable because leaf decomposer fungi maintained P concentration (Lodge 1987). In pastures, the total amount of immobilized P in litter significantly decreased because of less fungal biomass in the pasture litters.

Mycorrhizal associations have been intensively studied over the past several decades and increased understanding of the important role of this symbiosis in the functioning and performance of plants in tropical ecosystems. Besides mycorrhizal fungi providing phosphorus from host plants and reciprocal C provision from host to fungus, additional effects of mycorrhizal fungi on the functioning of their host plants include increased disease resistance, improved water relations, alternation in other soil properties, and acquisition of soil nutrients. Mycorrhizal fungi are of particularly importance in the tropical ecosystems because of their widely distribution with a higher proportion of total fungal biomass. In nutrient-poor soils of humid tropics, a number of late-successional woody species are obligately dependent on arbuscular mycorrhizae (Picone 2000). Previous studies have suggested that plant succession, restoration or reforestation could be inhibited by either the abundance or the diversity of arbuscular mycorrhizal fungi in old pastures or agricultural lands (Jenos 1988, van der Heijden et al. 1998). Because disturbances to the host plant community alter the mycorrhizal fungi community (Bever et al. 1996) and host plant diversity is much lower in pastures than in forests, mycorrhizal fungal diversity dramatically declines. Thus, unlike in forests, pastures are usually dominated by a few mycorrhizal species with each species produces more spores than the species in forests. Several hypotheses have been suggested to explain why most species of arbuscular mycorrhizal fungi produce more spores in pasture than in native forests (Picone 2000). For example, it is suggested by these hypotheses that higher fine root density and turnover rate, higher pH value, and more host specific sporulation in pasture favor producing more spores. Species richness of arbuscular mycorrhizal fungi also decreases in agroecosystems comparing to natural ecosystems. Frequently irrigated and fertilized agricultural lands reduce more arbuscular mycorrhizal fungi than un-intensive managed agricultural lands since high nutrient inputs diminish the soil colonization potential and the dependency of plants on mycorrhization (Höflich and Metz 1997). Phosphorous limitation to fungal processes has been demonstrated in forest ecosystems and such constraints are also present in the pastures. However, since conversion from forest to pasture dramatically reduces the abundance of spores and the diversity of arbuscular mycorrhizal fungi, forests tend to conserve limiting nutrients, such as P, more tightly than pastures because of the nutrient immobilization by fungi which maintain limiting nutrient concentration.

4.4.1.2 A Case Study: Fungal Biomass Dynamics Following Reforestation in Puerto Rico

Fungal biomass contains significant fractions of labile nutrients in the forest floors and surface soils of some tropical ecosystems (Yang and Insam 1991, Li et al. 2005). Nutrients can be immobilized and conserved by fungi and other microbes in their biomass during periods of high precipitation that cause leaching (Behera et al. 1991, Yang and Insam 1991). Therefore, fungal biomass and total biomass of all microorganisms are important indicators for the nutrient dynamics and availability. However, relatively few publications contain measurement of fungal or total microbial biomass in tropical primary forests and other forest ecosystems following reforestation, such secondary forests and plantations (Frankland et al. 1990, Korf 1997, Tufekcioglu et al. 2001, Myster and Schaefer 2003, Li et al. 2004). Secondary forests account for 40% of the total area of tropical forests and this percentage is still increasing throughout the tropics (Brown and Lugo 1990). The success of management of tropical forests in the future might well depend upon the adequacy of our ecological understanding of secondary forests (Cheng 1993). Tree plantations, covering 11 million ha of the tropics (Lanly 1982), are rapidly increasing in area in the tropics as a reforestation approach. Reforestation through plantation on abandoned and degraded agricultural lands in the tropics has been proposed as an effective C management approach (Montagnini and Porras 1998).

In this case study, total fungal biomass and active fungal biomass were measured in surface soil in subtropical wet forest zone at Guzmán site in Puerto Rico. Using a 20-year old pine plantation and a secondary forest originating from the same abandoned farmland in the wet tropics in Puerto Rico, we examined microbial biomass after 7-year litter exclusion, root exclusion, and both litter and root exclusion treatments in the plantation and the secondary forest. The study was conducted on two sites that were within 100 m of each other. One site was in a *Pinus caribaea* (Morelet)-dominated plantation and the other was in a secondary forest. Both sites were located in the Guzmán sector of the Luquillo Experimental Forest in northeastern Puerto Rico (18°18‘N, 65°50‘W). The plantation and the secondary forests originated from the same abandoned agricultural land with the same cropping system and management/disturbance history (Lugo 1992). The sites were characterized by a wet tropical climate with mean annual precipitation of 3,920 mm and mean annual air temperature of 22.3°C. The temperature was mild and stable with diurnal and seasonal temperature ranges of 3–4°C. Soils were classified as mixed isothermic tropohumult in both the plantation and the secondary forest. The sites were relatively flat with a slope of <5 degrees and an elevation of about 400 m above sea level.

The tree plantation was established on the abandoned cropland in 1976 as part of a reforestation program of the United States Forest Service (Lugo 1992). The secondary forest had naturally developed also on the same abandoned cropland since 1976. The plantation was dominated by *Pinus caribaea* (Morelet) with small trees and grass species underneath the canopy. When our study started in 1996, the average tree height was about 15 m and the average diameter at breast height (DBH) was 22 cm in the plantation. The secondary forest was characterized by a sparse overstory and a dense understory with abundant shrubs and grasses comparing to the plantation but there were still a lot of spaces among the vegetations. The dominant canopy species in the secondary forest include *Myrcia splendens* (Sw.), *Miconia prasina* (Sw.) and *Casearia arborea* (L C Richard). The major understory species include *Casearia sylvestris* (Sw.), *Miconia mirabilis* (Aubl.) and *Tabebuia heterophylla* (DC.). Total microbial biomass was measured using a fumigation-incubation procedure (Jenkinson and Powlson 1975) in August 1996 and March 1997, which represented a wet and a dry season, respectively. Biomass of active and total fungi was estimated using the agar film techniques (Lodge and Ingham 1991). Biomass of active and total bacteria was obtained using fluorescein isothiocyanate techniques (Zou and Bashkin 1998).

In this study, we found total soil microbial biomass in the litter-and-root exclusion plots decreased the most among all the treatments in both forests. Specifically, total fungal biomass was reduced 84.6 and 87.5% and total bacterial biomass was reduced 62.1 and 56.9%, respectively, in the plantation and the secondary forest. We found that microbial biomass demonstrated considerable seasonal variation in both the plantation and the secondary forest. Total microbial biomass, in general, was greater in the wet season than in the dry season with exceptions for the root exclusion and the litter-and-root exclusion plots in the secondary forest. Total and active fungal biomass was significantly ($P \leq 0.05$) higher in the plantation than in the secondary forest in all treatments during both the wet and dry seasons (Figs 4.3 and 4.4). Total and active fungal biomass was significantly greater in the wet season than in the dry season in both forests.

Microbial biomass was also significantly different among different treatments. The litter exclusion and the litter-and-root exclusion significantly reduced total fungal, total bacterial, active fungal, active bacterial biomass except for the active bacteria in the secondary forest during the wet season. In most plots, the litter-and-root exclusion reduced microbial biomass the most in comparison with the control and the other treatments. While root exclusion significantly reduced total fungal biomass in both forests (Fig. 4.3 a, b). Root exclusion plots, in general, had the least reduction in fungal biomass among all the treatment plots.

We also measured soil CO_2 efflux in the same site in another study (Li et al. 2005) and found microbial activity in the plantation was more sensitive to the litter input than the secondary forest. Soil CO_2 efflux had higher correlation with total fungal biomass but lower correlation with total bacterial biomass in the plantation than in the secondary forest, suggesting that the fungal communities may dominate soil heterotrophic respiration in the plantation, while bacterial communities may take the control in the secondary forest. By separating the total fungal biomass into active and inactive components, we found that active fungal biomass demonstrated higher correlations with soil CO_2 efflux than their corresponding total fungal and bacterial biomass.

In conclusion, fungal biomass and total microbial biomass are important variables in the reforestation processes. Although the plantations and the secondary forests originated from the same abandoned farm lands but they have different proportions of fungal communities and therefore might have different nutrient cycle

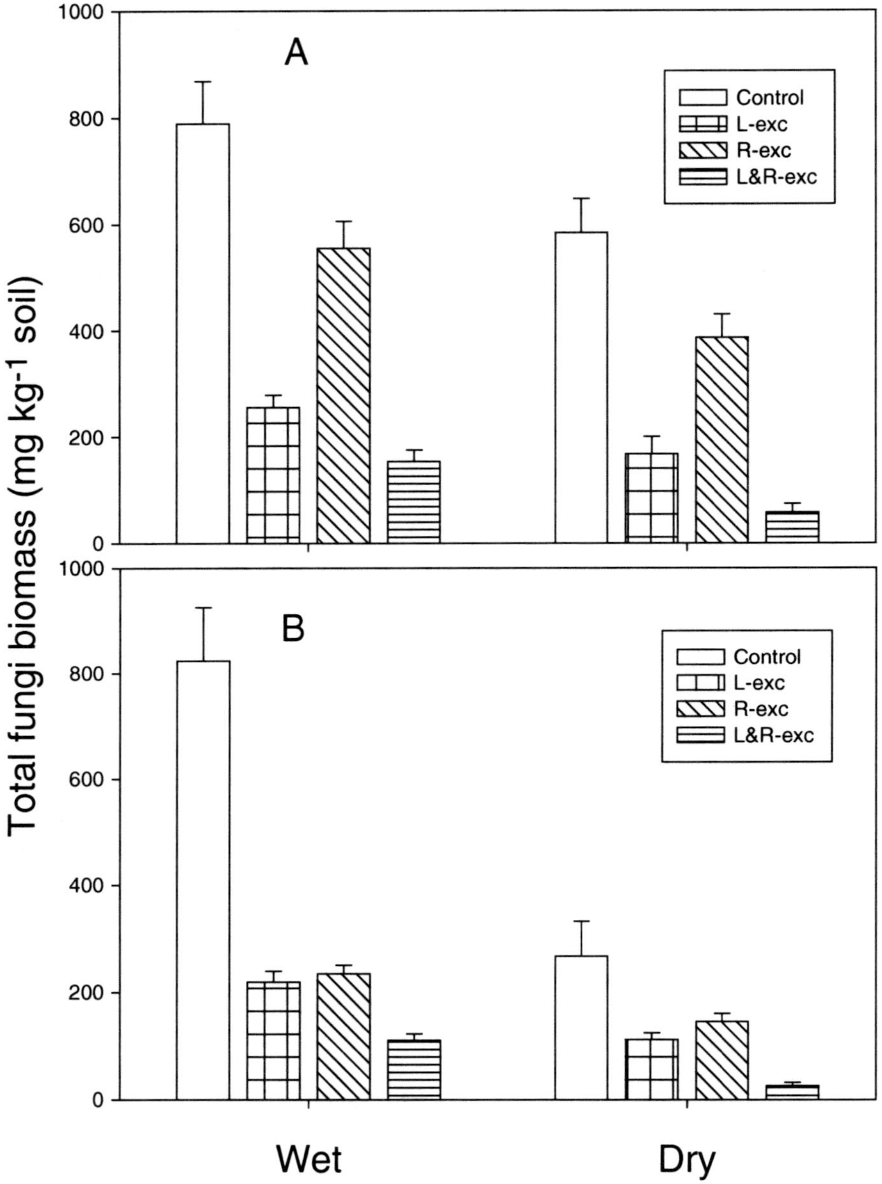

Fig. 4.3 Active fungal biomass (mg C kg^{-1} soil) in the plantation (A) and the secondary forest (B) and active bacterial biomass (mg C kg^{-1} soil) in the plantation (C) and the secondary forest (D) in the wet (August) and dry (March) seasons. Common letters on the bars indicate no significant difference between the treatments at 95% confidence level

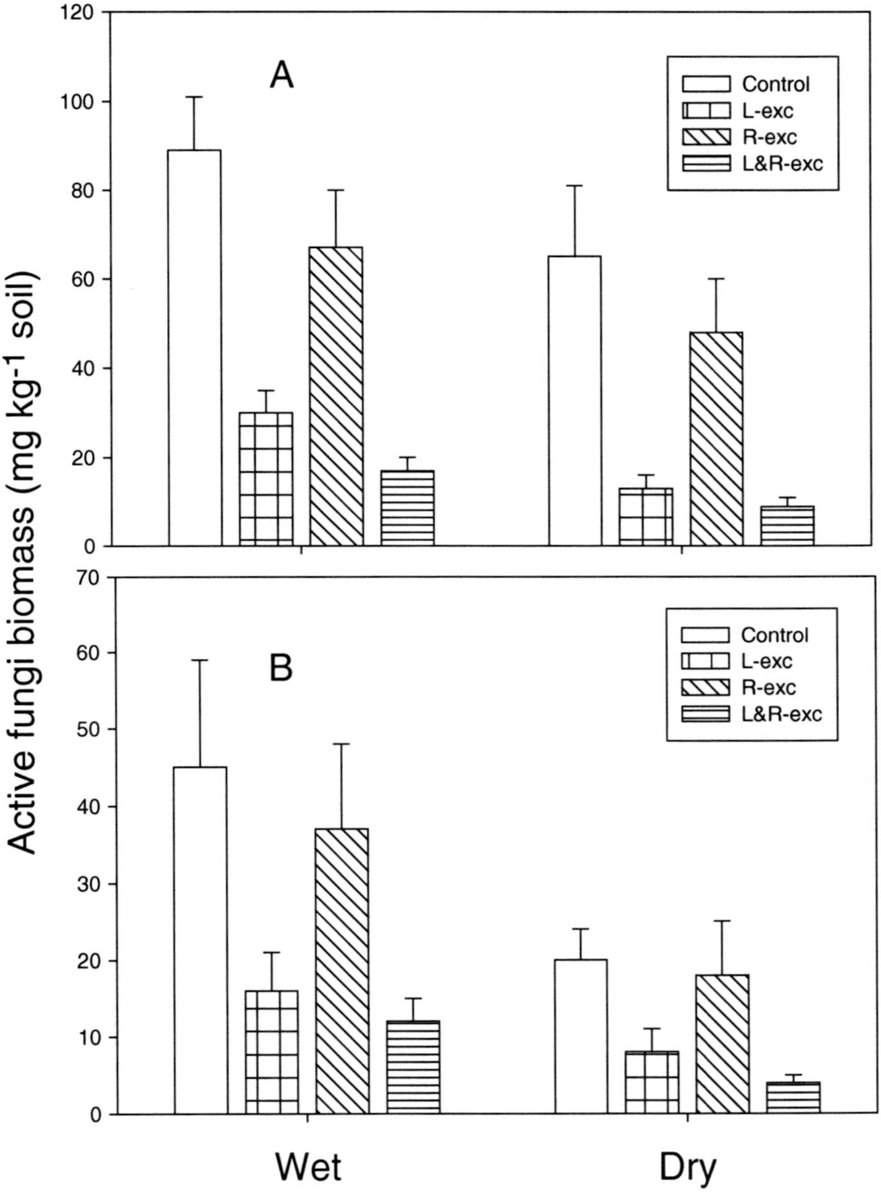

Fig. 4.4 Total fungal biomass (mg C kg^{-1} soil) in the plantation (A) and the secondary forest (B) and total bacterial biomass (mg C kg^{-1} soil) in the plantation (C) and the secondary forest (D) in the wet (August) and dry (March) seasons. Common letters on the bars indicate no significant difference between the treatments 95% confidence level

pathways. In addition, separation of active fungal biomass and total fungal biomass may enhance our understanding on the interactions between microbial communities and soil nutrient and organic C cycles.

4.4.2 Macrofauna

4.4.2.1 General Patterns of Macro-Fauna Composition Under Natural and Post-Agricultural Management Practices in the Neotropics

Earthworms, termites, ants, coleopterans, and myriapods constitute the largest component of the macro-fauna. Lavelle and Fragoso (1999) compared 12 communities from tropical rain forests, and estimated that earthworms and termites are the most important macro-fauna in these soils. Lavelle et al. (1994) analyzed the composition of 73 macro-invertebrate communities from 29 sites in the humid tropics and found termites, earthworms and litter arthropods (mostly millipedes and coleopterans) to account for the dominant proportion of the biomass. Further, Lavelle et al. (1997) argued that earthworms and termites are the most important engineers in terrestrial ecosystems based on the influence they exert on the diversity and activity of biota in subordinate trophic levels and the creation of organo-mineral structures that can affect soil physical and chemical properties. In general, earthworms are best represented in grasslands in humid areas as their abundance decreases towards humid forests and dry forests and grasslands (Lavelle et al. 1994). Termites are well adapted to dry environments where earthworms are not found (Menaut et al. 1985). Millipedes are important in litter dominated ecosystems; where their feeding activity is specialized on dead organic matter, or on saprophytic organism consumption (Bertrand and Lumaret 1992).

Disturbances linked to land use practices seem to severely affect the species richness and abundance of soil invertebrate communities world-wide (Lavelle et al. 1997). There are clear differences in soil macro-fauna between forested and grassland systems and other types of land use (Giller 1996). In the Neotropics, annual crops on sites where the natural vegetation has been cleared have depleted macro-invertebrate communities in terms of abundance and biomass (Lavelle et al. 1994); usually irrespective of the annual crop species but mostly associated to soil disturbance and the absence of a permanent soil cover (Barros et al. 2005. Termites seem thrived better than earthworms in annual crops and fallows and as compared to pastures, forests and agroforestry systems (e.g., Lavelle and Pashanasi 1989, Barros et al. 2002). Also for example, Lavelle and Pashanasi (1989) described the soil macrofauna communities of Yurimaguas (Peruvian Amazonia) and found the primary forest to have diverse and abundant fauna with a population density of 4,303 ind. m^{-2} and biomass of 53.9 g. m^{-2}—2 to 3 times higher than in similar environments in Mexico. Yet, clearing and cropping soon destroyed most of the fauna where half of the taxonomic units disappeared (see Table 4.1, Lavelle and Pashanasi 1989); where termites and ants become the dominant in terms of abundance. They also found pastures to have the highest biomass (83.3–159.2 g. m^{-2})

but the lowest population densities (922–2,347 ind. m^{-2}) and taxonomic richness (20–27) due to the dominant development of populations of *Pontoscolex corethrurus* (Oligochaeta: Glossoscolecidae) which represent 82.3–95% of the biomass (also see Chapter 5 on more information in earthworm dominance in pastures). Further, secondary forests had a soil invertebrate community with population densities on the same order (4,032 ind. m^{-2}) of the primary forest but biomass and taxonomic units were lower in the secondary forest as compared to the latter (see Table 4.1).

4.4.3 Crop Systems

Leaving soils to "rest" or "fallow" is a traditional management practice throughout the tropics for restoration of soil fertility during cropping (Sánchez 1995). Successful restoration of soil fertility normally requires a long fallow period for sufficient regeneration of the native vegetation and establishment of tree species (Young 1997). In this context, Thomas et al. (2004) studied the biodiversity of soil macrofauna along a chronosequence of abandoned rice paddy fields and plots without agricultural disturbance in Northern Argentina. They found that species diversity varied strongly along the chronosequence; it was highest in the natural grassland, lowest in the 2 years- fallow and increased progressively through the chronosequence, but was still lower in the fallow after 15 years of abandonment as compared to the natural grassland. Yet while working on Andean agroecosystems in Colombia, Barrios et al. (2005) found that improved fallows (as an alternative to the natural regeneration of the native flora) could help regenerate soil fertility in degraded volcanic-ash soils via changes in the soil macrofauna diversity. Similarly, in the

Table 4.1 Taxonomic richness, mean abundance and biomass of macroinvertebrates in different types of land-use at Yurimaguas, Peru. Modified from Lavelle and Pashanasi, 1989

Type of land use	Taxonomic units (no.)	Density (ind. m^{-2}) ± SE	Biomass (g. m^{-2}) ± SE
Forests			
Primary forest	41	4,304 ± 933	53.9 ± 8.54
Secondary forest (15 years)	27	4,099 ± 1,828	24.1 ± 5.85
Crops (1st crop)			
High input maize	20	730 ± 221	3.1 ± 1.04
Low input rice	24	3,683 ± 1,059	8.5 ± 2.75
Cassava	18	1,197 ± 283	7.6 ± 3.26
Pastures			
Improved (*Bracharia-Desmodium*)	27	922 ± 81	159.2 ± 16.4
Moist	23	1,768 ± 308	121.1 ± 20.8
Dry	20	2,347 ± 453	82.3 ± 15.5
Fallows			
6-mo. *Centrosema* after pasture	22	1,546 ± 253	111.9 ± 13.2
6-mo. Kudzu after high input crop	23	2,214 ± 969	15.5 ± 6.2
Peach-palm + Kudzu cover	32	1,858 ± 380	93.9 ± 25.2

above mentioned study from the Peruvian Amazonia, Lavelle and Pashanasi (1989) found that recent fallows (6 mo.) of Kudzu (Leguminosea) and *Centrosema* retained characteristics of the system from which they were originally started; continued with low taxonomic richness but then showed increases in the density of the macro-faunal population. Hence, they argued these fallows were on a slight development towards the original forest situation (Lavelle and Pashanasi 1989). Meanwhile, they found the association of palm-trees and Kudzu cover had macro-faunal elements of the primary forest and new species introduced after disturbance and the land was cleared. Thus, it seems that invertebrate communities are best conserved when the derived system has a structure similar to that of the original system, such as pastures planted in savanna areas and tree-based systems in forested areas (Decaëns et al. 1994, 2004, Fragoso et al. 1997, Barros 1999 and as presented by Barros et al. 2002; see Fig. 4.5). In Western Brazilian Amazonia (Barros et al. 2005) found that macrofauna communities in agroforestry systems were similar to the undisturbed

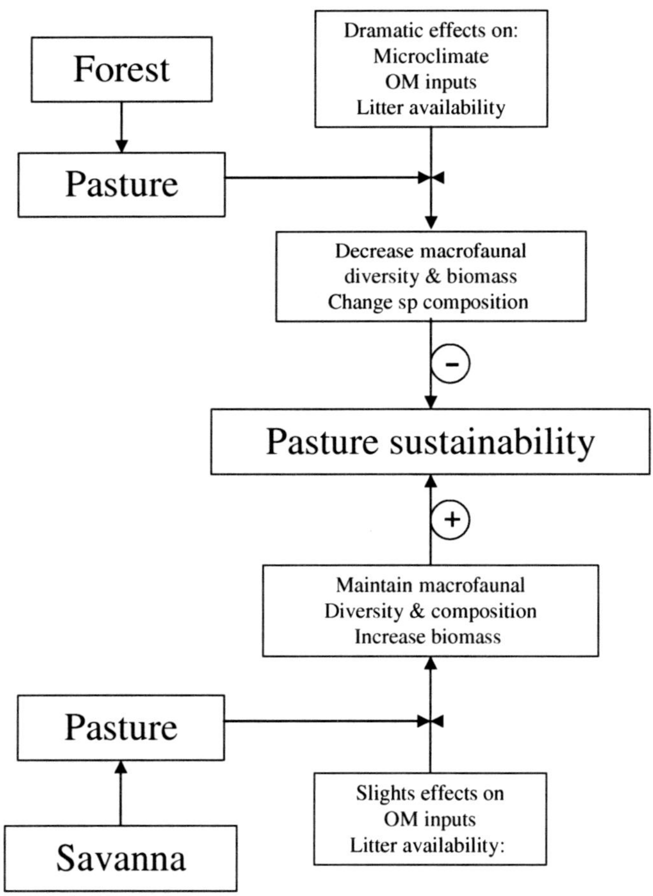

Fig. 4.5 A model of the response of macroinvertebrate communities to pasture establishment and the potential effects on agro-ecosystem sustainability (modified from Decaëns et al. 2004)

lowland rain forest, in spite of higher densities of social insects in the former. Yet in terms of abundance, they showed that land use practices (pastures, fallows, agroforestry and annual crop) were able to sustain sizeable macro-faunal communities similar to those of a disturbed forest.

4.4.4 Agroforestry

In general, pastures have been found to enhance earthworm abundance but reduce the overall macro-fauna diversity, whereas tree-based systems seem to be the best for conserving diversity and sustaining high levels of abundance (Mboukou-Kimbatsa et al. 1998, Barros et al. 2002). Alteration of the land use history plays a major role in determining the abundance and community structure of earthworms and the establishment of exotic earthworms in areas previously inhabited by worms (González et al. 2006). For example, in the tropics, conversion of forest to pastures has been associated to significant decreases in soil macro-invertebrate diversity (Lavelle and Pashanasi 1989) and the dominance of a few exotic earthworm species that can persist along different stages of plant succession after disturbance (e.g., Zou and González 1997, Sánchez-de León et al. 2003). Yet, considering the vast area under pastures and the different edapho-climatic and management parameters in the Neotropics, it is not surprising that the results on biological, physical and chemical properties may sometimes appear contradictory (Brossard et al. 2004). For example, Brown et al. (2004) studied soil macro-faunal communities in native and introduced pastures in Southeastern Mexico and suggested that if nutrient-poor savannas are converted to native pastures, these can maintain higher soil macrofauna populations and earthworm diversity than introduced pastures. Yet, Jiménez et al. (1998) studied earthworm communities in native savannas and man-made pastures of the Eastern Plains of Colombia and found a positive response of earthworm communities (higher density and biomass) to improved pastures; which is a type of land use being increasingly adopted in most Neotropical savannas. Further, Benito et al. (2004) compared macro-fauna communities from the native vegetation of a Brazilian Cerrado with renewed pastures and showed that renewed pastures had less diversity, population density and biomass of macro-fauna than the deforested Cerrado.

With the vast amount of abandoned tropical land due to non-sustainable farming practices, tropical tree plantations become an effective means in restoring soil productivity and preserving ecosystem biodiversity (Zou and González 2001). Tree planting accelerates the process of tree invasions and establishment (Myster 1993) in abandoned agricultural fields and it has been shown to mitigate barriers to secondary succession by re-establishing nutrient dynamics and improving the microclimate for plant understory species (Hagger et al. 1997, Binkley and Resh 1998). Forest management practices can influence earthworm communities through changing physical and chemical properties of soil, net primary productivity, or plant litter chemistry (Fig. 4.6, González et al. 1996, Zou and González 2001). Yet the response of the macro-faunal community to the forest management practices might differ depending among the various groups of fauna given their functionality. For

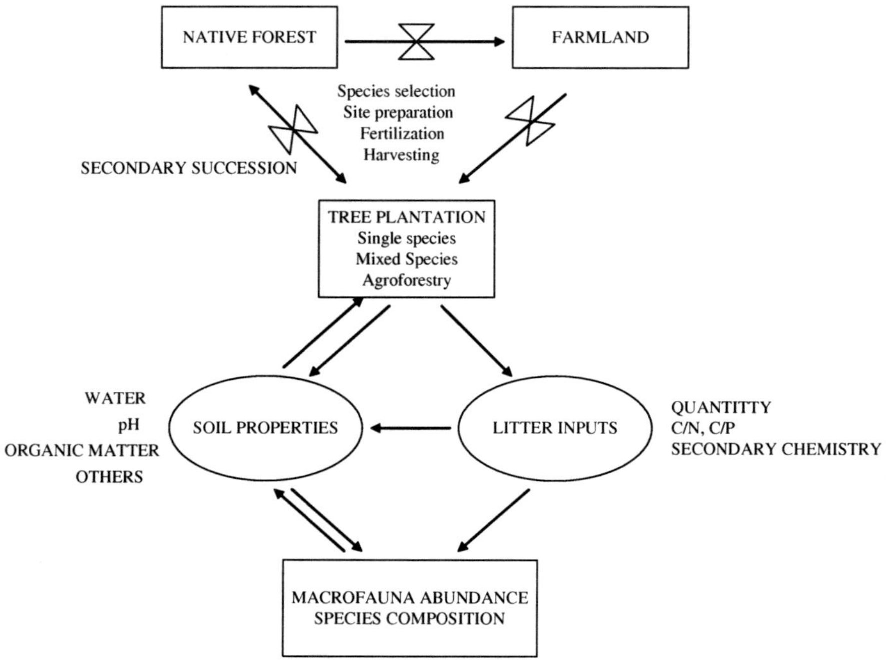

Fig. 4.6 Forest management practices (species selection, site preparation, fertilization and harvesting) that affect earthworm abundance and community structure in tropical tree plantations. Native forest refers to forests that are not disturbed by human activities. Farmland includes crop fields and pastures (modified from González et al. 1996)

example, Warren and Zou (2002) studied soil macro-fauna and litter inputs in three tropical plantations started from a disturbed site in Puerto Rico, and found that earthworm biomass did not differ among plantation treatments. However, they found the millipede biomass and abundance were significantly higher in *Leucaena* than in *Casuarina* and *Eucalyptus* plantations. Further, the millipede biomass was highly correlated to the N concentration and the C/N ratio in the incompletely decomposed organic layer (O_i). Warren and Zou (2002) suggested that in this study, millipedes may be influenced by litter chemistry at fairly small spatial scales, and earthworms are probably more affected by higher order controls such as soil physical conditions.

In the Neotropics and the world in general, soil fauna are important to consider if we are to sustain and maintain, functioning and healthy soils. As forests are converted to pastures, the effects of heavy machinery and trampling by cattle as a result from techniques used for deforestation and pasture management can have widespread and long lasting effects on the reduced abundance and diversity of macro-faunal communities in the newly created pastures (Chauvel et al. 1999); further instigating a cascade of deleterious effects which are linked to ecosystem health (see Chapter 5 for detailed examples on the effects of exotic earthworms on ecosystem processes). Land use alteration in the tropics has historically been dominated by a shift from forest to agriculture, but there are also trends towards increasing urbanization and reforestation. The role of exotic earthworms (the dominant group in tropical pastures) includes influence on current pasture ecosystems, their potential invasion into surrounding forests and consequent ecosystem effects, their influence on regenerating secondary forests in abandoned pasture, and their overall effects on tropical biodiversity. Many of these potential effects are yet unstudied and unknown (González et al. 2006).

In conclusion, the study of fungi and macrofauna, the major players in the decomposition processes of ecosystem functioning, is of particular importance since knowing the relationship between patterns and activities of fungal and macrofauna communities and biogeochemical changes following land conversion could greatly improve our understanding of the post-agricultural recovery mechanisms. In this chapter, we discussed general patterns of fungi and macrofauna in forests, pastures and agroforestry systems. In general, fungal and total microbial biomass significantly reduced following the conversion from forests to pastures or agricultural lands. In the case study of fungal biomass dynamics following reforestation in Puerto Rico, we found the plantations and the secondary forests have different proportions of fungal communities although they were developed originally from the same abandoned farm lands and therefore might have different nutrient cycle pathways. Soil macrofauna play a very important role in improving structure, organic matter content and distribution patterns of nutrient elements. Soil macrofauna may affect soil function and processes in a variety of ways, and could be used as indicators of nutrient status of soils. In general, earthworms are best represented in grasslands in humid areas as their abundance decreases towards humid forests and dry forests and grasslands. Termites are well adapted to dry environments where earthworms are not found. Millipedes are important in litter dominated ecosystems where their feeding activity is specialized on dead organic matter, or on saprophytic organism consumption. Fungal and macrofauna communities might be the most important variables controlling the biogeochemical changes following land conversion. However, most previous studies have focused on environmental or plant-driven factors over these biogeochemical changes. Future studies on post-agricultural recovery mechanisms need to pay more attention to the forms of trophic interactions among the functional groups, such as fungi and macrofauna, in soil.

Acknowledgment While writing this manuscript, Grizelle González received support of grant DEB-0218039 from the National Science Foundation to the Institute of Tropical Ecosystem Studies, University of Puerto Rico, and the USDA Forest Service, International Institute of Tropical Forestry as part of the Long-Term Ecological Research Program in the Luquillo Experimental Forest. Additional support was provided by the Forest Service (U.S. Department of Agriculture), and the University of Puerto Rico. María M. Rivera helped with graph reproduction. We are grateful to Dr. William A. Gould for valuable comments on early versions of the manuscript.

References

Amundson, R. 2001. The soil carbon cycles. Annual Reviews of Earth and Planetary Sciences 29: 535–562.

Anderson, J. M. 1977. The organization of soil animal communities. In Organisms as Components of Ecosystems: Proceedings of the VI International Soil Zoology Colloquium of the

International Society of Soil Science (SSSA), ed. U. Lohm and T. Persson, pp. 15–23. Stockholm: Swedish Natural Science Research Council.
Anderson, J. M. 1988. Spatiotemporal effects of invertebrates on soil processes. Biology and Fertility of Soils 6:216–227.
Barros, A. L. M. 1999. De capital, produtividade e crescimento da agricultura: O Brasil de 1970 a 1995. Piracicaba, SP: ESALQ/USP, 149 f. Tese (Doutorado) – Escola Superior de Agricultura "Luiz de Queiroz", Universidade de São Paulo, Piracicaba.
Bardgett, R. D. 2005. The Biology of Soils: A Community and Ecosystem Approach. Oxford University Press, Oxford, UK.
Barros, E., Pashanasi, B., Constantino, R., and Lavelle, P. 2005. Effects of land-use system on the soil macrofauna in western Brazilian Amazonia. Biology and Fertility of Soils 35: 338–347.
Behera, N., Pati, D. P., and Basu, S. 1991. Ecological studies of soil microfungi in a tropical forest soil of Orissa, India. Tropical Ecology 32: 136–143.
Benito, N. P., Brossard, M., Pasini, A., Guimarães, M. F., and Bobillier, B. 2004. Transformations of soil macroinvertebrate populations after native vegetation conversion to pasture cultivation (Brazilian Cerrado). European Journal of Soil Biology 40:147–154.
Bertrand, M., and Lumaret, J. P. 1992. The role of Diplopoda litter grazing activity on recycling processes in a Mediterranean climate. Vegetatio 99–100:289–297.
Bever, J. D., Morton, J., Antonovics, J., and Schultz, P. A. 1996. Host-dependent sporulation and species diversity of micorrhizal fungi in a mown grassland. Journal of Ecology 75:1965–1977.
Binkley, D., and Resh, S. C. 1998. Rapid changes in soils following Eucalyptus afforestation in Hawaii. Soil Sci. Soc. Am. J. 63:222–225.
Borneman, J., and Triplett, E. W. 1997 Molecular microbial diversity in soils from eastern Amazonia. Applied and Environmental Microbiology 63:2647–2653.
Brady, N. C., and Weil, R. W. 2002. The Nature and Properties of Soils, 13th edition. New Jersey: Prentice Hall.
Brossard, M., Boddey, R. M., and Blanchart, E. 2004. Soil processes under pastures in intertropical areas. Agriculture, Ecosystems and Environment 103:267–268.
Brown, S., and Lugo, A. E. 1990. Effects of forest clearing and succession on the carbon and nitrogen content of soils in Puerto Rico and US Virgin Islands. Plant and Soil 124:53–64.
Brown, G. G., Moreno, A. G., Barois, I., Fragoso, C., Rojas, P., Hernández, B. and Patrón, J. C. 2004. Soil macrofauna in SE Mexican pastures and the effect of conversion from native to introduced pastures. Agriculture, Ecosystems and Environment 103:313–327.
Buschbacher, R., Uhl, C., and Serrao, E. A. S. 1988. Abandoned pastures in eastern Amazonia. II. Nutrient stocks in the soil and vegetation. Journal of Ecology 776:682–699.
Chapin III, F. S., Matson, P. A., and Mooney, H. A. 2002. Principles of Terrestrial Ecosystem Ecology. New York: Springer-Verlag.
Chauvel, A., Grimaldi, M., Barros, E., Blanchart, E., Desjardins, T., Sarrazin, M., and Lavelle, P. 1999. Pasture damage by an Amazonian earthworm. Nature 398:32–33.
Cheng, L. 1993. Biodiversity in China. Beijing, China: Science Press.
Cleveland, C. C., Townsend, A. R., Schmidt, S. K., and Constance, B. C. 2003. Soil microbial dynamics and biogeochemistry in tropical forests and pastures, southern Costa Rica. Ecological Application 13:314–326.
Coleman, D. C. (1985). Through a ped darkly: an ecological assessment of root-soil-microbial-faunal interactions. In Ecological interactions in Soil. Plants, Microbes and Animals, ed. A. H. Fitter, pp. 1–21. Oxford, London etc.: Blackwell Scientific publications.
Decaëns, T., Lavelle, P., Jiménez, J. J., Escobar, G., and Rippstein, G. 1994. Impact of land management on soil macrofauna in the Oriental Llanos of Colombia. European Journal of Soil Biology 30:157–168.
Decaëns, T., Jiménez, J. J., Barros, E., Chauvel, A., Blanchart, E., and Lavelle, P. 2004. Soil macrofaunal communities in permanent pastures derived from tropical forest or savanna. Agriculture, Ecosystems and Environment 103:301–312.
Dodd, J. C., Burton, C. C., Burns R. G., and Jeffries P. 1987. Phosphatase activity associated with the roots and the rhizosphere of plants infected with vesicular-arbuscular mycorrhizal fungi. New Phytol. 107:163–172.

Doran, J. W., and Zeiss, M. R. 2000. Soil health and sustainability: managing the biotic components of soil quality. Appl Soil Ecol 15:3–11.
Fearnside, P. M. 1993. Deforestation in Brazilian Amazonia: the effect of population and land tenure. Ambio 22:537–545.
Fitter, A. H., and Garbaye, J. 1994. Interactions between mycorrhizal fungi and other soil organisms. Plant and Soil 159:123–132.
Fragoso, C., Brown, G. G., Patrón, J. C., Blanchart, E., Lavelle, P., Pashanasi, B., Senapati, B. K., and Kumar, T. 1997. Agricultural intensification, soil biodiversity and agroecosystem function in the tropics: the role of earthworms. Applied Soil Ecology 6:17–35.
Frankland, J. C., Dighton, J., and Boddy, L. 1990. Methods for studying fungi in soil and forest litter. In Methods in Microbiology, Vol. 22, ed. R. Grigorova and J. R. Norris, pp. 343–404. London: Academic Press.
Giller, P. S. 1996. The diversity of soil communities, the 'poor man's tropical rainforest'. Biodiversity and Conservation 5:135–168.
Gillison, A. W., Jones, D. T., Susilo, F. X., and Bignell, D. E. 2003. Vegetation indicates diversity of soil macroinvertebrates: a case study with termites along a land-use intensification gradient in lowland Sumatra. Organisms, Diversity and Evolution 3:111–126.
González, G., Zou, X., and Borges, S. 1996. Earthworm abundance and species composition in abandoned tropical croplands: comparison of tree plantations and secondary forests. Pedobiologia 40:385–391.
González, G., and Seastedt, T. R. 2001. Soil fauna and plant litter decomposition in tropical and subalpine forests. Ecology 82:955–964.
González, G., Huang, C. Y., Zou, X., and Rodríguez, C. 2006. Earthworm invasions in the tropics Biological Invasions 8(6):1247–1256.
Groffman, P. M., Driscoll, C.T., Fahey, T. J., Hardy, J. P., Fitzhugh, R. D., and Tierney, G. L. 2001. Colder soils in a warmer world: A snow manipulation study in a northern hardwood forest ecosystem. Biogeochemistry 56:135–150.
Hagger, J., Wightman, K., and Fisher, R. 1997. The potential of plantations to foster woody regeneration within a deforested landscape in lowland Costa Rica. Forest Ecology and Management 99(1–2):55–64.
Hansen, R. A. 1999. Red oak litter promotes a microarthropod functional group that accelerates its decomposition. Plant and Soil 209:37–45.
Hansen, R. A. 2000. Diversity in the decomposing landscape. In Invertebrates as Webmasters in Ecosystems, ed. D. C. Coleman and P. E. Hendrix, pp. 203–219. Wallingford, UK: CABI Press.
Heneghan, L., Coleman, D. C., Zou, X., Crossley, D. A., and Haines, B. L. 1999. Soil microarthropod contributions to decomposition dynamics: tropical-temperate comparisons of a single substrate. Ecology 80(6):1873–1882.
Henrot, J., and Robertson, G.P. 1994. Vegetation removal in two soils of the humid tropics: effect on microbial biomass, Soil biology and Biochemistry 26:111–116.
Höflich, G., and Metz, R. 1997. Nutrient inputs to soils and their uptake by alfalfa through long-term irrigation with untreated sewage effluent in Mexico. Soil Use and Management 14: 119–122.
Holl, K. D. 1999. Factors limiting tropical rain forest regeneration in abandoned pasture: seed rain, seed germination, microsite and soil. Biotropica 31:229–242.
Irmler, U. 2000. Changes in the fauna and its contribution to mass loss and N release during leaf litter decomposition in two deciduous forests. Pedobiologia 44:105–118.
Jayachandran K, Schwab, A. P., and Hetrick, B. A. D. 1992 Mineralization of organic phosphorus by vesicular-arbuscular mycorrhizal fungi. Soil Biol. Biochem. 24:897–903.
Jenkinson, D. S., and Powlson, D. S. 1975. The effects of biocidal treatments on metabolism in soil-5: a method for measuring soil biomass. Soil Biol. Biochem. 8:209–213.
Jenos, D. P. 1988. Mycorrhiza applications in tropical forestry: are temperate zone approaches appropriate? In Threes and Mycorrhiza, ed. F. P. Ng, pp. 133–188. Kuala Lumpur, Malaysia: Forest Research Institute.
Jiménez, J. J., Moreno, A. G., Decaëns, T., Lavelle, P., Fisher, M. J., and Thomas, R. J. 1998. Earthworm communities in native savannas and man-made pastures of the Eastern Plains of Colombia. Biology and Fertility of Soils 28(1):101–110.

Joner, E. J., and Jakobsen, I. 2005. Contribution by two arbuscular mycorrhizal fungi to P uptake by cucumber (*Cucumis sativus* L.) from^{32}P-labelled organic matter during mineralization in soil. Plant and Soil 163:203–209.

Korf, R. P. 1997. Tropical and subtropical discomycetes. In Biodiversity of Tropical Microfungi, ed. K. D. Hyde, pp. 229–240. Hong Kong: Hong Kong University Press.

Lanly, J. 1982. Tropical Forest Resources. Food and Agriculture Organization, Rome, FAO For. Pap. No. 30, p. 106.

Lavelle, P. 1997. Faunal activities and soil processes: Adaptive strategies that determine ecosystem function. Advances in Ecological Research 27:93–132.

Lavelle, P., and Pashanasi, B. 1989. Soil macrofauna and land management in Peruvian Amazonia (Yurimaguas, Loreto). Pedobiologia 33:283–291.

Lavelle, P., Bignell, D., Lepage, M., Wolters, V., Roger, P., Ineson, P., Heal, O. W., and Dhillion, S. 1997. Soil function in a changing world: the role of invertebrate ecosystem engineers. Soil Biology 33:159–193.

Lavelle, P., Barros, E., Blanchart, E., Brown, G. G., Desjardins, T., Mariani, L., and Rossi, J. P. 2001. SOM management in the tropics: Why feeding the soil macrofauna? Nutrient Cycling in Agroecosystems 61:53–61.

Lavelle, P., Dangerfield, J. M., Fragoso, C., Eschenbreener, V., Lopez-Hernandez, D., Pashanasi, B., and Brussaard, L. 1994. The relationship between soil macrofauna and tropical soil fertility. In Woomer, P. L., Swift, M. J., eds The Biological Management of Tropical Soil Fertility. Chichester (UK): Wiley 137–169.

Li, Y., Xu, M., Sun, O. J., and Cui, W. 2004. Effects of root and litter exclusion on soil CO_2 efflux and microbial biomass in wet tropical forests. Soil Biology & Biochemistry 36: 2111–2114.

Li, Y., Xu, M., Zou, X., and Xia, Y. 2005. Soil CO_2 efflux and fungal and bacterial biomass in a plantation and a secondary forest. Plant and Soil 268:151–160.

Li, Y., Xu, M., Zou, X., and Zhang, Y. 2005. Comparing of soil organic carbon dynamics in plantations and secondary forests in wet tropics in Puerto Rico. Global Change Biology 11, 239–248.

Lodge, D. J. 1987. Nutrient concentrations, percentage moisture and density of field-collected fungal mycelia. Soil Biology and Biochemistry 19:727–733.

Lodge, D. J. 1993. Nutrient cycling by fungi in wet tropical forests. In Aspects of tropical mycology, ed. S. Isaac, J. C. Frankland, R. Watling, and A. J. S. Whalley. British Mycological Society Symposium Series. Vol. 19, pp. 37–58. Cambridge: Cambridge University Press.

Lodge, D. J., and Cantrell, S. 1995. Fungal communities in wet tropical forests: variation in time and space. Canadian Journal of Botany 73:1391–1398.

Lodge, D. J., and Ingham, E. R. 1991. A comparison of agar film techniques for estimating fungal biovolumes in litter and soil. Agriculture, Ecosystems and Environment 34: 131–144.

Lugo, A. E. 1992. Comparison of tropical tree plantations with secondary forests of similar age. Ecol. Monogr. 62:1–41.

Luizao, R. C. C., Bonde, T. A., and Rosswall, T. 1992. Seasonal variation of soil microbial biomass: the effects of clearfelling a tropical rainforest and establishment of pasture in the Central Amazon. Soil Biology and Biochemistry 24:805–813

Macfadyen, A. 1963. The contribution of the microfauna to total soil metabolism. In Soil Organisms, ed. J. Doeksen and Van der Drift, pp 3–15. Amsterdam: J. North Holland.

Martius, C., Tiessen, H., and Vlek, P. L.G. 2001. The management of organic matter in tropical soils: What are the properties? Nutr. Cycling Agroecosyst. 61:1–6.

Mboukou-Kimbatsa, I. M. C., Bernhard-Reversat, F., and Loumeto, J. J. 1998. Change in soil macrofauna and vegetation when fast-growing trees are planted on savanna soils. Forest Ecology and Management 110:1–12.

Menaut, J. C., Barbault, R., Lavelle, P., and Lepage, M. 1985. African savannas: biological systems of humification and mineralization. In Management of the world's savannas, ed. I. C. Tothill and J. J. Mott. Canberra, Australia: Australian Academy of Sciences.

Moguel, P., and Toledo, V. M. 1999. Biodiversity conservation in traditional coffee systems of Mexico. Conservation Biology 13(1):11–21.

Montagnini, F., and Porras, C. 1998. Evaluating the role of plantations as C sinks: an example of integrative approach from the humid tropics. Environ. Manage. 22:459–470.
Myster, R. W. 1993. Tree invasion and establishment in oldfields at Hutcheson Memorial Forest. The Botanical Review 59:251–272.
Myster, R. W. 2004. Post-agricultural invasion, establishment, and growth of neotropical trees. The Botanical Review 70:381–402.
Myster, R. W. 2006. Light and nutrient effects on growth and uptake strategies of Inga vera, a landslide tree in Puerto Rico. Canadian Journal of Forest Research 13:1–8.
Myster, R. W., and Schaefer, D. A. 2003. Species and microsite effects on litter decomposition in a Puerto Rican landslide. Community Ecology 4:157–162
Neill, C., Melillo, J. M., Steudler, P. A., Cerri, C. C., de Moraes, J. F. L., Piccolo, M. C., and Brito, M. 1997. Soil C and nitrogen stocks following forest clearing for pasture in the southwestern Brazilian Amazon. Ecological Applications 7:1216–1225.
Nepstad, D. C., Verissimo, A., Alencar, A., Nobre, C., Lima, E., Lefebvre, P., Schlesinger, P., Potter, C., Moutinho, P., Mendoza, E., Cochrane, M., and Brooks, V. 1999. Large-scale impoverishment of Amazonian forests by logging and fire. Nature 398:505–508.
Nüsslein, K., and Tiedje, J. M. 1999. Soil bacterial community shift correlated with change from forest to pasture vegetation in a tropical soil. Applied and Environmental Microbiology 65:3622–3626.
Paul, E. A., and Clark, F. E. 1996. Components of soil biota. In Soil Microbiology and Biochemistry, ed. E. A. Paul and F. E. Clark, pp. 71–106. New York: Academic Press.
Picone, C. 2000. Diversity and abundance of arbuscular-mycorrhizal fungus spores in tropical forest and pasture. Biotropica 32:734–750.
Post, W. M., and Kwon, K. C. 2000. Soil C sequestration and land-use change: processes and potential. Global Change Biology 6:317–327.
Rao, M. R, Niang, A., Kwesiga, F. R., Duguma, B., Franzel, S., Jama, B., and Buresh, R. 1998. Soil fertility replenishment in sub-Saharan Africa: new techniques and the spread of their use on farms. Agroforestry Today 10:3–8.
Read, D. D., and Perez-Moreno, J. 2003. Mycorrhizas and nutrient cycling in ecosystems: a journey towards relevance? The New Phytologist 157:475–492.
Salas, A. M., Elliott, E. T., Westfall, D. G., Cole, C. V., and Six, J. 2003. The role of particulate organic matter in phosphorus cycling. Soil Science Society of America Journal 67:181–189.
Sánchez, P. 1995. Science in agroforestry. Agrofor. Syst. 30:5–55.
Sánchez-De León, Y., Zou, X., Borges, S., and Ruan, H. 2003. Recovery of native earthworms in abandoned tropical pastures. Conservation Biology 17:1–8.
Sanders, F. E., and Tinker, P. B. 1971 Mechanism of absorption of phosphate from Soil by *Endogone* mycorrhizas. Nature 233:278–279.
Spaans, E. J. A., Baltissen, G. A. M., Miedema, R., Lansu, A. L. E., Schooderbeek, D., and Wielemaker, W. G. 1989. Changes in physical properties of young and old volcanic surface soils in Costa Rica after clearing of tropical rain forest. Hydrol. Proc. 3:383–392.
Swift, M. J., Heal, O. W., and Anderson, J. W. 1979. Decomposition in terrestrial ecosystems. Oxford: Blackwell.
Thomas, T., Folgarait, P., Lavelle, P., and Rossi, J. P. 2004. Soil macrofaunal communities along an abandoned rice field chronosequence in Northern Argentina. Applied Soil Ecology 27:23–29.
Thomlinson, J. R., Serrano, M. I., Del, M., Lopez, T., Aide, T. M., and Zimmerman, J. K. 1996. Land-use dynamics in a post-agricultural Puerto Rican landscape (1936–1988). Biotropica 28:525–536.
Townsend, A. R., Asner, G. P., Cleveland, C. C., Lefer, M. E., and Bustamante, M. C. 2002. Unexpected changes in soil phosphorus dynamics following tropical deforestation to cattle pasture. Journal of Geophysical Research 107(D20):8067, doi: 10.1029/2001 JD000650, 2002.
Tufekcioglu, A., Raich, J. W., Isenhart, T., and Schultz, R. C. 2001 Soil respiration within riparian buffers and adjacent crop fields. Plant Soil 299:117–124.
van der Heijden, M. G. A., Klironomos, J. N., Ursic, M., Moutoglis, P., Streitwole-Engel, R., Boller, T., Weimken, A., and Sanders, I. R. 1998. Mycorrhizal fungal diversity determines plant biodiversity, ecosystem variability, productivity. Nature 396:69–72.

Vanlauwe, B., Sanginga, N., and Merckz, R. 1997. Decomposition of four Leucaena and Senna prunings in alley cropping systems under subhumid tropical conditions: the process and its modifiers. Soil Biology & Biochemistry 29:131–137.

Wallwork, J. A. 1970. Ecology of Soil Animals. London: McGraw-Hill.

Warren, M., and, Zou, X. 2002. Soil macrofauna and litter nutrients in three tropical tree plantations on a disturbed site in Puerto Rico. Forest Ecology and Management 170:161–171.

Yang, J. C., and Insam, H. 1991. Microbial biomass and relative contributions of bacteria and fungi in soil beneath tropical rain forest, Hainan Island, China. Journal of Tropical Ecology 7:385–395.

Young, A. 1997. Agroforestry for Soil Management, 2nd edition, p. 320. New York, USA: CAB International and ICRAF.

Zou, X., and González, G. 1997. Changes in earthworm density and community structure during secondary succession in abandoned tropical pastures. Soil Biology and Biochemistry 29: 627–629.

Zou, X. M., and Bashkin, M. 1998. Soil carbon accretion and earthworm recovery following revegetation in abandoned sugarcane fields. Soil Biol. Biochem. 30:825–830.

Zou, X., and González, G. 2001. Earthworms in tropical tree plantations: effects of management and relations with soil carbon and nutrient use efficiency. In Management of Tropical Plantation Forests and Their Soil Litter System, ed. M. V. Reddy, pp. 283–295. New Delhi, India: Oxford University Press.

Chapter 5
Earthworms and Post-Agricultural Succession

Grizelle González, Ching-Yu Huang and Shu-Chun Chuang

5.1 Introduction

Earthworms are the best known and often the most important animals influencing the functioning of soil ecosystems (Hendrix and Bohlen 2002). In mature forests of the Neotropics, earthworms usually dominate the soil food web in terms of biomass (e.g., Odum and Pigeon 1970, Lee 1985, Lavelle, Brussaard and Hendrix 1999). Consequently, they have significant influence on soil structure, nutrient cycling and crop productivity as their activities can increase soil porosity, stimulate microbial activities, and accelerate litter decay and the release of nutrients into the soil (Lee 1985, Lavelle et al. 1999, González and Seastedt 2001, González 2002, Liu and Zou 2002).

Earthworms are classified into endogeic, anecic, and epigeic species to represent soil, soil and litter, and litter feeders, respectively (Bouché 1977). Earthworms can alter soil physical properties and biogeochemical processes (e.g., Edwards and Bohlen 1996) according to their functionality. Through borrowing activities, earthworm may alter soil structure and soil water infiltration rates, thus alter the environmental conditions for microbial growth and biogeochemical fluxes (González et al. 2006). Earthworms also act as inoculators of soil microbes for the freshly fallen plant materials, and consequently accelerate the decomposition of plant materials (González 2002). Endogeic earthworms alter soil properties primarily through changing soil physical and chemical properties and epigeic earthworms mainly affect microbial inoculation and the associated decomposition processes. Anecic earthworm influence soils through both borrowing and inoculation activities (González et al. 2006).

Plant community succession alters the quantity and chemistry of organic inputs to soils (Zou and González 1997). These differences in the organic input influence the patterns of response of the soil biota as do tree species during the process of change from pastures to post-agricultural forests. In addition, land use history and invasions by earthworm exotic species can affect the relative abundance and species composition of a local earthworm fauna distinctly from soil, climate, vegetation, and topography (Hendrix and Bohlen 2002). Regularly in the Neotropics, cosmopolitan earthworm species dominate the pastures at the expense of native ones (Table 5.1). With the conversion of pastures to secondary forests, the richness of earthworm

R. W. Myster, *Post-Agricultural Succession in the Neotropics.*

Table 5.1 Changes of native and exotic earthworm communities along post-agricultural successional stages in the Neotropics

Location	Pattern	Native earthworm	Exotic earthworm	Reference
Brazil				
Manaus	Abandoned plantation of rubber trees	*Andiorrhinus sp. 1*	*Pontoscolex corethrurus*	Römbke et al. (1999)
		An. amazonicus	*Dichogaster bolaui*	
		An. venezuelanus tarumanis		
		Rhinodrilus contornus		
		R. priollii		
		R. brasiliensis		
		Tuiba dianaea		
	Secondary forest	Data not available	None	
	Primary forest	Data not available	None	
Mexico				
Tabasco	*Brachiaria decumbens* pasture	None	*P. corethrurus*	Geissen and Gúzman (2006)
	Cynodon plectostachyus pasture	None	*P. corethrurus*	
	Successional forest	None	*P. corethrurus*	

(continued)

Table 5.1 (continued)

Location	Pattern	Native earthworm	Exotic earthworm	Reference
Puerto Rico				
Cayey Mt.	Abandon pasture	None	*P. corethrurus* *Amynthas gracilis*	Sánchez-de León et al. (2003)
	Secondary forest	None	*P. corethrurus*	
	Mature forest	*Borgesia sedecimesetae* *Estherella* sp. *Onychochaeta borincana* *Neotrigaster rufa* *Trigaster longissimus*	*P. corethrurus*	
Luquillo Mt.	*Pinus caribaea* plantation	None	*P. corethrurus* *A. rodericensis*	González et al. (1996)
	Swietenia macrophylla plantation	None	*P. corethrurus*	
	Secondary forest	*P. spiralis* *E. montana* *E. gatesi*	*P. corethrurus,* *A. rodericensis*	
Sabana, Luquillo Mt.	Active pasture	None	*P. corethrurus*	Zou and González (1997)
	Grass-vine-fern	None	*P. corethrurus*	
	Shrub-small tree	*E. gatesi*	*P. corethrurus* *A. rodericensis*	
	Forest	*E. gatesi*	*P. corethrurus* *A. rodericensis*	

species increases due to the presence of native species (Zou and González 1997). Yet, the exotic earthworms are still dominant in terms of biomass and density (Zou and González 1997, Sánchez-de León et al. 2003). In Puerto Rico, González et al. (1996) found that *Pontoscolex corethrurus* (an exotic earthworm) dominated both plantations and secondary forests but native earthworms occurred only in secondary forests; suggesting that naturally regenerated secondary forests are preferable to plantations for maintaining a high level of earthworm density, biomass and native species in post-agricultural forests. The restoration of secondary forests through natural regeneration on abandoned pastures can promote the recovery of both anecic earthworms and native species (González et al. 1996, Sánchez-de León et al. 2003). Within this context then, in this chapter, we discuss (1) factors that affect the composition of earthworms (soil physico-chemical properties, soil nutrient status, and biotic interactions), (2) habitat disturbance and the invasions of exotic earthworm species, and (3) the physiological mechanisms by which exotic earthworms can successfully invade pastures and disturbed environments. Finally, we highlight gaps in knowledge that might stimulate further studying of native and non native earthworms in natural, agricultural and post-agricultural sites in the Neotropics.

5.2 Discussion

5.2.1 Factors that Affect the Composition of Earthworms

Earthworm communities can be described in terms of numbers, biomass, composition, and diversity of species. These components of earthworm communities are regulated by a combination of environmental factors and biotic interactions. Beyond landscape and regional scales, phylogenetic constraints and geological history determine the distribution and composition of earthworm assemblages. In the North Neotropical region, earthworm communities are mainly constituted of Megascolecidae (Acanthodrilinae and Dichogastrini), then followed by Ocnerodrilidae and some Glossoscolecidae (Fragoso et al. 1995). Glossoscolecidae becomes dominant at South America, especially in Brazil, Perú and Colombia (Fragoso et al. 1999a). Earthworm biogeography is also related to past geological history. For example, plate tectonic movements in Caribbean islands might partly explain the differences in earthworm communities and abundance among the Greater Antilles, the Lesser Antilles, and the Bahamas archipelago (Fragoso et al. 1995, James 2004). At the local scale, earthworm community structure may be controlled by a variety of factors such as climate, vegetation, edaphic properties and the biotic interactions in the soil. It has been suggested that these factors affect earthworm communities hierarchically; climate (air temperature) being the predominant factor explaining the variation of earthworm communities, followed by soil nutrient status and seasonal fluctuations of humidity (Fragoso et al. 1999a). However, the predominant factors determining earthworm communities can vary depending on local habitat characteristics. Also, it is difficult to separate the independent effects of environmental factors as they are often interrelated (e.g., soil temperature and moisture retention) (Lee 1985). Biotic factors may become determinant for earthworm communities, especially in areas invaded by exotic species. Predatory, competitive, or facilitative interactions between native and exotic earthworm species may change the dynamics and composition of the original earthworm community in invaded areas. The community structure we observe in the field is a result of the combined effects from the aforementioned and other factors. We can divide the regulating factors of earthworm communities into two main categories, abiotic and biotic; and the following sections discuss the potential impacts of each on earthworm communities.

5.2.1.1 Abiotic Factors

Soil temperature and moisture. – Edaphic properties have strong controls on earthworm community structure (Fragoso and Lavelle 1992), primarily soil temperature and moisture content, which govern basic requirements for maintaining earthworm activities. Few field studies describe the preferred and/or optimum soil temperatures in relation to earthworm populations due to the difficulty of standardization and concurrently measuring soil temperatures and earthworm activities in natural settings. Generally, the temperature tolerance of earthworms varies depending on the species, collecting site, and the observed conditions (laboratory vs field observation, and methods of experimentation employed) (Lee 1985). However, it has been determined that the optimum temperature range for tropical earthworm growth and activity is between 20–30°C at the surface or sub-surface of the soil in the field (Lee 1985).

Among all environmental factors within the Neotropics, moisture seems to be the principal regulator of earthworm dynamics. Soil moisture content also regulates soil water tension, gas transfer, and other soil fauna activities (i.e. microbial populations), which indirectly affect earthworm communities (Araujo and López-Hernández 1999). In general, total earthworm numbers and biomass increase with an increase in soil moisture content (Garnsey 1994, Araujo and López-Hernández 1999, Schmidt and Curry 2001). Fragoso and Lavelle (1992) found maximum earthworm density and biomass in wet tropical forests with 2,000–4,000 mm annual rainfall. Beyond this range, the soils were either too wet or too dry for sustaining an optimum earthworm community. Nevertheless, once soil moisture content meets the basic requirement for earthworm activities, this factor may have less influence on earthworm dynamics. In Puerto Rico, González et al. (1999) found that soil water content, which ranged 40 to over 100%, in wet (tabonuco) forests had no influence on earthworm communities and *Pontoscolex corethrurus* densities.

Soil moisture content usually closely relates to soil temperature fluctuations. The rise of soil temperature due to removal of litter floor or loss of canopy may also result in an increase in soil evapotranspiration rates; which in turn decrease the soil water content. Both soil moisture and temperature fluctuations cause spatial and temporal (seasonal and annual) heterogeneity in earthworm communities at a variety of ecosystems within the Neotropics. Araujo and López-Hernández (1999) found that soil temperature and moisture were positively correlated to earthworm densities in a natural savanna of Venezuela. In the east Mexican tropics, soil temperature and moisture significantly affected the relative dominance of endogeic and epigeic groups of earthworm communities in different types of forests (Fragoso et al. 1995). In the tropics, earthworms can either become inactive or migrate into deeper soil when confronting extremely low soil moisture conditions like during droughts (Lavelle 1988). In Colombian pastures and savannas, earthworms showed uniform vertical distribution in the top 20 cm of the soil profile during the rainy season; while during the dry season, most worms moved to deeper soils responding to the decline of soil moisture at the top 20 cm of soil (Jiménez and Decaëns 2000). Seasonal or frequent droughts may convert earthworm communities from mixed species with different ecological strategies to the dominance of some specific species, which are more likely to be opportunistic poly-humic species (Lavelle 1988). Earthworm species richness is negatively correlated to the quiescent period of earthworm population in tropical regions (Lavelle 1988). However, other edaphic characteristics and biotic regulating factors may become more predominant than soil temperature and moisture when the climate factors remain relatively constant, such as in some tropical wet and rain forests.

Soil pH and texture. – Soil properties may be useful predictive variables of earthworm communities, particularly soil pH and texture. Some studies have showed a strong correlation between soil texture and earthworm communities in the Neotropics. Hubers et al. (2003) suggested that "severe" soil physicochemical conditions in Nipe soils of Maricao State Forest, Puerto Rico, resulted in lower diversity and abundance of earthworm communities as compared to Mexico, Venezuela, and other sites in Puerto Rico. These factors included lower soil pH, lower organic matter, and higher exchangeable iron (Fe), magnesium (Mg) and aluminum (Al) concentrations in Nipe soils. However, in the Luquillo Mountains of Puerto Rico, soil pH alone could not explain the distribution and biomass of earthworm communities (González et al. 1996, González et al. 1999, González and Zou, 1999b). The correlation between soil texture and earthworm communities should be explained with caution, since the casting and burrowing activities of earthworms result in the modification of soil texture (silt, clay, and sand contents), soil pH, and bulk density (Lee 1985, Edwards 2004).

Soil nutrients. – Depending on the feeding strategies of earthworms (epigeic, endogeic and anecic groups), earthworms can utilize litter, soil organic matter, plant roots, root exudates and microbes as food resources. Litterfall input (quantity) and quality are more important determinants of the presence and abundance of epigeic and anecic than endogeic earthworm species because of direct consumption of litter by the former groups. Fragoso and Lavelle (1992) showed that earthworm biomass increased with an increase in litter quantity among 14 localities of tropical rain forests. Further, the reduction of litter inputs in most agroecosystems via forest clearing and aboveground crop harvest, which is detrimental to epigeic and anecic species, have resulted in the dominance of endogeic earthworm species in the tropics (Fragoso et al. 1999a). Earthworm abundance is often positively correlated to litter quality (such as litter nitrogen, or phosphorus) and is negatively related to litter polyphenol content (Lee 1985). In the Luquillo Experimental Forest of

Puerto Rico, González and Zou (1999) described a correlation between litter N and phosphorus (P) contents and the biomass of anecic earthworms in both *Dacryodes* and *Heliconia* tree communities. Fragoso and Lavelle (1987) found that litter quality and soil organic matter explained the aggregated distribution of earthworm communities in a Mexican tropical rainforest.

Soil organic matter is considered an important food source for endogeic earthworm species (Lavelle 1988). Earthworms are always found in soils with higher soil organic matter (Lee 1985, Hubers et al. 2003). Hubers et al. (2003) found the density of *P. corethrurus* was higher in sites with soil organic matter above 6.5%. Fragoso and Lavelle (1992) showed that the soil nutrients status, including N, calcium (Ca), and Mg, accounts for 28% of the variation of earthworm communities, which is the second most important factor after soil moisture (45%), in tropical rainforests of Central America, South America, South Africa and Asia. The improved soil fertility status in the soil, either due to increased litterfall inputs and/or the applications of organic and inorganic fertilizers, can enhance earthworm biomass, density and diversity (Zou and González 1997, López-Hernández et al. 2004). Plant living and dead roots, root exudates and associated microbes which form the "rhizosphere", provide alternative food resources for earthworms (Lee 1985, Brown et al. 2000). Aggregated earthworm communities are usually found near the rhizosphere and dominated by endogeic species, because of higher soil carbon input in this zone as compared to the surrounding soil. For example, direct and indirect evidence showed that endogeic earthworm, *P. corethrurus*, fed in and aggregated around the rhizosphere (Brown et al. 2000, Sánchez-de León et al. 2003). *P. corethrurus* was shown to assimilate root-derived carbon of maize plants and sugar cane by directly observing the ^{13}C shift in worm tissues from an originally lower ^{13}C value (from C3 vegetation) to a higher ^{13}C content (from C4 plants) (Brown et al. 2000). Indirectly, Sánchez-de León et al. (2003) suggested that grass roots can be an important food resources for endogeic earthworms, *P. corethrurus*, because root biomass affected the earthworm community and distribution in a chronosequence of pastures in the Cayey Mountains of Puerto Rico. Moreover, Zou and González (1997) suggested root quality as an important factor explaining the difference on earthworm communities between pastures and woody plantations in Puerto Rico. Still, few studies deal with the potential effects that root exudates or microbial-earthworm interactions in the rhizosphere might have on earthworm community patterns. More research is needed in this area; as both microorganisms and root exudates are believed to be favorable food for earthworms. Generally, energy (food quality) and nutrient availability in the soil are good indicators of earthworm communities and their distributions.

5.2.1.2 Biotic Factors

Decaëns and Rossi (2001) investigated the spatio-temporal structure of earthworm communities and its relationship with soil heterogeneity in Colombian pastures; and found that the spatial and temporal variations of soil heterogeneity at regional scales could not completely explain the variability of earthworm distributions (Decaëns and Rossi 2001). Further, they argued that soil properties explained a small portion of the spatial and temporal structure of earthworm distribution, suggesting that earthworm communities themselves were at a non-equilibrium state, which was possibly caused by biotic interaction, such as competition exclusion (Decaëns and Rossi 2001) or predatory pressure.

Predatory pressure. – The earthworm tissue is enriched with proteins and thus earthworms are favored food for many animal species, which include birds, mammals, amphibians, reptiles, beetles, flatworms and some other invertebrates (Macdonald 1983, Lee 1985, Judas 1989, Blackshaw 1995). Most studies on earthworm predation focused either on the effects of predators on a single earthworm species or on the predator behavior. Lee (1985) argued that few studies have gathered data on amounts of earthworms captured or eaten by the predators, how predators affect earthworm populations, or the significance of earthworms in the energy flow of the food web. Yet, some studies had been done on the potential impacts of predatory pressure (mostly birds) on the structure of earthworm communities in temperate forests (Bengtson et al. 1976, Barnard and Thompson 1985, Judas 1989, Blackshaw 1995). For example, Bengtson et al. (1976) found twice the abundance and biomass of earthworms in protected plots than in the plots exposed to predation of golden plovers (*Pluvialis apricaria*) in an Icelandic hayfield. Judas (1989) found the average mortality of earthworms that were exposed to the predators (chilopod, *Strigamia acuminata*) was two times higher (64%) than the control (chilopod-excluded; 31%) in a 2-month field microcosm experiment. However, he did not find any significant effects of predatory pressure from macro-predators (birds, carabids, rodents, and shrews) on earthworm populations and vertical distribution in exclusion and control plots in another 7.5 mo field exclusion experiment in a beech forest (Judas 1989). The predation pressure may not always play a significant role on earthworm communities, and its inconsistent influence may depend on the seasons and certain situations (such as changes in vegetation and invasion of new predators). Shifts of vegetation types due to anthropogenic factors can result in the presence and/or absence of litter floor and canopy, which in turn may change the intensity of predatory pressure on earthworm communities though predatory efficiency and/or as related to different assemblages of earthworm predators. Blackshaw (1995) investigated the population sizes of invasive predatory flatworm, *Artioposthia triangulata*, and its impacts on earthworm communities in grasslands of Ireland. With the presence of predatory flatworms, he found a reduction in earthworm species richness (Blackshaw 1995). However, potential relationships between earthworms and their predators due to vegetation changes have not been intensively explored. In the Neotropics, many tropical pastures have been gradually converted to grasslands or secondary forests following the abandonment of agricultural practices due to soil degradation and changes in economic strategies (e.g., Brazil and Puerto Rico, see Buschbacher 1986, Helmer 2004). This post-agricultural succession provides great opportunities to investigate the dynamics of earthworms and their potential predators along the successional stages.

Intra- and inter-populations interactions. – Three ecological groups of earthworms (epigeic, endogeic, and anecic groups) represent different abilities to utilize resources (niche partitioning), including space and nutrients (Bouché 1977). Epigeic species are litter-dwelling and mostly feed on surface litter, while endogeic species live in the soil and consume organic residues and soil. Anecic species, which live in permanent vertical burrows, eat and bury surface litter (Lee 1985). When the niche of different earthworm species overlap, significant competition would be expected due to the use of similar resources. Since 1980s, several researchers have been interested in the competitive interaction among earthworms. Competition intensities are found to vary among different earthworm species from many laboratory studies (Abbott 1980, Hamilton et al. 1988, Butt 1998, Dalby et al. 1998, Baker et al. 2002, Lowe and Butt 2002). For example, Abbott (1980) showed persistent competition for food between *Eisenia foetida* and *Microscolex dubius*, but not between *E. foetida* and *Allolobophora trapezoides*. In a laboratory study by Butt (1998), the decrease of growth rates and cocoon production of *Allolobophora longa* resulted from both intra-specific competition and inter-specific competition with *Lumbricus terrestris*. Winsome et al. (2006) described that *Argilophilus marmoratus* had lower growth and reproduction rates in the presence of *A. trapezoides* in different habitats in a California grassland. They attributed this to the competition for exploiting similar food resources (particularly microbes) (Winsome et al. 2006). In contrast to the abovementioned studies from temperate ecosystems, few studies have focused on the competitive relationships between native and exotic earthworm species in the tropics (Hendrix et al. 1999, Lachnicht et al. 2002). Hendrix et al. (1999) examined earthworm resource utilization under field conditions in three different forest ecosystems (elfin forest, tabonuco forest, and a pasture) in Puerto Rico by using the natural abundance of ^{13}C and ^{15}N. The differences in ^{15}N fractions within native and exotic earthworm species when they co-exist in the soil indicated that there might be resource competition between native and exotic species (Hendrix et al. 1999). Also in Puerto Rico, Lachnicht et al. (2002) set up a 19-day incubation experiment to evaluate the interactive effects of *P. corethrurus* (an exotic earthworm) with native *Estherella* sp. by using stable isotopes. *P. corethrurus* and *Estherella* sp. were found to inhabit different soil layers when incubated together. In addition, *P. corethrurus* assimilated different nitrogen resources in the presence of *Estherella* sp. (Lachnicht et al. 2002). In the field, habitat disturbance and invasion of exotic earthworm species usually trigger or enhance competitive exclusion within earthworm communities due to the shortage of food resources. Nevertheless, it is still not clear how competitive exclusion and its intensity affect the structure of earthworm communities in different ecosystems; especially in the tropics.

Although environmental (abiotic) and biotic factors are determinants on earthworm communities, earthworm activities can reversibly affect their environments. Earthworms, as ecological engineers, have significant impacts on the soil abiotic and biotic environment and can regulate the availability of resources for other species by directly or indirectly modifying their habitats (Lee 1985, Edwards 2004). Soil moisture, soil texture, organic matter distribution, and soil nutrient cycling can be modified by earthworm activities, e.g. casting and building burrows (Lee 1985, Edwards 2004). Hence, unique earthworm assemblages are generally believed to reflect not only their specific adaptations or preferences to the local surroundings, but also the habitat they build.

5.2.2 Habitat Disturbance and the Invasion of Exotic Earthworm Species

Habitat disturbance can modify biotic and abiotic factors, which in turn can have significant effects on the composition of earthworm communities as discussed earlier. Natural disturbances, such as flooding and landslides caused by hurricanes, can result in temporal changes of earthworm communities through the alteration of litter inputs and soil properties in localized areas. However, anthropogenic disturbances, like forest clearing, human inhabitation and land-use management, can occur more frequently than natural disturbances and have more permanent and devastating influences on earthworm communities. For instance, large areas of tropical moist forests had been converted into savannas and pastures in Asia, Africa, South America, and Central America (Buschbacher 1986). Tropical trees in the original forests of Central America have been replaced with shrubs, annual/perennial crops or grasses (Nations and Komer 1983). This transformation of vegetation types causes a reduction of litter quality/quantity, alteration of root distribution and availability, and modification of soil properties. As a result, these abrupt fluctuations on soil physical/chemical conditions and food resource availability from habitat disturbance can modify the original composition and distribution of earthworms taxonomically and functionally (Abbott 1985, González et al. 1996, Sánchez-de León et al. 2003, Sánchez-de León and Zou 2004). Often, there is a significant decline of earthworm diversity, with endogeic earthworm species becoming dominant after vegetation conversion from forests to agroecosystems in most tropical countries (Blanchart and Julka 1997, Fragoso et al. 1999b, González et al. 2006). However, the alteration of earthworm communities will vary with the nature of land-use management and as it relates to the intensity and frequency of the disturbance (Abbott 1985, Fragoso et al. 1999a). For example, the dominance of earthworm ecological groups can be modified from the endogeic species in natural savannas (*Glossodrilus* n. sp.; 80%) to anecic species (88%) in the man-made pastures in Eastern Colombia (Jiménez et al. 1998).

Recently, it has been found that anthropogenic habitat disturbance can enhance the introduction of exotic earthworm species (Fragoso and Lavelle 1992, González et al. 2006). The invasion of exotic earthworm species can alter local native earthworm communities (Kalisz 1993, Kalisz and Wood 1995). The success for exotic earthworms to invade novel areas or their potential impacts on native earthworm communities depends greatly on population characteristics of both exotic and native earthworms and the local environmental and soil properties. Hendrix et al. (2006) illustrated the possible sequences of invasion depending on degree of habitat disturbance and invasion success by exotic earthworms invading ecosystems inhabited by native earthworms (Figure 5.1). They hypothesized that the intensity of habitat disturbance (severe, moderate or minimal) would influence native earthworm communities in the natural ecosystems differentially. Exotic species can occupy disturbed areas, exclusively or by coexisting with native earthworm species where competition displacement could happen in the later case (Hendrix et al. 2006). Hendrix et al. (2006) also proposed three main mechanisms (propagule pressure, habitat match

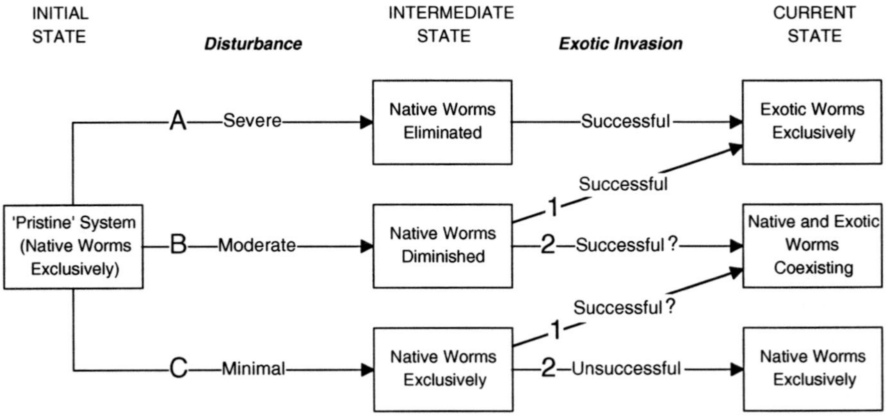

Fig. 5.1 Hypothesized sequences of invasion depending on degree of habitat disturbance and invasion success by exotic earthworms invading ecosystems inhabited by native earthworms (reproduced with kind permission of Springer Science and Business Media from Fig. 1 of Hendrix et al. 2006)

and biotic resistance) to determine invasion success or failure of exotic earthworm species during the invasion processes. Propagule pressure relates to the magnitude of the exotic earthworm introduction into the invaded soils, in terms of the frequency of the introduction and size of the introduced population (Hendrix et al. 2006). Repeated introductions and more individuals involved during each introduction can enhance the success of exotic earthworm populations to establish, proliferate and disperse. Hence, population characteristics (such as life history and reproduction strategy) of the invasive species become determinant elements at this initial stage of invasion. For example, high fecundity and parthenogenesis can boost certain exotic earthworm species to build their populations in a relative short time (Hendrix et al. 2006). Nevertheless, even with the advantage of the propagule pressure, exotic earthworm species still have to adapt to the local environment that they attempt to invade, which may encompass different plant communities, food resource availability, and soil properties. As noted earlier, these abiotic environmental factors are basic requirements for earthworm populations and significantly affect earthworm communities and distributions. Hendrix et al. (2006) called this "adaptation" step to the invaded environment as the "habitat matching" mechanism, which is relevant for distribution, growth and reproduction of exotic earthworms. The characteristics of higher tolerance and flexible plasticity to the environment can make exotic earthworm species more competitive than other earthworm species, especially when they invade into the disturbed areas. Superior adaptation ability to new environments and reproduction biology (e.g., parthenogenesis and *r*-selected strategy) are generally attributed to the success of exotic earthworm species, e.g. *Pontoscolex corethrurus*, a pan-tropical distributed invasive earthworm species (Fragoso et al. 1999a, González et al. 2006, Hendrix 2006). The third mechanism, "biotic resistance", hypothesized the potentially resistant forces from local native communities to impede the invasion of exotic earthworms. Predatory pressure, parasitism, and competition with indigenous earthworm species for resources can retard the establishment of exotic earthworms even when they are able to overcome the challenges of propagule pressure and habitat matching (Hendrix et al. 2006). Biotic resistance gives a reasonable explanation for the lack of invasion of exotic earthworms in some undisturbed forests adjacent to areas they have invaded (Lavelle and Pashanasi 1989, Kalisz 1993). In addition, recognition systems on exotic earthworms by new predators and parasites may not develop in a relative short time at the beginning of the invasion. Competition relationships with local native earthworm communities may play a more important role for invasion success of exotic earthworm at the early stage of invasion. The resistance to the invasion from the native earthworm communities can be weakened by habitat disturbance. The shortage of food availability and unsuitable soil conditions from habitat disturbance are usually detrimental to the local native earthworms. As a result, habitat disturbances devastate the integrity and equilibrium of local earthworm communities and can curtail their resistance strengths to impede the invasion of exotic species. In disturbed areas, the mechanisms for exotic earthworms to become the dominant species include not only the two advantages discussed above (adaptation plasticity and reproduction biology), but also the lack of competition stress from local native earthworm communities (Fragoso et al. 1999a, González et al. 2006).

Consistent with Hendrix et al. (2006), González et al. (2006) showed different pathways by which exotic earthworms can establish populations in natural and disturbed ecosystems (Figure 5.2). In natural ecosystems with no or minor disturbance,

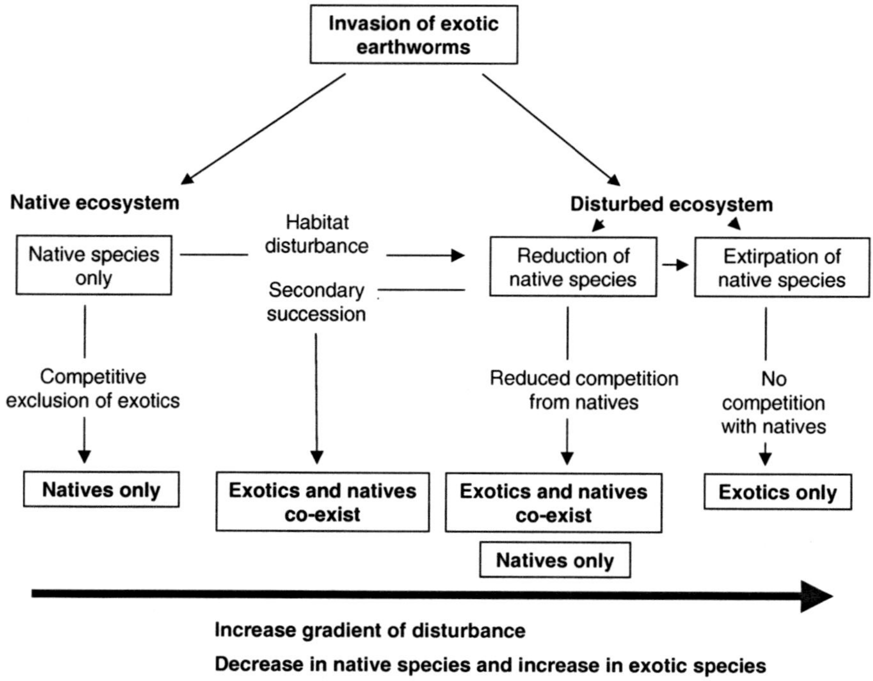

Fig. 5.2 Model illustrating the paths by which invasion of exotic earthworms affect native earthworm species in undisturbed and disturbed ecosystems (reproduced with kind permission of Springer Science and Business Media from Fig. 1 of González et al. 2006)

the integrity of local native earthworm communities gives few chances for exotic earthworm species to invade. Once native earthworm communities are excluded and/or disrupted by habitat disturbance in the disturbed areas, exotic earthworm species can occupy the empty niches released by native earthworms to build their populations. The intensity and frequency of disturbance will significantly affect the completeness of native earthworm communities (biotic resistance), therefore the invasion success for exotic earthworm species (González et al. 2006). In reality, these three mechanisms (propagule pressure, habitat matching, and biotic resistance) correlate to one another when explaining the invasion success of exotic earthworms. More efforts are needed to test and clarify each of these mechanisms. The relative importance of each mechanism may be diverse in the different ecosystems.

5.2.3 The Physiology of Exotic Earthworms: Mechanisms for Successful Introductions in Pastures and Disturbed Environments

The reproductive biology of exotic species is an important characteristic to deem in the context of invasion (González et al. 2006). Tropical peregrine earthworms (e.g., *P. corethrurus*, *Perionex excavatus*, *Dichogaster modigliani* and *Polypheretima elongata*) are often considered to be continuous breeders with high fecundity (Bhattacharjee and Chaudhuri 2002). So that on the basis of response to selection pressure, high fecundity, short incubation period with high hatching success are probably adaptative strategies of *r*-selected organisms that enables them to survive drastic environmental changes, especially heat, drought and predation in the soil (Pianka 1970, Bhattacharjee and Chaudhuri 2002). Up until now, endogeic earthworms are more frequent invaders of disturbed tropical pastures than epigeic species. Interestingly, exotic endogeic species (e.g., *P. corethrurus*, *P. elongata* and *Drawida nepalis*) have been shown to increase their rate of cocoon production and incubation period with increased temperature (Figure 5.3) while epigeics decreased their reproductive capabilities (Bhattacharjee and Chaudhuri 2002). Thus, the interactions between the characteristics of the exotic earthworm species—particularly their functionality and activities—could determine their potential for establishment (González et al. 2006).

Animals have physiological adaptations to the environment, and earthworms are no exception. Efficient adaptations to the environment by earthworms ensure the survival of species, particularly that of non-natives. How can *P. corethrurus* adapt and distribute widely in tropical areas around the world? The physiological adaptations of this species may be very important to consider when answering this question. For example, some earthworm species, such as *Lumbricus terrestris*, *Amynthas gracilis*, *A. robustus*, and *Metaphire schmardae*, crawl out of the soil after a heavy rain exposing themselves to predation (Chuang et al. 2004, Darwin 1881, Tsai 1964). Other earthworm species do not. In tropical and subtropical Taiwan, *P. corethrurus* is a common exotic worm; where it has

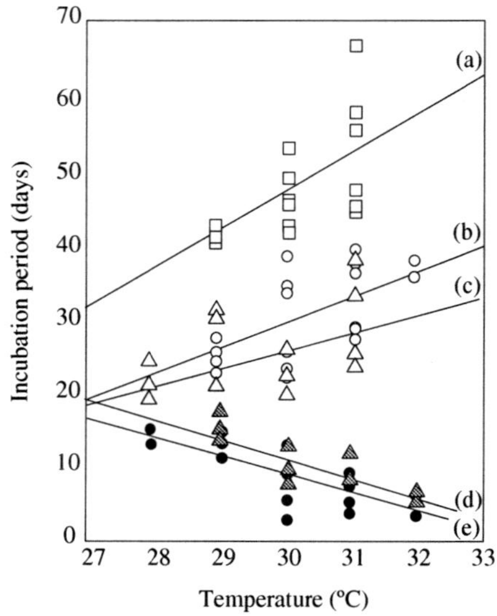

Fig. 5.3 Relationship between temperature and incubation period in different earthworm species (a) *Polypheretima elongata*, (b) *Drawida nepalensis*, (c) *Pontoscolex corethrurus* , (d) *Dichogaster modiglianii*, and (e) *Peryonyx excavatus*. Endogeic species are represented with open symbols and epigeic species are represented with filled symbols (modified from Bhattacharjee and Chaudhuri 2002)

not been observed to crawl out of the soil after a heavy rain. To understand why some earthworms crawl out of the soil, Chuang et al. (2004) and Chuang and Chen (in press) investigated the respiration physiology of *A. gracilis* (a native earthworm to Taiwan) and *P. corethrurus* by placing earthworms in open and sealed bottles filled with air-saturated water. They found the mean survival time of *A. gracilis* in water was 5.4 ± 1.34 h in a sealed bottle and 13.4 ± 6.95 h in an open bottle. A control earthworm remained alive for more than 72 h (end of testing) in a humidity box. However, *P. corethrurus* survived five to ten times longer than *A. gracilis* under the same conditions, surviving in water for 45.4 ± 3.58 h in a sealed bottle and to the end of the test (76 h) in an open bottle and for at least 96 h (end of test) in a humidity box (Figure 5.4). When *A. gracilis* was classified as dead, the residual oxygen concentration in the sealed and open bottle was 1.87 or 1.5 μg/ml, respectively. However, *P. corethrurus* could survive until the oxygen concentration in the sealed and open bottles fell to 0.73 or 0.64 μg/ml, respectively (Figure 5.5). Therefore, it is reasonable to infer that *P. corethrurus* might use anaerobic respiration in oxygen-shortage conditions. *Amynthas gracilis* was found to have a diurnal rhythm of oxygen consumption, using more oxygen at night than during the day. However, *P. corethrurus* lacked a diurnal rhythm (Figure 5.6). *P. corethrurus* consumed similar amounts of oxygen at different time periods and temperatures. In addition, the cocoon and juveniles of *P. corethrurus* are easily found on the soil surface throughout the entire year unless the soil is extremely dry (Chuang et al. 2004). If the soil humidity is too low, then *P. corethrurus* will go into torpor. They twist their body into a coil and secrete mucus to maintain moisture (Chuang et al. 2004). This phenomenon was also described for *Glossoscolex paulistus* (Abe 1985), and they are both classified as Glossoscolecidae. Thus, *P. corethrurus* has lower oxygen consumption, tolerates poor oxygen concentrations in the environment and adapts to a wide range of soil temperatures. These physiological adaptations could well

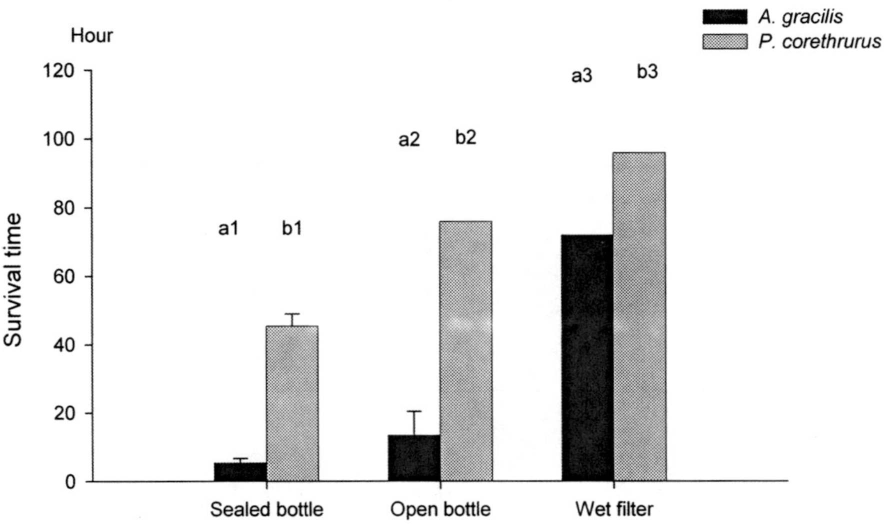

Fig. 5.4 Survival times of submerged earthworms. Earthworms were kept in an open or sealed bottle filled with artificial spring water (ASW) at 25°C. Different letters indicate significant differences. The data are the mean±SD ($n = 20$) for the survival time (Chuang et al. in press).

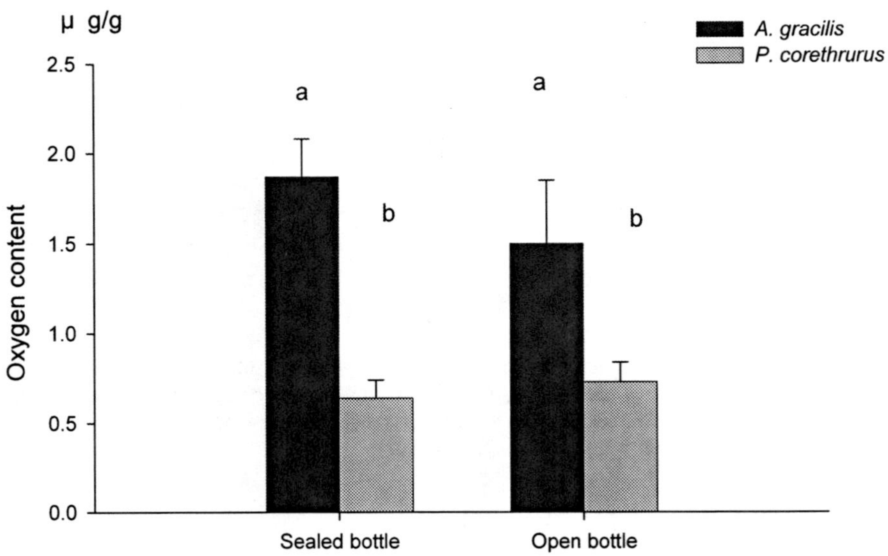

Fig. 5.5 Oxygen remaining in artificial spring water-containing bottle at 25°C. The value was measured immediately after the worm was found to be dead. Different letters for different treatments indicate significant differences. The data are the mean±SD ($n = 20$) for the residual oxygen content (Chuang et al. in press)

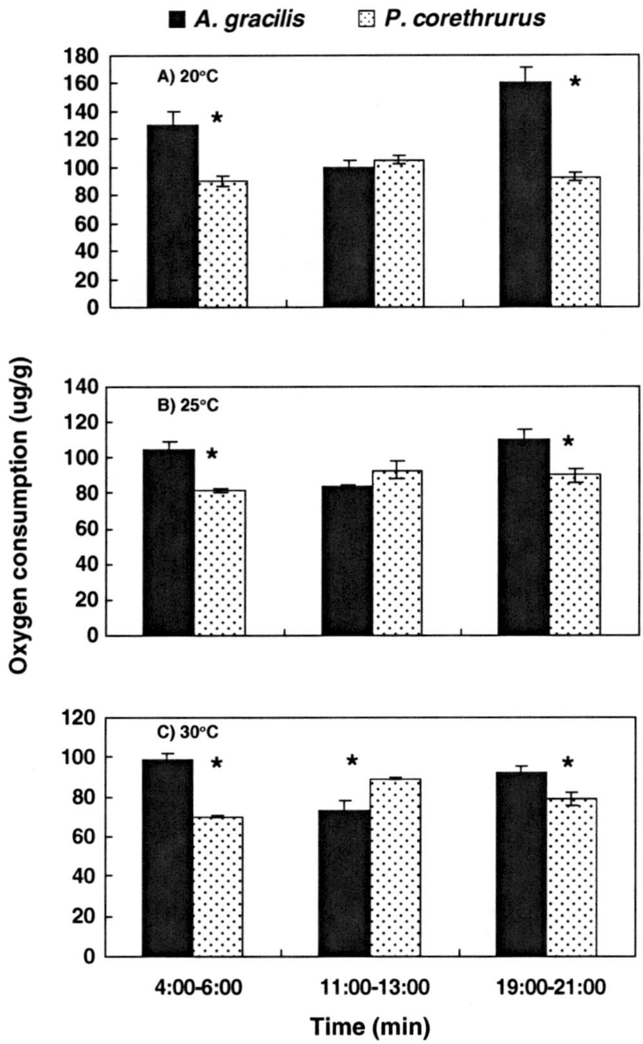

Fig. 5.6 Total oxygen consumption of two species of earthworms at different temperatures. The oxygen consumption of *A. gracilis* was significantly higher than *P. corethrurus*, irregardless of being at the different temperature ($P < 0.001$). The asterisks indicate significant difference (from Chuang et al. 2004)

explain why *P. corethrurus* is widely distributed in the tropics around the world. On the other hand, the high oxygen demand of *A. gracilis* (especially at night), causes it to expose its body on the soil surface to exchange for air, consequently making them more susceptible to predation. This behavior is not observed in *P. corethrurus* (as it is an endogeic species), which might be another explanation as to why *P. corethrurus* can be a successful exotic earthworm (Chuang et al. 2004) in pastures and disturbed sites with a non continuous canopy cover. Another characteristic that makes *P. corethrurus* a hard-hitting invader is the ability that the juveniles have to enter diapause and regenerate after amputation independent of soil moisture (Fragoso and Lozano 1992). This phenomenon and the fact that parthenogenesis is common for most tropical exotics (Fragoso et al. 1999b) could well be adaptive strategies of tropical exotic earthworms to avoid predation pressure in a climatically harsh environment; making them without doubt strong invaders and competitors (González et al. 2006).

In addition to the respiration adaptation to water-saturated environments, *P. corethrurus* exhibits other traits that help them survive in severe environments. *P. corethrurus* has high tolerance to ultraviolet radiation (UV) (Chuang et al. 2006). Chuang et al. (2006) tested UV effects on the behavior and physiology of three species of earthworms: *A gracilis*, *M. posthuma* and *P. corethrurus*. They found that *P. corethrurus* would crawl slowly after UV (Figure 5.7), but surprisingly, they would not die or have tissue damage even after 24 h of ultraviolet radiation (Figure 5.8). The influence of UV to *P. corethrurus* may be temporary, as UV influences their crawling activity, but these earthworms may have some mechanism to protect themselves. It is known that the development of pigment is an important

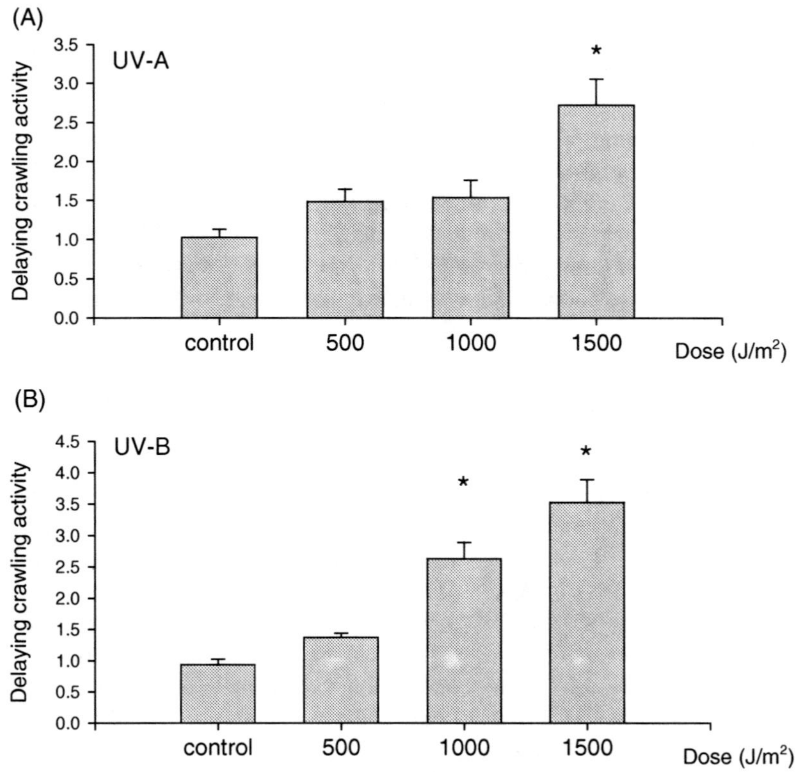

Fig. 5.7 Relative crawling time of earthworms after exposure to either UV-A and UV-B. A) The crawling time of *Pontoscolex corethrurus* exposed to $1{,}500\cdot J/m^{--2}$ of UV-A was significantly increased compared to controls (ANOVA, N=6, *P<0.01) B) After UV-B exposure, *P. corethrurus* show a significantly increased crawling time compared to controls (ANOVA, N=6, *P<0.01) (from Chuang et al. 2006)

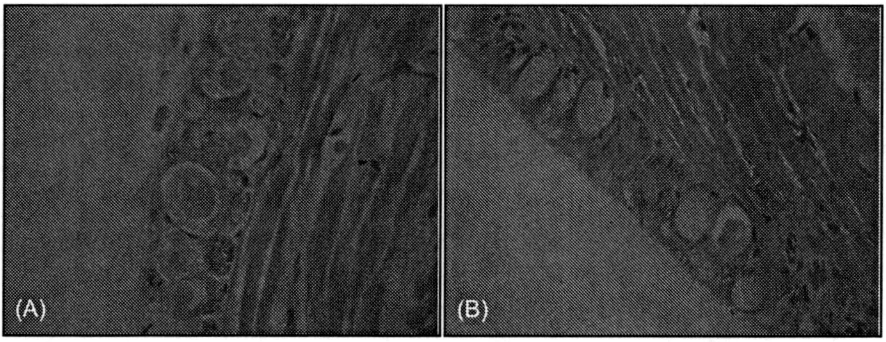

Fig. 5.8 Cross section photographs of *Pontoscolex corethrurus* tissue A) with no UV radiation (control) and B) as exposed to UV radiation at a dose of 1,500 J/m^2 after 24 h. No changes in the epidermis were observed after 2, 12 and 24 h (as shown in B) of UV radiation. Also note the many granular cells in the epidermis by which *P. corethrurus* can secrete more mucus than *A. gracilis* or *M. posthuma* (photographs taken by Shu-Chun Chuang)

adaptation against solar radiation. However, *P. corethrurus* does not show much pigmentation in the epidermis. Yet, the mucus and epidermis of *P. corethrurus* is thicker than other species of earthworms (Chuang, personal observation). In addition, *P. corethrurus* mucus may contain chemicals that help absorb UV and protect their skin. Experimentation with earthworm mucus is problematical given difficulties in the extraction of mucus from the earthworm skin surface (Chuang, personal communication). Accordingly, Chuang et al. (2006) proposed that *P. corethrurus* might utilize substances such as mycosporine-like amino acid (MAA) or flavonol quercetin, and/or some enzymes like photolyase to protect itself. Chuang et al. (unpublished data, not shown), showed that *P. corethrurus* has a high activity of Protein kinase before and after UV radiation. Protein kinase activity is related to the release of acetylcholine, which is an important hormone in neuromuscular junction. Earthworms can have better movement behavior with higher acetylcholine activity. On the other hand, oxidative stress and lipid peroxidation after ultraviolet radiation may be lethal for earthworms because the final product of lipid peroxidation is lipid peroxide (LOOH). LOOH will lead to the accumulation of malondialdehye (MDA) in the earthworm muscle/skin tissue. When MDA reacts with thiobarbituric acid (TBA), it forms TBA-reactive substances (TBARS). Therefore, we can extract MDA from earthworm skin/muscle tissue to react with TBA and test the level of lipid peroxide. The higher TBARS means the higher LOOH. Therefore, TBARS can be good indicators of lipid peroxidation. For example, when *P. corethrurus* is exposed to 1,000 J/m^2 UV-B, the concentration of TBARS increased after 2 hr, however, at other doses of UV-B radiation, the concentration of TBARS did not change; meaning that *P. corethrurus* did not have much lipid peroxidation after UV-B exposure (Table 5.2, unpublished data).

Chuang et al. (unpublished data) also found that the anti-oxidant enzyme activity of other earthworm species decreased after UV radiation, but in *P. corethrurus*, it remained high after UV radiation (Table 5.2, unpublished data). Girotti (2001) classified the photo-oxidative damage into four degrees and according to these

Table 5.2 MDA level and catalase activity in muscle/skin tissue (Chuang et al. unpublished data)

Dosage (J/m^2)	Time (min)	TBARS (mmol/g)[1]	Catalase activity (mU/mg)[2]
Control		15.029±1.91	4.67±0.94
500	15	17.57±1.38	5.62±0.36
	120	15.07±3.92	5.52±0.62
	240	14.88±2.24	5.04±0.52
1,000	15	20.12±2.85	5.65±0.38
	120	23.45±1.22 (*[3])	5.07±1.13
	240	14.53±1.01	4.94±0.24
1,500	15	22.12±3.97	5.95±0.27
	120	23.31±3.58	4.79±0.87
	240	17.74±2.68	5.30±0.37
3,000	15	16.06±4.27	5.69±0.38
	120	21.40±2.52	5.53±0.38
	240	14.78±1.12	4.87±0.21

[1]The MDA level of muscle/skin tissue depended on minutes expose and doses of UV-B radiation. The concentration of MDA was used to estimate the extent of lipid peroxidation.
[2]Effect of UV-B radiation on catalase activity in muscle/skin tissue. The activity meant the ability of anti-oxidative stress.
[3]Asterisk (*) meant significant differences compare to control treatment (Two way ANOVA test, $p<0.05$).

results from *P. corethrurus*; it should be classified into secondary degree. It means that epidermal cells of *P. corethrurus* have been influenced slightly after ultraviolet radiation, but the increase of lipid peroxidation also turned on the anti-oxidative mechanism and further protected the epidermal cells. In the aforementioned study, the anti-oxidative enzyme of *P. corethrurus* was also found to be kept at a high activity after UV radiation and the reasons for this are unknown. Therefore, *P. corethrurus* still had enough protection from UV and did not die. The oxidative stress relates not only to the effects of ultraviolet radiation but also to other stresses such as heavy metal pollution and an adversely compromised physiology. Therefore, if *P. corethrurus* has high tolerance to oxidative stress, they would better survive in a harsh environment. Certainly, we also argue that this mechanism can help *P. corethrurus* invade successfully disturbed sites throughout in the tropics.

5.3 Summary

In the Neotropics, large areas of pasture have been gradually converted to grasslands or secondary forests following the abandonment of agricultural practices. These areas covered with the secondary succession are expected to expand in the near future because of the adjustment of economic strategies and soil degradation problems in these tropical countries (e.g., Brazil and Puerto Rico, Buschbacher 1986, Birdsey and Weaver 1987, Helmer 2004). For now at least, these post-agricultural succession processes boost the recovery of natural landscapes and vegetations and help the restoration of natural ecosystems. Natural succession (no attempted management) evolved from these post-agricultural lands provides an opportunity to observe the dynamics of vegetation communities, the re-colonization sequence of soil fauna, and further the interactive relationships among plants, soil fauna, and microbes within considerable areas along the continuum of successional forests. These areas also represent an excellent opportunity to explore the linkage between aboveground communities (forest types, vegetation cover) and belowground soil fauna (especially earthworm communities) in post-agricultural lands. In addition, the variety of exotic and native earthworm communities observed in post-agricultural fields provide for good opportunities to test the mechanisms of invasion biology. The case study from post-agricultural pastures and secondary forests of Puerto Rico showed that native earthworm communities are gradually re-colonizing the forests as aboveground plant communities mature. We can test the three mechanisms (propagule pressure, habitat matching, and biotic resistance) proposed by Hendrix et al. (2006) by monitoring dynamics of exotic and native earthworm populations and/or conducting field experiments along post-agricultural/succession gradients. Comparisons between different ecosystems and geographic regions are also possible by using earthworms as an experimental model for invasion biology.

5.4 Future Research Directions

The history of the introductions of non-native flora and fauna is much more complex in the Neotropics than in temperate North America; as it is related to the complex human history of migration and use of the landscape, water barriers and island ecosystems. An interdisciplinary approach (i.e., history and ecology) can help elucidate the spreading of non-native species in the tropics and the development of sustainable land management practices (González et al. 2006). Landscape scale dynamics of earthworm distributions within changing landscapes; and physiological and other mechanisms controlling community dynamics are important areas of future study on earthworm ecology in the Neotropics. As discussed above, *P. corethrurus* shows better physiological adaptations than other earthworms in terms of oxygen utilization, oxygen consumption, ultraviolet tolerance and high anti-oxidative enzyme activities. These physiological traits can be powerful and effective competitive adaptations when non native earthworms invade new areas. Yet, many questions related to the physiological adaptation of exotic earthworms like *P. corethrurus* remained unanswered. For example, why can *P. corethrurus* survive in oxygen-shortage conditions? Why does *P. corethrurus* have high tolerance to ultraviolet radiation? Why does it have higher anti-oxidative stress in its body? The hemoglobin of earthworms (mega-hemoglobin) is different from vertebrates. Thus, is the high oxygen binding affinity of *P. corethrurus* hemoglobin that can help worms survive in anoxia condition? Also, although we do know that *P. corethrurus* is tolerant to UV; we still do not understand the mechanism by which it can protect itself from the radiation. *P. corethrurus* mucus is thick and sticky and we infer it has some protective chemicals. MAA is a chemical which can absorb ultraviolet in sea hare (Carefoot et al. 1998) and photolyase is an enzyme known to repair DNA damage in frogs (Blaustein et al. 1998). To date, it is unknown whether or not earthworms have these chemicals, but these topics merit study. On the other hand, heavy metals, pesticides or chemical fertilizers are also harmful to earthworms, but their toxicity to *P. corethrurus* is not well understood. Exactly how species such as *P. corethrurus* can survive in pastures and / or agricultural fields requires further investigation.

Acknowledgment While writing this manuscript, Grizelle González received support of grant DEB-0218039 from the National Science Foundation to the Institute of Tropical Ecosystem Studies, University of Puerto Rico, and the USDA Forest Service, International Institute of Tropical Forestry as part of the Long-Term Ecological Research Program in the Luquillo Experimental Forest. Additional support was provided by the Forest Service (U.S. Department of Agriculture), and the University of Puerto Rico. María M. Rivera helped with graph reproduction. We are grateful to Drs. William A. Gould, Timothy R. Seastedt, and Ariel E. Lugo for kindly commenting on earlier versions of the manuscript.

References

Abbott, I. 1980. Do earthworms compete for food? Soil Biol. Biochem. 12:523–530.

Abbott, I. 1985. Distribution of introduced earthworms in the Northern Jarrah Forest of Western Australia. Aust. J. Soil Res. 23:263–270.

Abe, A. S. 1985. Oxygen uptake of active and aestivating earthworm Glossoscolex paulistus (Oligochaeta, Glossoscolecidae). Comp Biochem Physiol 81A(1):63–66.

Araujo, Y., and López-Hernández, D. 1999. Earthworm populations in a savanna-agroforestry system of Venezuelan Amazonia. Biol. Fertil. Soils 29:413–418.

Baker, G., Carter, P., Barrett, V., Hirth, J., Mele, P., and Gourley, C. 2002. Does the deep-burrowing earthworm, *Aporrectodea longa*, compete with resident earthworm communities when introduced to pastures in south-eastern Australia? Eur. J. Soil Biol. 38:39–42.

Barnard, C. I., and Thompson, B. A. 1985. Gulls and Plovers: The Ecology of Mixed-species Feeding Groups. New York: Columbia University Press.

Bengtson, S.-A., Nilsson, A., Nordström, S. and Rundgren, S. 1976. Effects of bird predation on lumbricid populations. Oikos 27:9–12.

Bhattacharjee, F., and Chaudhuri, P. S. 2002. Cocoon production, morphology, hatching pattern and fecundity in seven tropical earthworm species: a laboratory-based investigation. Journal of Biosciences 27:283–294

Birdsey, R. A., and Weaver, R. W. 1987. Forest area trends in Puerto Rico. Research Note SO-331, USDA Forest Service, Washington, D.C., USA.

Blackshaw, R. P. 1995. Changes in populations of the predatory flatworm *Artioposthia triangulata* and its earthworm prey in grassland. Acta Zoolog. Fennica 196:107–110.

Blanchart, E., and Julka, J. M. 1997. Influence of forest disturbance on earthworm (Oligochaeta) communities in the Western Ghats (South India). Soil Biol. Biochem. 29:303–306.

Blaustein, A. R., Kiesecker, J. M. D., Chivers, P., Hokit, D. G., Marco, A., Beloden, L. K., and Hatch, A. 1998. Effect of ultraviolet radiation on amphibians: field experiments. Am. Zool. 38:799–812.

Bouché, M. B. 1977. Ecologie et paraecologie: peut-on apprecier le role de la faune dans les cycles biogeochimiques? Soil Organism as Components of Ecosystems 25:157–163.

Brown, G., Barois, I., and Lavelle, P. 2000. Regulation of soil organic matter dynamics and microbial activity in the drilosphere and the role of interactions with other edaphic functional domains. Eur. J. Soil Biol. 36:177–198.

Buschbacher, R. 1986. Tropical deforestation and pasture development. BioScience 36:22–28.

Butt, K. R. 1998. Interactions between selected earthworm species: a preliminary, laboratory-based study. Appl. Soil Ecol. 9:75–79.

Carefoot, T. H., Harris, M., Taylor, B. E., Donovan, D., and Karentz, D. 1998. Mycosporine-like amino acids: possible UV protection in eggs of the sea hare *Aplysia dactylomela*. Mar. Biol. 130:389–396.

Chuang, S. C., Lee, H., and Chen, J. H. 2004. Diurnal rhythm and effect of temperature on oxygen consumption in earthworms, *Amynthas gracilis* and *Pontoscolex corethrurus*. J. Exp. Zoolog. A Comp. Exp. Biol. 301:731–744.

Chuang, S. C., Lai, W. S., and Chen, J. H. 2006. Influence of ultraviolet radiation on selected physiological responses of Earthworms. J. Exp. Biol. 209:4304–4312.

Chuang, S. C., and Chen, J. H. In press. Diurnal rhythm of oxygen consumption may cause earthworms to crawl out of the soil at night after heavy rain. Inverte. Biol. (in press).

Dalby, P. R., Baker, G. H., and Smith, S. E. 1998. Competition and cocoon consumption by the earthworm *Aporrectodea longa*. Appl Soil Ecol. 10:127–136.

Darwin, C. 1881. The Formation of Vegetable Mould Through the Action of Worms, with Observation of Their Habits. London: John Murray.

Decaëns, T., and Rossi, J.-P. 2001. Spatio-temporal structure of earthworm community and soil heterogeneity in a tropical pasture. Ecography 24:671–682.

Edwards, A. L. 2004. Earthworm Ecology. Boca Raton: CRC Press.

Edwards, C. A., and Bohlen, P. J. 1996. Biology and Ecology of Earthworms. New York: Chapman & Hall.

Fragoso, C., and Lavelle, P. 1987. The earthworm community of a Mexican tropical rainforest (Chajul, Chiapas). In On Earthworms, ed. A. M. Bonvicini-Pagliai and P. Omodeo, pp. 281–295. Mucchi: Modena.

Fragoso, C., and Lavelle, P. 1992. Earthworm communities of tropical rain forest. Soil Biol. Biochem. 24:1397–1408.

Fragoso, C., and Lozano, N. 1992. Resource allocation strategies imposed by caudal amputation and soil moisture in the tropical earthworm *Pontoscolex corethrurus*. Soil Biology and Biochemistry 24:1237–1240.

Fragoso, C., James, C., and Borges, S. 1995. Native earthworms of the North Neotropical Region: current status and controversies. In Earthworm Ecology and Biogeography in North America, ed. P. F. Hendrix, pp. 67–115. Boca Raton, FL: Lewis Publishers.

Fragoso, C., Lavelle, P., Blanchart, E., Senapati, B. K., Jiménez, J. J., Martínez, M. Decaëns, T., and Tondoh, J. 1999a. Earthworm communities of tropical agroecosystems: origin, structure and influence of management practices. In Earthworm Management in Tropical Agroecosystems, ed. P. Lavelle, L. Brussaard and P. Hendrix, pp. 27–56. New York: CAB International.

Fragoso C., Kanyonyo, J., Moreno, A., Senapati, B. K., Blanchart, E., and Rodríguez, C. 1999b. A survey of tropical earthworms: taxonomy, biogeography and environmental plasticity. In Earthworm Management in Tropical Agroecosystems, ed. P. Lavelle, L. Brussaard, and P. Hendrix, pp. 1–26. New York: CABI Publishing.

Garnsey, R. B. 1994. Seasonal activity and aestivation of Lumbricid earthworms in the Midlands of Tasmania. Aust. J. Soil Res. 32:1355–1367.

Geissen, V., and Gúzman, G. M. 2006. Fertility of tropical soils under different land use systems - a case study of soil sin Tabasco, Mexico. Appl. Soil Ecol. 31:169–178.

Girotti, A. W. 2001. Photosensitized oxidation of membrane lipids: reaction pathways, cytotoxic effects and cytoprotective mechanisms. J. Photochem. Photobiol. B Biol. 63:103–113.

González, G. 2002. Soil organisms and litter decomposition. In Modern Trends in Applied Terrestrial Ecology, ed. R. S. Ambasht and N. K. Ambasht, pp. 315–329, New York: Kluwer Academic / Plenum Publishers.

González, G., and Seastedt, T. R. 2001. Soil fauna and plant litter decomposition in tropical and subalpine forests. Ecology 82:955–964.

González, G., Zou, X., and Borges, S. 1996. Earthworm abundance and species composition in abandoned tropical croplands: comparison of tree plantations and secondary forests. Pedobiologia 40:385–391.

González, G., Zou, X., Sabat, A., and Fetcher, N. 1999. Earthworm abundance and distribution pattern in contrasting plant communities within a tropical wet forest in Puerto Rico. Caribb. J. Sci. 35:93–100.

González, G., and Zou, X. 1999. Plant and litter influences in earthworm abundance and community structure in a tropical wet forest. Biotropica 31:486–493.

González, G., Huang, C.-Y., Zou, X., and Rodríguez, C. 2006. Earthworm invasions in the tropics. Biological Invasions 8:1247–1256.

Hamilton, W. E., Dindal, D. L., Parkinson, C. M., and Mitchell, M. J. 1988. Interaction of earthworm species in sewage sludge-amended soil microcosms: *Lumbricus terrestris* and *Eisenia fetida*. J. Appl. Ecol. 25:847–852.

Helmer, E. H. 2004. Forest conservation and land development in Puerto Rico. Landsc. Ecol. 19:29–40.

Hendrix, P. F. 2006. Biological invasions belowground- earthworms as invasive species. Biological Invasions 8:1201–1204.

Hendrix, P. F., and Bohlen, P. J. 2002. Exotic earthworm invasions in North America: ecological and policy implications. BioScience 52:801–811

Hendrix, P. F., Lachnicht, S. L., Callaham, M. A., and Zou, X. 1999. Stable isotopic studies of earthworm feeding ecology in tropical ecosystems of Puerto Rico. Rapid Comm. in Mass Spectrom. 13:1295–1299.

Hendrix, P. F., Baker, G. H., Callaham, M. A., Damoff, G. A., Fragoso, C., González, G., Winsome, T., and Zou, X. 2006. Invasion of exotic earthworms into ecosystems inhabited by native earthworms. Biological Invasions 8:1287–1300.

Hubers, H., Borges, S., and Alfaro, M. 2003. The Oligochaetofauna of the Nipe soils in the Maricao State Forest, Puerto Rico. Pedobiologia 47:475–478.

James, S. W. 2004. Planetary processes and their interactions with earthworm distributions and ecology. In Earthworm Ecology, ed. C. A. Edwards. Boca Raton: CRC press.

Jiménez, J. J., and Decaëns, T. 2000. Vertical distribution of earthworms in grassland soils of the Colombian Llanos. Biol. Fertil. Soils 32:463–473.

Jiménez, J. J., Moreno, A. G., Decaëns, T., Lavelle, P., Fisher, M. J., and Thomas, R. J. 1998. Earthworm communities in native savannas and man-made pastures of the Eastern Plains of Colombia. Biol. Fertil. Soils 28:101–110.

Judas, M. 1989. Predator-pressure on earthworms: field experiments in a beechwood. Pedobiologia 33:339–354.

Kalisz, P. J. 1993. Native and exotic earthworms in deciduous forest soils of Eastern North America. Biological Pollution: the control and impact of invasive exotic species. Proceedings of a Symposium held at the University Place Conference Center, Indiana University-Purdue University, Indianapolis, October 25 & 26, 1991:93–100.

Kalisz, P. J., and Wood, H. B. 1995. Native and exotic earthworms in wildland ecosystems. In Earthworm Ecology and Biogeography in North America, ed. P. F. Hendrix, pp. 117–126. Boca Raton, FL: Lewis Publishers.

Lachnicht, S. L., Hendrix, P. F., and Zou, X. 2002. Interactive effects of native and exotic earthworms on resource use and nutrient mineralization in a tropical wet forest soil of Puerto Rico. Biol. Fertil. Soils 36:43–52.

Lavelle, P. 1988. Earthworm activities and the soil system. Biol. Fertil. Soils 6:237–251.

Lavelle, P., and Pashanasi, B. 1989. Soil macrofauna and land management in Peruvian Amazonia. Pedobiologia 33:283–291.

Lavelle, P., Brussaard, L., and Hendrix, P. 1999. Earthworm Management in Tropical Agroecosystems, p. 300. New York: CABI Publishing.

Lee, K. E. 1985. Earthworms, Their Ecology and Relationships with Soils and Land Use. New York: Academic Press.

Liu, Z. G., and Zou, X. M. 2002. Exotic earthworms accelerate plant litter decomposition in a Puerto Rican pasture and a wet forest. Ecological Applications 12:1406–1417

López-Hernández, D., Araujo, Y., López, A., Hernández-Valencia, I., and Hernández, C. 2004. Changes in soil properties and earthworm populations induced by long-term organic fertilization of a sandy soil in the Venezuelan Amazonia. Soil Sci. 169:188–194.

Lowe, C. N., and Butt, K. R. 2002. Growth of hatchling earthworms in the presence of adults: interactions in laboratory culture. Biol. Fertil. Soils 35:204–209.

Macdonald, D. W. 1983. Predation on earthworms by terrestrial vertebrates. In Earthworm Ecology: From Darwin to Vermiculture, ed. J. E. Satchell, pp. 393–414. London: Chapman and Hall.

Nations, J. D., and D. I. Komer. 1983. Central America's tropical rainforests- positive steps for survival. Ambio 12:232–238.

Odum, H. T., and Pigeon, R. F. 1970. A tropical rain forest: a study of irradiation and ecology at El Verde, Puerto Rico. S.S. Atomic Energy Commission, National Technical Information Service, Springfield, VA.

Pianka, E. R. 1970. On r and K-selection. American Naturalist 104:459–466.

Römbke, J., Meller, M., and García, M. 1999. Earthworm densities in central Amazonia primary and secondary forests and a polyculture forestry plantation. Pedobiologia 43:518–522.

Sánchez-de León, Y., and Zou, X. 2004. Plant influences on native and exotic earthworms during secondary succession in old tropical pastures. Pedobiologia 48:215–226.

Sánchez-de León, Y., Zou, X., S. Borges, and Ruan, H. 2003. Recovery of native earthworms in abandoned tropical pastures. Conservation Biology 17:999-1006.

Schmidt, O., and Curry, J. P. 2001. Population dynamics of earthworms (Lumbricidae) and their role in nitrogen turnover in wheat and wheat-clover cropping systems. Pedobiologia 45: 174–187.

Tsai, C. F. 1964. On some earthworms belonging to the genus *Pheretima* Kinberg collected from Taipei area in North Taiwan. Q. J. Taiwan Mus. 17:1–35.

Winsome, T., Epstein, L., Hendrix, P. F., and Horwath, W. R. 2006. Competitive interactions between native and exotic earthworm species as influenced by habitat quality in a California grassland. Appl. Soil Ecol. 32:38–53.

Zou, X., and G. González. 1997. Changes in earthworm density and community structure during secondary succession in abandoned tropical pastures. Soil Biol. Biochem. 29:627–629.

Part II
Mechanisms and Tolerances that Cause Plant-Plant Replacements, Producing Old Field Vegetation Patterns

Chapter 6
Factors Affecting the Species Richness and Composition of Neotropical Secondary Succession: A Case Study of Abandoned Agricultural Land in Panama

Elaine R. Hooper

6.1 Neotropical Secondary Succession

Deforestation of extensive areas throughout the tropics and consequent biodiversity loss is of worldwide concern. Often, unsustainable land use follows deforestation, producing vast areas of degraded land in various stages of secondary succession (Brown and Lugo 1990, FAO 2001). Increasing fragmentation of mature tropical forests is resulting in landscapes where primary forest appears as islands in a sea of secondary vegetation and anthropogenic grasslands (Uhl 1987). While species richness may recover rapidly during secondary forest succession, species composition of secondary forest remains very different from that of mature forest for many decades (Brown and Lugo 1990, Finegan 1996). The processes determining the species composition during neotropical secondary succession remain poorly understood (Wijdeven and Kuzee 2000). Successional research in temperate areas has been used as the basis for modeling neotropical secondary succession, however, temperate-based models do not include stochastic factors such as seed availability and dispersal (Finegan 1984, McCook 1994). In tropical areas, most tree species are animal-dispersed (Howe and Smallwood 1982) suggesting potential limitations in characterizing tropical succession using temperate models. For example, some of the major seed dispersers in the tropics (frugivorous primates, large frugivorous birds and bats) are absent from temperate areas (Chapman and Chapman 1999).

The classical description of neotropical secondary forest succession involves four phases, each characterized by a different suite of species. Changes in environmental conditions and differences among species' growth rates, shade-tolerance, longevity and size at maturity are thought to drive the sequential physiognomic dominance of the site by various suites of species as succession proceeds (Finegan 1984, 1996). The first phase of succession is dominated by herbs, shrubs and climbers. These disappear under the shade of short-lived pioneers trees, which rapidly develop a closed canopy and dominate the second phase for 10–30 years. Long-lived pioneers predominate in the third phase of succession, which can last for 75–150 years.

R. W. Myster, *Post-Agricultural Succession in the Neotropics.*

Colonization by shade-tolerant, mature forest species occurs continuously, and eventual dominance by these species characterizes the final, mature forest stage (Finegan 1996, Guariguata and Ostertag 2001). While life history variation and competition among species are thought to drive successional change, factors like initial site conditions and seed dispersal also play a role, and must be factored in for a more complete description of succession (Finegan 1996).

6.2 Factors Constraining Secondary Succession in the Neotropics

The successional sequence following deforestation, agricultural use and subsequent land abandonment often follows less predictable successional trajectories (Guariguata and Ostertag 2001) because land degradation can delay forest regeneration in proportion to the intensity of prior land use (Uhl et al. 1988, Aide et al. 1995). Site-specific factors also add to the unpredictability of successional species composition. Interactions between site-specific factors and land-use may facilitate, inhibit (sensu Connell and Slatyer 1977), or even divert successional sequences predicted by the general successional model, depending upon the type and intensity of land use before abandonment (Mesquita et al. 2001). While old-growth forests are high priorities for conservation because of their complex structure and diverse floristic composition, the conservation value of secondary forests is becoming increasingly important as mature forests are rapidly lost through logging and conversion to agriculture (DeWalt et al. 2003). Successional processes may restore the diversity and species composition to that of old-growth forests, however research throughout the neotropics has shown that a number of factors, often interacting, may impede tropical forest regeneration at deforested, abandoned sites. These include biotic factors such as seed dispersal limitations and seed predation, and abiotic factors including grass competition, fire, drought and low soil nutrients.

6.2.1 Biotic Factors

When intensive land use, often coupled with fire eliminates tree regeneration from the seed bank, root-sprouts, and advance regeneration (Uhl et al. 1988, Nepstad et al. 1990), the recovery of mature-forest species composition can be very slow (Guariguata and Ostertag 2001) and dependent on seed dispersal as the critical mechanism for determining the template of species recruitment. When seed dispersal is the only possibility for regeneration, succession can potentially be constrained. While over 80% of tropical tree species have animal-dispersed fruits (Howe and Smallwood 1982), most animals will not venture into abandoned old fields because of limited caloric and structural resources, such as perching sites (Cardosa da Silva et al. 1996) and they often do not venture far from forest edge (Aide and Cavelier 1994). Large-seeded species are often the most severely dispersal-limited because the predominant frugivores in abandoned areas (small birds and bats) carry small seeds (Nepstad et al. 1996). Consequently, insufficient seed dispersal is a major limitation to natural regeneration throughout the neotropics (Janzen 1988, Uhl et al. 1988, Aide and Cavelier 1994, Thomlinson et al. 1996, Holl 1999, Holl et al. 2000, Wijdeven and Kuzee 2000, Zimmerman et al. 2000). In extremely degraded and fragmented landscapes, such as the state of São Paulo, Brazil, where remaining forested areas are estimated to be only 9% of the original native vegetation, even areas adjacent to remaining forest fragments have limited floristic development because of the small fraction of regenerating species originating from dispersed seeds (Maluf de Souza and Batista 2004).

The effects of remnant, or residual vegetation (Chazdon 2003) can also play a critical role in structuring species richness and composition following disturbance. For example remnant trees (Guevara et al. 1986, Guevara and Laborde 1993), shrubs (Vieira et al. 1994, Holl 2002, but see Zahawi and Augspurger 1999) and large-leaved monocots (Duncan and Chapman 1999) can attract seed dispersers and facilitate native rainforest regeneration under their canopies. They may also attract seed predators and herbivores, whose activity can modify the original seed dispersal shadows centered around this remnant vegetation (Myster and Pickett 1992). The ability of some tree and shrub species to resprout following disturbance is another important successional mechanism. For example, trees damaged during hurricanes can resprout, promoting a rapid return of the species composition of mature forest by circumventing secondary succession (Yih et al. 1991). Species differ in their ability to resprout following cutting and fire. In areas where fire is a recurrent disturbance phenomena, this differential resprouting ability can have an unparalleled impact on forest regeneration, extinguishing those species which can not resprout and consequently impoverishing and altering successional trajectories (Cochrane and Schulze 1999).

Post-dispersal biotic factors such as seed and seedling predation may play a significant role in determining the species richness and composition of regenerating communities, as well as convergence characteristics between secondary and primary forest. Research in primary forest suggests that post-dispersal factors are a major factor increasing diversity from dispersed seeds to recruits (Harms et al. 2000, Wills et al. 2006). Whether these post-dispersal factors contribute to the diversity and species composition of forests regenerating after agriculture deserves research attention. In temperate areas, differential palatability to seed and seedling predators can influence species composition during succession by modifying seed and seedling distributions characteristic of dispersal shadows (leptokurtic seeding distributions centered around adult trees) (Myster and Pickett 1992). In the neotropics, research suggests that seed predation has the potential to make significant impacts upon the diversity of floristic recovery, for example losses due to seed predators can be the predominant mechanism of seed mortality in some neotropical old-fields (Nepstad et al. 1996, Myster 2003a, 2004). Seed predation may play a pivotal role in structuring species composition during succession if seed removal is species specific, as has been found in old fields in Bolivia (Peña-Claros and De Boo 2002) and Puerto Rico (Myster 2003b). At one site in Costa Rica, dominance of the species *Pentaclethra macroloba* in mature-forest canopy is mirrored by dominance of the same species in the seedling communities in secondary forests 16–18 years after pasture abandonment. In contrast, other mature-forest trees in the region, for example species of *Dipteryx, Tetragastris and Pouteria* that are present as remnant overstory individuals in secondary stands have low regeneration. The authors (Guariguata et al. 1997) attribute this dominance by *Pentaclethra* to the fact that its seedlings lack significant predation, while *Dipteryx*, *Tetragastris* and *Pouteria* have minimum juvenile recruitment because their fleshy fruits are prone to heavy seed predation.

Abandoned agricultural lands are often invaded by exotic grasses, which can constrain or arrest succession of tree species. Grass invasions limit tree regeneration at many ecological levels (D'Antonio and Vitousek 1992). At the population level, alien grasses successfully compete with tree seedlings for water and nutrients because the grass has dense, shallow root systems (Nepstad 1989, Nepstad et al. 1996, Gonzales-Montagut 1996). Grass competition can affect plant community composition by preventing many tree species from regenerating, especially small-seeded species (Nepstad et al. 1990, 1996, Hooper et al. 2002).

6.2.2 Abiotic Factors

The harsh physical conditions resulting from the presence of grasses including a high irradiance regime, desiccating microclimate, and exposure to drought and fire may also prevent many tree species from regenerating (Nepstad et al. 1990). At the ecosystem level, grasses are more flammable and able to recover more rapidly after fire in comparison to trees. Fire arrests natural forest regeneration in abandoned, deforested lands and encroaches on uncut forest (Janzen 1988, Nepstad et al. 1991). Fire can also impoverish soils, consequently reducing seedling growth (Aide and Cavelier 1994), and exacerbating the problem that total nutrient stocks (especially N) decline with increasing land-use intensity prior to abandonment (Buschbacher et al. 1988, Reiners et al. 1994). At some sites, soil nutrient losses following tropical deforestation can limit forest regeneration (Buschbacher et al. 1988). For example, in abandoned Amazonian pastures, total nutrient stocks and forest recovery were negatively related to land use intensity prior to abandonment; lightly-used sites were N depleted while heavily-used sites were N, K and Mg depleted (Buschbacher et al. 1988). At other sites, research suggests that a lack of nutrients reduces seedling growth in abandoned agricultural lands (Aide and Cavelier 1994). This combination of factors can make these exotic grassland systems too extreme for successful establishment of many rain forest tree species throughout the tropics (Ashton et al. 2001). Those species that do dominate extreme post-agricultural landscapes are often a fire resistant, drought resistant subset of the original native forest species complement with fast growth in full sunlight and the ability to resprout repeatedly when burned or cut with a machete (Boucher et al. 2001).

6.3 Measuring Floristic Recovery During Succession

Forest recovery during succession can be measured using a number of variables, including structural (basal area, biomass, tree height or stem density), functional (soil variables including soil carbon, nitrogen mineralization, soil water availability, cation exchange capacity) or floristic. Floristic variables, including species density, richness and composition have been used by researchers to assess forest recovery and are the subject of this comparison as they have a high relevance to conservation and restoration efforts. Research throughout the neotropics suggests that species richness and composition vary independently (Finegan 1996). Secondary forest succession can recover species richness rapidly, however virtually all studies of post-agricultural succession suggest that the recovery of mature-forest species composition is comparatively slow (Brown and Lugo 1990, Chazdon 2003, Howorth and Pendry 2006). Chronosequence studies suggest that while species richness may recover to mature-forest conditions over time in neotropical forests, species composition may not converge. Instead, initial site conditions, proximity to mature forest, type and degree of disturbance may interact to prevent successional forests from achieving the species composition of the original forest (Capers et al. 2005). Yet, more information is needed on plant community floristics during secondary succession to assess their potential for maintenance of plant species diversity, stand compositional dynamics, and to delineate guidelines for their management and conservation (Guariguata et al. 1997).

6.3.1 Multivariate Techniques- Indirect Gradient Analysis

Multivariate techniques are effective at summarizing data that relates changes in species composition over time, and these methods have advantages over traditional descriptive studies (Austin 1977). Multivariate analyses have been successfully applied to analyze successional data and have allowed researchers to dissect temporal trends from environmental patterns and randomness and to understand successional trajectories of species composition. While both direct and indirect methods of gradient analysis have been used, most studies have utilized indirect gradient analysis, which allows a description of changes in vegetation structure in relation to differences at sites. Indirect gradient analysis has been used to confirm the general successional model (Peña-Claros 2003) proposed by Finegan (1984, 1996), to show a floristic separation between secondary forest regenerating following different land-use histories (Mesquita et al. 2001, Healey and Gara 2003), hurricane compared to agricultural disturbance (Boucher et al. 2000), on land with different soil and topography (Finegan and Delgado 2000), as well as a chronosequence of sites with different times since disturbance (Knight 1975, Sheil 1999, Aide et al. 2000, Howorth and Pendry 2006).

6.3.2 Multivariate Techniques- Direct Gradient Analysis

Direct gradient analysis is advantageous because it can be used to determine the mechanisms driving successional changes. Direct gradient analysis can be used with chronosequence data to determine how much of the floristic variation is explained by successional trend and to illustrate changing patterns of species composition in relation to time since disturbance. Additionally, it can show the mechanisms underlying successional change (ex Purata 1986, Aide et al. 1996), perhaps contributing to a recent increase in the use of these direct gradient multivariate analyses by scientists throughout the neotropics to test hypotheses regarding the effect of different variables on vegetation patterns. For example, direct gradient analysis has been used to examine how patterns of species composition are affected by factors thought to affect floristic development such as land use history, soils and elevation (ex. Aide et al. 1996, Pascarella et al. 2000). Purata (1986) used direct gradient analysis to show how the slow recovery of species composition in secondary forest can be attributed to inadequate seed dispersal and the extent of site degradation. Throughout the neotropics, chronosequence analysis using multivariate analysis illustrates some of the complexities and barriers to achieving community compositional convergence. Multivariate techniques are advantageous in that they can show the interaction of time since disturbance (succession) with mechanisms that may alter successional pathways, and allow a quantification of the amount of floristic variation that is explained by these various interacting factors. In the following overview, I will present results from a range of floristic studies that use multivariate analyses to characterize forest recovery following deforestation and agricultural land use, beginning with sites that present a minimum of barriers to succession (nutrient rich sites close to seed sources) and progressing through those where barriers to forest regeneration can be severe.

6.3.3 Lightly-Disturbed Sites

A recent study in Costa Rica (Capers et al. 2005) on lightly-used sites suggests that the species richness of secondary forest seedling communities 13–26 years following pasture abandonment was similar to that of mature forest. However, community composition of chronosequence plots showed no evidence of increasing similarity over time. Detrended Correspondence analysis (DCA) showed an initial site distinctness in seedling community composition; these sites remained distinct over the period of study (5 years) and did not show any trend of convergence (Capers et al. 2005). This result is underscored by examining the DCA biplot; the primary axis of variation, time since disturbance explained over 50% of the variation in floristic composition, whereas the secondary axis explained only approximately 10% of the variation. Another study in Costa Rica of secondary stands 16–18 years after relatively light pasture use (Guariguata et al. 1997) showed that tree and treelet species richness was lower in secondary stands than in mature forest stands, yet these differences were much less pronounced in seedling and sapling communities. Sapling community composition in these secondary forests was floristically more similar to mature forest stands than was the overstory, mainly because of dominance of *Pentaclethra macroloba* in both (Guariguata et al. 1997). The ubiquity of *Pentaclethra macroloba* in the study area and its ability to survive seed predation and shade conditions may allow for a greater convergence in floristic composition than at most other neotropical sites where species dominance is much less pronounced.

A chronosequence study in Panama following abandonment of lightly-used sites near mature forest suggests that secondary succession on such sites can proceed rapidly, with species richness approaching that of mature forest within 20 years of regrowth (Denslow and Guzman 2000, DeWalt et al. 2003). However, even on these sites where successional processes are uninterrupted (DeWalt et al. 2003), floristic differences were pronounced 100 years following abandonment. Secondary forest stands differed in tree and liana species composition among themselves, and with old-growth stands. While similarity to old-growth composition increased with secondary forest age, 100 year old secondary stands and old growth forest had only a 41% similarity (Sorenson's index) in tree species composition, suggesting that species composition converges on old-growth more slowly than does forest structure (DeWalt et al. 2003).

In Northeast Puerto Rico, conditions for forest recovery are also considered favourable because of small pasture size, short distances to nearby forest, lack of fire, and nutrient rich soils. In these areas, direct gradient analysis suggests that successional trend is not deflected by factors related to anthropogenic use, but is related to natural site factors, specifically elevation (Aide et al. 1996). Species distributions were significantly affected by age since abandonment and elevation; these effects were nearly perpendicular on the ordination and explained 14% of the variation. No effect of distance to the forest was detected, suggesting that the relatively short distances to forest in this landscape do not limit forest regeneration.

6.3.4 Moderately-Disturbed Sites

In a subsequent study in Puerto Rico at sites with more severe disturbance, direct gradient analysis similarly identified the importance of age since abandonment and elevation in structuring species community composition, and additionally showed that substrate type and bulldozing significantly contributed to species composition (Chinea 2002). In total 26% of the variance in species community composition was explained, with over half the variance explained by the two environmental variables (elevation, substrate), while slightly less variation was explained by land-use history variables (site age, bulldozing). Very little variance was shared by the two sets of variables; this effect is underscored by the fact that their effect on species composition was perpendicular in ordination space. The first ordination axis was positively correlated with bulldozing and negatively correlated with age since abandonment, suggesting that bulldozed sites have a species composition similar to early-successional communities. Site elevation was positively correlated with the second axis, a result similar to sites studied by Aide et al. (1996).

6.3.5 Heavily-Disturbed Sites

At sites where previous land use has been more severe, multivariate analyses suggest that not only does time since disturbance play a primary role in structuring species composition (as was shown for lightly-used sites), but land-use type use also explains much of the floristic variation, and these two factors interact. Correspondence analysis (CA) of a chronosequence (11–50 years after abandonment) in Northwest Argentinian neotropical montane forest comparing abandoned herbaceous crops and abandoned citrus orchard shows that time since abandonment and previous land use were both important influences on species composition (Grau et al. 1997). The first axis of the CA accounted for 41% of the floristic variation, and was related to chronosequence age, while the second CA axis also explained a significant proportion of the variation (23%) and was related to the difference in land use (herbaceous vs. citrus). The floristic trend suggests that dispersal mechanisms mediated by crop structure play an important role in structuring species composition; animal-dispersed species predominated in the abandoned citrus orchards, while wind-dispersed species dominated early successional stages. Species richness at these sites reached similar values to old-growth forest within 50 years (Grau et al. 1997).

In abandoned fields in the Mexican humid tropics, research also suggests the importance of land use and seed dispersal; Purata (1986) studied how intensity of land-use following abandonment and proximity to forest affect chronosequence results. She used a direct gradient analysis (Canonical Correspondence Analysis – CCA) to illustrate how regrowth age (successional trend) interacts with length of cropping period and proximity to forest to determine floristic composition. The two main CCA axes of variation explained a high proportion of the floristic variation (Axis 1–50%; Axis 2–22%) and interestingly all three factors were significantly related to the first axis, while the second axis was significantly related to regrowth age and forested perimeter. In contrast to previously-discussed results from lightly-used sites (ex Capers et al. 2005) where the major axis of floristic variation is explained solely be time since disturbance, at these more heavily-used sites, time since disturbance no longer primarily determines floristic development. Instead it interacts with intensity of land use and proximity to forest showing how these factors can permanently deflect succession and prevent convergence. Floristic development was positively related to regrowth age and amount of forested perimeter and negatively related to length of cropping period. The latter result suggests that intense land use inhibits succession; at sites where the length of the cropping period was low, diversity was higher and floristic composition richer than when it was high. The positive effect of forested perimeter on floristic development stresses the importance of nearby forest propagules to community recovery; without the addition of forest species floristic trend is deflected and does not converge with mature forest composition (Purata 1986).

While species composition may not follow predictable trajectories because of interacting factors such as unique site and disturbance characteristics, predictable trends in life-form composition of the seedling community with time have been shown at a number of sites. In Costa Rica, canopy palms, understory palms and the proportion of canopy tree seedlings increased in abundance and species richness over time, while shrub and liana relative abundance declined; these trends correlated to declining light levels with succession (Capers et al. 2005). The proportion of rare species also increased over time because of the increasing recruitment of canopy tree species whose propagules originated outside the study plots. The authors (Capers et al. 2005) suggest that these rare species have large propagules that are animal-dispersed and therefore seed dispersal by animals may be a driver of this diversity-generating process. In Panama, palm importance values similarly increased with stand age (DeWalt et al. 2003). Understory shrub assemblages in Costa Rican premontane tropical wet forest (La Selva Biological station) were more diverse, richer and had a higher density in secondary forest than in old-growth forest (Laska 1997), suggesting an inverse correlation between diversity trends in trees and shrubs. While these predictable trends in life form may suggest that species composition will return to original conditions through the process of secondary succession, recent research in Puerto Rico shows irreversible changes in species composition due to human activity (Lugo and Helmer 2004). These changes include persistence of alien species in mature phases of forest development and a low complement of endemic species compared to native forests.

6.4 Species Richness and Composition on Sites with Heavy and Prolonged Disturbance: a Case History from Panama

While previous floristic studies have shown the relationship between time since disturbance, anthropogenic and landscape factors which can affect species composition on lightly to heavily disturbed sites, no multivariate studies have shown how species composition can be modified on heavily disturbed sites invaded by exotic grasses and modified repeatedly by fire. In central Panama, many abandoned sites are invaded by an exotic grass species, *Saccharum spontaneum* L. ssp. *spontaneum* , that grows in tall (2.5 m) dense, impenetrable, fire-prone stands. In the following case study, I discuss results from experiments (Hooper et al. 2002, 2004, 2005) designed to investigate factors affecting the floristics of early stages of secondary forest succession on abandoned agricultural lands in Panama invaded by *Saccharum spontaneum.* At five study sites located near the Panama Canal (9°06 ′N, 79°53 ′W), 4 km southwest of the Barro Colorado Nature Monument (BCI) I used a multifactorial experimental design to test the effect of limited seed dispersal, *Saccharum* competition, fire, and remnant vegetation (trees, large-leaved monocots and shrubs) on the diversity and floristic composition of regenerating tree and shrub species [see Hooper et al. (2004, 2005) for a detailed description of the experiments]. The effect of distance from the forest edge (10, 35, and 85 m) was assessed to evaluate the role of potential limitations to seed dispersal. Competition with *Saccharum* was studied by comparing mown (once and thrice mown yearly) and unmown treatments. The effect of fire was investigated by performing a prescribed burn. Soil nutrients were compared in *Saccharum*-dominated sites and adjacent forest. Tree and shrub sprouts, seedlings and saplings were monitored over a 1-year period beginning in August, 1996 to assess the effects of these factors on natural regeneration. To determine the effect of trees, shrubs and large-leaved monocots (*Musa* and *Heliconia* spp.) on natural tree regeneration I documented their location and measured their proximity to each regenerating tree and shrub seedling. In another experiment (Hooper et al. 2002), I planted 15,000 seeds of 20 native tree species varying in seed size and shade tolerance characteristics under combinations of *Saccharum* shading and mowing treatments and followed their germination and survival to further assess the effect of *Saccharum* competiton on tree regeneration. ANOVA, linear and nonlinear regression analyses and various multivariate analyses including distance-based redundancy analysis (Legendre and Anderson 1999) forward selection using RDA (following appropriate transformation) and fourth-corner analysis (Legendre et al. 1997) were used to evaluate the effect of the experimental factors.

In total, I documented 4,984 individuals of 80 species (mean diameter 6 mm) naturally regenerating in the *Saccharum spontaneum*. Seedling density was surprisingly high underneath the dense *Saccharum*, and depended on the season (dry season: 29 $\pm$6 10 m^{-2}; wet season: 35 $\pm$ 8 10 m^{-2}). The most common species (total number of individuals) were *Gustavia superba* (1,060), *Piper marginatum* (522), *Cochlospermum vitifolium* (400), *Spondias mombin* (290) and *Cordia alliodora* (241); a full species list (Appendix C) is included in Hooper et al. (2004). Direct gradient analysis results show that *Saccharum* competition (P = 0.001), distance from the forest (P = 0.038) and time since fire (P = 0.016), significantly affect species community composition of regenerating tree and shrub seedlings (Hooper et al. 2004). Fire and distance from the forest significantly affect species diversity. Remnant vegetation in the *Saccharum* also significantly affects the density, species richness, and species composition of forest regeneration.

6.4.1 Saccharum *competition*

Results from the experiment where I planted 15,000 seeds of 20 native tree species in four *Saccharum* mowing and shading treatments and unshaded, unmown controls (Hooper et al. 2002) shows that the invasive grass, *Saccharum spontaneum*, negatively affects germination, survival, and growth of native tree seedlings in abandoned Panamanian farmland. Most tree species had higher germination, survival and growth when the *Saccharum* was shaded in contrast to unshaded control conditions. As shading treatments essentially eliminated the *Saccharum*, consequently reducing below-ground competition, while understory light levels were similar when the *Saccharum* was 95% shaded as compared to the control, I attribute high seedling performance in shaded treatments to a decrease in below-ground competition with the *Saccharum*. The enhancement of seedling performance in shaded treatments as compared to the *Saccharum* control was pronounced in the dry season, leading to a further conclusion that dry-season water stress limits the regeneration in the *Saccharum*. This conclusion concurs with results in Amazonian pastures, where seedling establishment and survival was similarly limited by high temperatures and lower moisture availability in tropical pastures as compared to forest because the dense grass root mass in the upper 50 cm of soil significantly decreases soil moisture availability (Uhl 1987, Nepstad et al. 1990).

Species responded differently to shading and mowing conditions, suggesting that species community composition can be affected by conditions in the *Saccharum*.

These species differences in response to experimental treatments were consistently predicted by their seed size and shade tolerance characteristics. Small-seeded, light-dependent species (*Trema*, *Annona*, *Vochysia* and *Jacaranda* spp.) had the lowest overall germination and survival in the *Saccharum* control and their performance in the treatments suggest that they can tolerate neither above nor below-ground constraints imposed by the *Saccharum*. Light-demanding gap-colonists (*Byrsonima*, *Spondias* and *Sterculia* spp.) with larger seeds had high performance in the thrice mown *Saccharum* treatment and low performance in the *Saccharum* control, suggesting that these species will colonize after a fire, but will not persist once overtopped by the *Saccharum*. Small-seeded, shade-tolerant species (*Posoqueria*, *Genipa*, and *Heisteri* spp.) had high performance in shaded treatments, but low performance in the mown and control *Saccharum*, suggesting that these species could colonize sites where the *Saccharum* is shaded out, such as below remnant vegetation but will not successfully regenerate in open conditions after fire, or in dense *Saccharum*. Large- to very large-seeded (2.9 to 50.4g), shade-tolerant species (*Dipteryx*, *Calophyllum*, *Carapa*, and *Virola*) characteristic of mature-forest had the highest performance in the *Saccharum* control, combining high germination and high survival.

These results suggest that the potential for succession to be constrained in *Saccharum*-dominated sites is high. At sites in close proximity to ours (< 4 km) on Barro Colorado Island (BCI) that are not invaded by *Saccharum spontaneum*, successional trajectories follow Finegan's (1996) general neotropical successional model. For example, Kenoyer (1929) reports a four-stage, post-disturbance succession on BCI: (1) grasses and weedy plants, large monocots (*Heliconia* spp.), and shrubs of the genus *Piper*; (2) dominance by short-lived pioneers; (3) mixed secondary forest of longer lived pioneers; (4) climax forest stage. In contrast, experimental results at *Saccharum*-dominated sites (Hooper et al. 2002) suggest that this successional sequence may be constrained. Small-seeded, light-demanding tree species of the second phase of succession (pioneers) can not invade because they do not successfully compete with the *Saccharum*. The larger-seeded, light-dependent species which would form the third successional phase (long-lived pioneers) are also precluded because they need more light. Their high performance in mown *Saccharum* suggests that these species may potentially regenerate directly after fire before the *Saccharum* has re-established. Small-seeded, shade-tolerant species of the mature-forest which would continuously colonize the understory in the typical successional sequence (Finegan 1996) are also absent because they can not compete with the *Saccharum*. In contrast, large-seeded, shade-tolerant species typical of mature forest may regenerate in the *Saccharum* as they have the highest potential performance in the *Saccharum* controls, and were the only species able to resprout after recurrent fires.

6.4.2 Seed Dispersal

A comparison of the above results from experimentally-dispersed species to those naturally-regenerating in the *Saccharum* suggests insufficient seed dispersal is a major barrier to forest regeneration. When seed input was equalized, large-seeded shade-tolerant species had the highest germination and survival but most individuals naturally regenerating were small-seeded gap species (Hooper et al. 2005). The high germination and survival of the large-seeded shade-tolerant species suggest that if they were dispersed into the *Saccharum*, they would be found in the highest proportions. Yet, they are rarely found naturally regenerating. These findings suggest similar regeneration constraints in Panamanian and Amazonian secondary succession. In Amazonian pastures, the small-seeded bat- bird-dispersed species most likely to be dispersed into abandoned Amazonian pastures had the lowest survival whereas large-seeded species were constrained by the fact that their mammalian dispersers rarely moved into abandoned pastures (Nepstad et al. 1990, 1996). My research results suggest that these dispersal limitations depend on distance from the forest, distance from remnant vegetation, and time since fire.

6.4.3 Distance from the Forest

Distance from the forest significantly affected species richness (ANOVA; $P = 0.036$) and species community composition (db-RDA; $P = 0.001$). Species richness was almost threefold lower at 35 m as compared to 10 m from the forest. Similar declines in species richness away from forest edge throughout the neotropics (Uhl et al. 1988, Aide and Cavelier 1994, Holl 1999, Cubiña and Aide 2001, Rodrigues et al. 2004) suggest that if distance to the nearest seed source is large, this will limit the speed, diversity, and species composition of natural forest regeneration. The effect of distance from the forest on species composition depended on time since fire (db-RDA; effect of site × distance interaction on community composition: $P = 0.001$, 12.2% of the variance explained) (Hooper et al. 2004). Sites with early post-fire communities (1–2 years post-fire) had significant differences in community composition as a result of distance from the forest, while sites surveyed 3 and 4 years after fire showed no significant effect ($P = 0.119$, $P = 0.116$) of distance from the forest on community composition. Purata (1986) similarly reports that the interaction of time since disturbance and distance to the forest was important in determining floristic composition in abandoned Mexican fields. Finegan (1996) also reports how initial site conditions affect propagule dispersal, which determines species richness and composition during neotropical rainforest succession.

At *Saccharum*-dominated sites, community composition of tree seedling regeneration 1 year after fire is dominated by light-demanding, wind-dispersed species, with some small-bird-dispersed species present; their abundance declines exponentially with increasing distance from the forest. This decline is steep enough that few individuals are found 35 m from the forest. In contrast to sites close to the forest, the density of the bat-dispersed (*Piper marginatum*) increases with distance from the forest, peaking at 35 m and dominating the community there. At other neotropical sites, bats have a similar effect on tree regeneration because they defecate in flight (Charles-Dominique 1986) and thus disperse more seeds to locations farther from perches than birds (Gorchov et al. 1993, Cardosa da Silva et al. 1996, Medellin and

Gaona 1999). At distances furthest from the forest, few individuals are found except for species that can recruit from seed after fire or are able to reproduce in these *Saccharum* grasslands. No large monocots, trees, saplings, or shrubs are present 1 year after fire. In contrast, as time since fire increases, naturally-regenerating wind-dispersed species become progressively less abundant, large-animal-dispersed species become more abundant, remnant vegetation become established and reaches a height great enough to overtop the *Saccharum*, and the importance of distance from the forest on seedling community composition declines. Distance from remnant vegetation becomes more important with time. Two years post-fire, significant differences in species composition result from the interaction of distance to remnant vegetation and forest. Three and 4 years post-fire, no significant differences in community composition are found because all sites are dominated by species regenerating close to the remnant vegetation (Hooper et al. 2004).

6.4.4 Remnant Vegetation

I utilized non-linear regression analysis to model the effect of distance from the forest and distance from various types of remnant vegetation (isolated trees, shrubs, and large-leaved monocots) on seedling density and species richness and used a backwards selection procedure to determine the relative importance of each (Table 6.1). Results suggest that remnant vegetation, specifically large-leaved monocots is the single most important factor affecting seedling density. Distance from the large monocots accounts for 56.6% of the variation in seedling density, with a fourfold increase in seedling density in plots with large monocots present as compared to those with no large monocots. Three of the most common species (accounting for 42% of the total density of all naturally-regenerating individuals) had densities that declined exponentially with distance from these monocots including a bat-dispersed species, an agouti-dispersed species and a species dispersed by both birds and bats (Dalling et al. 1998b). Other remnant vegetation in these Panamanian old-fields also influenced different species of tree regeneration. Isolated trees and shrubs had more small-bird-dispersed species, and more *Byrsonima crassifolia*, a species dispersed by large birds and mammals. These results concur with research throughout the tropics, where trees (Guevara et al. 1986, Guevara and Laborde 1993, Cardosa da Silva et al. 1996, Duncan and Chapman 1999, Toh et al. 1999, Wijdeven and Kuzee 2000), shrubs (Vieira et al. 1994, Holl 2002, but see Zahawi and Augspurger 1999) and large monocots (Duncan and Chapman 1999) act as recruitment foci for tropical tree seedlings at abandoned sites because they attract seed dispersers. Differences in the relative density of recruitment below different types of remnant vegetation may result from differential attraction of frugivores.

In contrast, the density of naturally-regenerating trees and shrubs did not correlate with distance to the forest. Duncan and Duncan (2000) and Slocum and Horvitz (2000) similarly report that in areas containing remnant vegetation, seedling density does not correlate with distance from the forest, and Guevara et al. (1986) and Willson and Crome (1989) record little effect of distance from the forest on

Table 6.1 Summary of results of nonlinear regression analyses, modeling the effect of distance from various types of old-field vegetation (biotic) and abiotic factors on the density of naturally-regenerating tree and shrub species per 10 m^2. Only the variables that were significant ($P < 0.05$) following backward elimination are presented. For old-field vegetation types, both the negative exponential (Ex) and linear functions (Lin) were used in the model. All reported values in the 10 left-hand columns are R^2 statistics resulting from regression against a single explanatory variable of distance from the various old-field vegetation types or abiotic factors. Positive exponent: increasing density with distance from source; negative exponent: decreasing density with distance from source. Source includes: distance from the forest (Forest), distance from the nearest large-leaved monocot (Monocot), distance from the nearest old-field tree larger than 5 m in height (Tree), and distance from the nearest shrub or sapling 2.5–5 m in height (Shrub). Habitat: see note[a]. Biotic = partial R^2 for all biotic variables. Abiotic = partial R^2 for all abiotic variables. Total = total R^2 for the (multiple) nonlinear regression model. F = F-statistic; P = P-value; $n = 90$ (reported in *Hooper* et al. 2004; digital Appendix D)

	Forest		Monocot		Tree		Shrub		Habitat[a]		Biotic	Abiotic			
Species[b]	Ex	Lin[c]	Ex	Lin[d]	Ex	Lin[e]	Ex	Lin[f]	linear	linear	partial	partial	Total	F	P
Total density of all individuals															
	–	–	0.566^{-}	–	–	–	–	–	–	–	0.566	0.000	0.566	$F_{3,87} = 22.65$	<0.0001
Total number of species															
	–	0.611^{-}	0.662^{-}	–	–	–	–	–	–	–	0.679	0.000	0.679	$F_{4,86} = 7.24$	0.0002
1. Wind-dispersed species (not reproducing)															
Tric ga	–	0.091^{-}	–	–	–	–	–	–	Fire 0.108^{-}	–	0.091	0.108	0.161	$F_{3,87} = 6.15$	0.0032
Lueh se	–	0.133^{-}	–	–	–	–	–	–	–	–	0.133	0.000	0.133	$F_{2,88} = 6.80$	0.0107
Cord al	–	–	–	–	–	–	–	–	Fire 0.164^{+}	–	0.000	0.164	0.164	$F_{1,89} = 17.47$	<0.0001
2. Wind-dispersed species (reproducing)															
Coch vi	–	–	–	–	–	–	–	–	Fire 0.111^{-}	Dead 0.120^{-}	0.000	0.171	0.171	$F_{3,87} = 4.49$	0.0139
3. Ballistic-dispersed species															
Hyba pr	–	0.113^{-}	0.335^{-}	–	–	0.155^{-}	–	–	–	–	0.464	0.000	0.464	$F_{5,85} = 15.85$	<0.0001

4. Small-bird-dispersed species															
Mico ar	–	–	–	–	0.173^{-}	–	–	–	–	–	0.173	0.000	0.173	$F_{2,88} = 9.21$	0.0002
Cecr in	–	–	0.312^{-}	–	–	0.168^{+}	–	–	Mid 0.136^{+}	Low 0.091^{-}	0.440	0.149	0.487	$F_{6,84} = 12.03$	<0.0001
Bana gu	–	–	–	–	–	–	–	–	–	–	0.000	0.000	0.000		
Trem mi	–	–	–	–	–	–	–	–	–	–	0.000	0.000	0.000		
Psyc gr	–	–	–	–	0.147^{-}	–	–	–	–	–	0.147	0.000	0.147	$F_{2,88} = 7.59$	0.0009
5. Bat-dispersed species															
Pipe ma	–	–	0.577^{-}	–	–	–	–	–	–	–	0.577	0.000	0.577	$F_{3,87} = 47.97$	<0.0001
6. Large-arboreal-animal-dispersed species															
Thev ah	–	0.096^{-}	–	–	–	–	–	–	–	–	0.096	0.000	0.096	$F_{2,88} = 4.50$	0.0368
Quas am	–	–	–	–	–	–	–	–	–	–	0.000	0.000	0.000		
Byrs cr	–	–	–	–	–	–	–	0.112^{-}	–	–	0.112	0.000	0.112	$F_{2,88} = 5.18$	0.0252
Swar s2	–	0.250^{+}	–	–	–	–	–	–	Dead 0.284^{+}	–	0.250	0.284	0.317	$F_{3,87} = 6.48$	0.0024
Inga ve	–	0.223^{-}	–	–	–	–	–	–	–	–	0.223	0.000	0.223	$F_{2,88} = 4.76$	0.0318
Spon mo	–	–	–	–	–	–	–	–	–	–	0.000	0.000	0.000		
7. Agouti-dispersed species															
Gust su	–	–	0.449^{-}	–	–	–	–	–	Dead 0.283^{+}	–	0.449	0.283	0.479	$F_{4,86} = 17.73$	<0.0001

[a]Abiotic variables include: Mid = Light (PAR) at 0.5 m (μ moles m^{-2} sec^{-1}), Low = Light (PAR) at 0.1 m (μ moles m^{-2} sec^{-1}), Live = biomass of live *Saccharum* (g), Dead = biomass of *Saccharum* litter (g), Fire = time since wildfire (months), Slope = slope (angle above horizontal) . All abiotic variables were ln $(x + 1)$ transformed, except fire and slope. Slope was square root $(x + 0.1)$ transformed; time since the last wildfire was untransformed

[b]Species: Bana gu = *Banara guianensis*, Byrs cr = *Byrsonima crassifolia*, Cecr in = *Cecropia insignis*, Coch vi = *Cochlospermum vitifolium*, Cord al = *Cordia alliodora*, Gust su = *Gustavia superba*, Hyba pr = *Hybanthus prunifolius*, Inga ve = *Inga vera*, Lueh se = *Luehea seemannii*, Mico ar = *Miconia argentea*, Pipe ma = *Piper marginatum*, Psyc gr = *Psychotria grandis*, Quas am = *Quassia amara*, Spon mo = *Spondias mombin*, Swar s2 = *Swartzia simplex* var. *ochnacea* , Thev ah = *Thevetia ahouai*, Trem mi = *Trema micrantha*, Tric ga = *Trichospermum galeotti*

[c]Square root of distance from the forest

[d]Distance from the nearest large-leaved monocot (*Musa* sp., *Heliconia* sp.)

[e]Square root of distance from the nearest isolated old fields tree larger than 5 m in height

[f]Ln (distance from the nearest shrub or sapling between 2.5 and 5 m in height)

animal-dispersed seed rain near isolated trees and shrubs present in old-fields. Slocum and Horvitz (2000) conclude that distance to the forest is most important when remnant vegetation is not found in degraded lands, like very early in succession. Seedling distributions at my Panamanian sites support this hypothesis. While I attribute these results to the effect of different seed dispersal vectors, it is possible that post-dispersal seed predation modifies the original leptokurtic distributions of seeds dispersed near forest. For example, Myster and Pickett (1992) similarly found that the negative exponential decline in density of dispersed species from vegetation became weaker as succession proceeded, and attribute this result to herbivore pressure. My results can not distinguish between the effects of seed dispersal and post-dispersal factors as I measured the distributions of seedlings, however the impact of seed predators and herbivores deserves research attention.

Species richness was significantly affected by both distance from the forest and distance from the large-leaved monocots, accounting for 68.1% of the variation (Table 6.1). Many species were significantly affected by proximity to the forest, with the density of larger-arboreal-animal-dispersed, wind-dispersed, and ballistic-dispersed species declining linearly with increasing distance from the forest. Holl (1999) and Aide and Cavelier (1994) record similar declines in seedling species richness with distance from the forest edge at other neotropical sites, which may result from dispersal limitations. For example, Zimmerman et al. (2000) document a decline in the species richness of seed rain with increasing distance from the forest. In contrast to these biotic factors, abiotic factors in Panamanian old-fields did not significantly affect the overall density or species richness of naturally-regenerating trees and shrubs. Similarly, soil nutrient levels were not a significant factor in influencing natural regeneration (Hooper et al. 2005). These results suggest that biotic factors are much more important than abiotic post-dispersal factors in affecting species richness and composition. However, plots with remnant vegetation, and those close to the forest may also provide superior abiotic conditions for growth and survival compared to grass-dominated sites (Guevara et al. 1992, Vieira et al. 1994, Nepstad et al. 1996, Holl 2002, Hooper et al. 2002) because of lowered grass competition and a more favourable microclimate (Vieira et al. 1994) These results underscore the importance of remaining forest and remnant vegetation in attracting seed dispersers (and potentially seed predators) and consequently increasing diversity and driving succession in these *Saccharum*-dominated old-fields. These results suggest that in their absence, forest regeneration will be slow and limited to locations close to pre-existing forest. Proximity to forest and different remnant vegetation in the grasslands differentially affects seed dispersal vectors, resulting in spatially distinct variations in community composition in relation to this vegetation.

6.4.5 Fire

Fires burn yearly in the dry season in these *Saccharum*-dominated abandoned pastures. My data show that these wildfires are a major barrier to tree regeneration because they decrease species richness and change community composition. Fire significantly (P = 0.025) decreased the number of species from 6 ± 0.1 10 m^{-2} pre-fire to 3 ± 0.6 10 m^{-2}post-fire. In contrast, the density of regenerating tree and shrub species was not significantly affected by burning because a subset of the original species resprouted in high densities after fire. Species that were able to sprout after fire formed a large part of the initial floristic recovery. The percentage of seedlings originating as rootsprouts significantly (ANOVA; P = 0.036) increased in burned plots (unburned: 31 ± 7 %, burned: 52 ± 10 %) (Hooper et al. 2005). Species differed in the effect that fire had on the amount of regeneration from resprouts and recruits (from seed); these differences significantly impacted upon community composition (db-RDA; P = 0.001) (Hooper et al. 2004). Many more species showed a decline in one or both mechanisms of regeneration compared to those that increased in response to fire, leading to a low-diversity community following fire, as has been similarly documented at other tropical sites (Uhl et al. 1988, Nepstad et al. 1990). For example, 43 species had decreased recruitment from seed, while only 11 had enhanced recruitment after fire (Hooper et al. 2004). Half the species able to resprout had lowered or no resprouting following fire. Other authors similarly found that burning reduces the number of species resprouting at neotropical sites (Uhl et al. 1990, Sampaio et al. 1993). The few species that had increased resprouting after fire, especially *Gustavia superba*, *Cochlospermum vitifolium*, and *Cordia alliodora* dominated post-fire community composition.

6.4.6 Succession Following Fire

To characterize species differences in relation to time since fire, I used a series of 4th corner correlations (Legendre et al. 1997) relating time since fire with species' dispersal mechanism (Table 6.2), species' regeneration characteristics (shade-tolerance, seed size, resprouting ability) (Table 6.3) and life-form (Table 6.4). The species characteristics negatively correlating with time since fire include wind dispersal and ability to reproduce in *Saccharum*-dominated old-fields (Table 6.2), shade-intolerance (Table 6.3), and mid-sized tree life-form (Table 6.4). In contrast, the species characteristics positively correlating with time since fire include large-animal-dispersal (Table 6.2), regeneration from root sprouts (Tables 6.2 and 6.3),

Table 6.2 Dispersal vector / mechanism (see text) compared to time since fire. The table reports correlation fourth-corner statistics r(i,j), and probabilities (Prob) tested using 999 permutations followed by adjustment using Holm's procedure (reported in *Hooper* et al. 2004; digital Appendix E)

	1. Wind, not reproducing	2. Wind, reproducing	3. Ballistic	4. Small birds	5. Bats	6. Large arboreal animals	7. Agouti	Resprout index[a]
r(i,j)	–0.0435	–0.1329	0.0806	–0.0390	–0.0380	0.1231	–0.0364	0.1382
Prob	0.5400	0.0080*	0.1200	0.5400	0.5400	0.0080*	0.5400	0.0080*

*P < 0.01

[a]Resprout index, calculated as the number of individuals originating from root sprouts divided by the total number of individuals, indicates the ability of each species to resprout

Table 6.3 Species characteristics including light index, seed size and resprout index compared to abiotic variables. Only those abiotic variables that significantly correlated with species characteristics are presented. The table reports correlation fourth-corner statistics *r*(i,j), and probabilites (Prob) tested using 999 permutations (reported in Hooper et al. 2004; digital Appendix F)

	Light index[a]	Seed size (mm)	Resprout index[b]
Time since fire (months)			
r(i,j)	–0.15507	0.04682	0.16238
Prob	0.00100*	0.08200	0.00100*

*$P < 0.01$

[a]Light index was calculated in Condit et al. (1996). Species with a high light index (> 3.0) are pioneers and light demanding, those with a low light index (< 0) are shade tolerant, and those between 0 and 3.0 are gap-dependent

[b]Resprout index (calculated as the number of individuals originating from root sprouts divided by the total number of individuals) indicates the ability of each species to resprout

shade-tolerance (Table 6.3), and shrub life-form (Table 6.4). These results suggest a very general characterization of succession following fire begins with light-dependent species regenerating from small-seeded, wind-dispersed seeds; these mid-sized tree species such as *Cochlospermum vitifolium* are able to reproduce quickly within the old-fields. As succession proceeds, larger-seeded, shade-tolerant species, especially shrubs but also large trees (result marginally non-significant; $P = 0.07$) that can resprout and that are dispersed by large arboreal animals predominate.

This data generally supports predictions by the general successional model of Finegan (1996) that species composition will change from small-seeded pioneer species dominating early in succession, to more shade-tolerant, mature forest species dominating late-successional communities. However, the chronosequence at my Panamanian sites differs fundamentally from Finegan's successional model (Hooper et al. 2004). While I similarly record a dominance of the site by grasses (*Saccharum*) as a first successional stage, the second successional stage of short-lived, light-demanding species is virtually absent. Short-lived pioneers of the genera *Cecropia*, *Trema* and *Ochroma* predominate the 3rd to 15th years following disturbance at nearby sites that are not invaded by *Saccharum* (ex: Kenoyer 1929). In contrast, light-demanding species significantly declined in abundance after only a few years following fire, suggesting that pioneer species do not persist in *Saccharum* grasslands. *Saccharum* sites lack dominance by core pioneers (especially of the genus *Cecropia*) that dominate the seed bank at unburned sites less than 4 km away (Dalling et al. 1998a). Research at Amazonian sites also demonstrates that *Cecropia*

Table 6.4 Adult size of individuals, as defined in Condit et al. (1995), regenerating in the old fields, related to time since fire (months). The table reports correlation fourth-corner correlation statistics r(i,j), and probabilities tested using 999 permutations (reported in Hooper et al. 2004; digital Appendix G)

	Large tree	Mid-sized tree	Treelet	Shrub
r(i,j)	0.0711	–0.1842	0.0441	0.1201
Probability	0.0700	0.0040**	0.1560	0.0150*

*$P < 0.05$;
**$P < 0.01$

species are extinguished as a component of the regeneration (Didham and Lawton 1999, Mesquita et al. 2001) when fire is present. The small-seeded pioneers that do survive in the seed bank and germinate following fire have very low performance in the *Saccharum* because they can tolerate neither above-ground nor below-ground *Saccharum* competition (Hooper et al. 2002).

Species of the long-lived pioneer genera *Luehea*, *Apeiba*, *Miconia*, and *Cordia*, characteristic of the third phase of the general neotropical successional model are the major component of the regeneration at *Saccharum*-dominated sites. Results from the planting experiment (Hooper et al. 2002) suggest that these species can tolerate below-ground *Saccharum* competition because of their larger seed size. Their abundance supports the hypothesis of Finegan (1996) that regeneration of the long-lived pioneers is abundant in the first years of succession. Interestingly, many species regenerating in the *Saccharum* are characteristic of mature-forest (Hooper et al. 2004); for example 54 of the 80 species regenerating in *Saccharum* are also present in mature-phase forest on Barro Colorado Island nearby (Condit et al. 1996). The spatial distribution of these species is positively affected by proximity to the forest and remnant vegetation, while species composition of these communities is significantly affected by the identity of the vegetation (ex forest, large monocot) found in close proximity; these results follow the nucleation model of Yarranton and Morrison (1974). Therefore, while species composition is similar to the third and fourth stages of the general successional model (Finegan 1996), their spatial distribution follows the nucleation model of succession, highlighting the importance of seed dispersal and initial site conditions on propagule dispersal, seedling establishment, and survival. Results at *Saccharum*-dominated sites suggest that both temporal and spatial factors must be considered for a complete description of succession.

In the continual presence of fire, however, this successional sequence does not have a chance to proceed, and a *Saccharum*-dominated grassland will continue to effectively preclude woody species regeneration. Authors throughout the neotropics report how grass dominance on repeatedly-burned lands can suppress tree seedling recruitment for years (Cochrane and Schulze 1998) or permanently deflect succession to a savanna and scrub vegetation (D'Antonio and Vitousek 1992). This could lead to impoverishment of vast expanses of tropical lands as fire becomes more widespread (Uhl 1998). This irreversible degradation is extremely likely in the presence of invasive exotic grasses, as demonstrated in much of the Old World tropics (Richards 1964, Otsamo et al. 1995). At neotropical sites invaded by exotic grasses, results in Panama and Brazil suggest that the inability to resprout can drive local species extinction, lower biodiversity and modifying species composition if seed dispersal is limited, and the seed bank becomes degraded (Miller and Kauffmann 1998, Hooper et al. 2004).

In conclusion, results from heavily-disturbed sites in Panama concur with research at other moderately- to heavily-disturbed neotropical sites (Purata 1986, Grau et al. 1997, Chinea 2002) that illustrate how floristic composition following land abandonment differs from that predicted by the general successional model because of the interaction of previous land use and the spatial dynamics of seed dispersal. In abandoned, *Saccharum*-dominated sites in Panama, as at other neotropical locations (Cochrane and Schulze 1999), fire is a recurrent disturbance phenomena. Fire permanently deflects succession and lowers biodiversity by extinguishing many species from the seed bank and favouring a limited subset of fire-adapted species that can resprout. Competition with the *Saccharum spontaneum* further modifies predicted successional trajectories by preventing small-seeded species from regenerating, in effect eliminating the pioneer phase of succession. Larger-seeded species are able to survive *Saccharum* competition, however these species are mostly animal-dispersed and consequently dispersal-limited, a result consistent at many neotropical sites (Janzen 1988, Uhl et al. 1988, Aide and Cavelier 1994, Cardosa da Silva et al. 1996, Thomlinson et al. 1996, Holl 1999, Holl et al. 2000, Wijdeven and Kuzee 2000, Zimmerman et al. 2000). The density and spatial distribution of these vertebrate-dispersed species are influenced by the location of pre-existing forest and also remnant vegetation, such as trees, shrubs, or large-leaved monocots which attract seed dispersal vectors and ameliorate conditions for seedling regeneration. Direct gradient analysis confirms that at neotropical sites with moderate to heavy disturbance histories (Purata 1986, Grau et al. 1997), and Panamanian *Saccharum*-dominated sites, distance from this vegetation explains much of the variation in species composition, and is consequently a major driver of the floristics during succession. These results concur with McCook (1994) and Finegan (1984) who suggest that spatial and stochastic factors, such as seed availability and dispersal should be included for a more complete description of neotropical succession.

Acknowledgment This work was supported by a grant from the Fundación AVINA. Richard Condit and Pierre Legendre provided invaluable support throughout. The Smithsonian Tropical Research Institute in Panama provided generous logistical assistance. I thank members of the Center for Tropical Forest Science, including Jorge Aranda, Salomón Aguilar and Rolando Pérez for tree seedling identification, Jorge Aranda for help with vegetation mapping, Suzanne Loo de Lao for help with initial database management, and Lidia Valencia for logistical support. I am grateful to M.S. Ashton for valuable comments on later versions of this Manuscript. Special thanks to the people of Las Pavas for their dedicated field assistance and Alejandro Sanchez for sharing his knowledge of the local flora.

References

Aide, T. M., and Cavelier, J. 1994. Barriers to lowland tropical forest restoration in the Sierra Nevada de Santa Marta, Columbia. Restoration Ecology 2:219–229.

Aide, T. M., Zimmerman, J. K., Herrera, L., Rosario, M., and Serrano, M. 1995. Forest recovery in abandoned tropical pastures in Puerto Rico. Forest Ecology and Management 77:77–86.

Aide, T. M., Zimmerman, J. K., Rosario, M., and Marcano, H. 1996. Forest recovery in abandoned cattle pastures along an elevational gradient in northeastern Puerto Rico. Biotropica 28: 537–548.

Aide, T. M., Zimmerman, J. K., Pascarella, B., Rivera, L., and Marcano-Vega, H. 2000. Forest regeneration in a chronosequence of tropical abandoned pastures: implications for restoration ecology. Restoration Ecology 8:328–338.

Ashton, M. S., Gunatilleke, C. V. S., Singhakumura, B. M. P., and Gunatilleke, I. A. U. N. 2001. Restoration pathways for rain forest in southwest Sri Lanka: a review of concepts and models. Forest Ecology and Management 154:409–430.

Austin, M. P. 1977. Use of ordination and other multivariate descriptive methods to study succession. Vegetation 35:165–175.

Boucher, D. H., Vandermeer, J. H., Granzow de la Cerda, I., Mallona, M. A., Perfecto, I., and Zamora, N. 2001. Post-agriculture versus post-hurricane succession in southeastern Nicaraguan rain forest. Plant Ecology 156(2):131–137.

Brown , S., and Lugo, A. E. 1990. Tropical secondary forests. Journal of Tropical Ecology 6:1–32.

Buschbacher, R., Uhl, C., and Serrão, E. A. S. 1988. Abandoned pastures in Eastern Amazonia. II. Nutrient stocks in the soil and vegetation. Journal of Ecology 76:682–699.

Capers, R. S., Chazdon, R. L., Brenes, A. R., and Alvarado, B. V. 2005. Successional dynamics of woody seedling communities in wet tropical secondary forests. Journal of Ecology 93:1071–1084.

Cardosa da Silva, J. M., Uhl, C., and Murray, G. 1996. Plant succession, landscape management, and the ecology of frugivorous birds in abandoned Amazonian pastures. Conservation Biology 10:491–503.

Cavelier, J. 1992. Fine-root biomass and soil properties in a semideciduous and a lower montane rain forest in Panama. Plant and Soil 142:187–201.

Chapman, C. A., and Chapman, L. J. 1999. Forest restoration in abandoned agricultural land: a case study from East Africa. Conservation Biology 13:1301–1311.

Charles-Dominique, P. 1986. Inter-relations between frugivorous vertebrates and pioneer plants: *Cecropia*, birds, and bats in French Guyana. In Frugivores and Seed Dispersal, ed. A. Estrada and T. H. Fleming, Dordrecht, The Netherlands: Dr. Junk Publishers.

Chazdon, R. L. 2003. Tropical forest recovery: legacies of human impact and natural disturbances. Perspectives in Plant Ecology, Evolution and Systematics 6(1,2):51–71.

Chinea, J. D. 2002. Tropical forest succession on abandoned farms in the Humacao municipality of eastern Puerto Rico. Forest Ecology and Management 167:195–207.

Cochrane, M. A., and Schulze, M. D. 1998. Forest fires in the Brazilian Amazon. Conservation Biology 12:948–949.

Cochrane, M. A., and Schulze, M. D. 1999. Fire as a recurrent event in tropical forests of the Eastern Amazon: Effects on forest structure, biomass, and species composition. Biotropica 31(1):2–16.

Condit, R., Hubbell, S. P., and Foster, R. B. 1996. Assessing the response of plant functional types to climatic change in tropical forests. Journal of Vegetation Science 7:405–416.

Connell, J. H., and Slatyer, R. O. 1977. Mechanisms of natural succession in natural communities and their role in community stability and organization. American Naturalist 111:1119–1144.

Cubiña, A., and Aide, T. M. 2001. The effect of distance from forest edge on seed rain and soil seed bank in a tropical pasture. Biotropica 33:260–267.

D'Antonio, C. M., and Vitousek, P. M. 1992. Biological invasions by exotic grasses, the grass/fire cycle, and global change. Annual Review of Ecology and Systematics 23:63–87.

Dalling, J. W., Hubbell, S. P., and Silvera, K. 1998a. Seed dispersal, seedling establishment and gap partitioning among tropical pioneer trees. Journal of Ecology 86:674–689.

Dalling, J. W., Swaine, M. D., and Garwood, N. 1998b. Dispersal patterns and seed bank dynamics of pioneer trees in moist tropical forest. Ecology 79:564–578.

Denslow, J. S., and Guzman, S. G. 2000. Variation in stand structure, light, and seedling abundance across a tropical moist forest chronosequence, Panama. Journal of Vegetation Science 11:201–212.

DeWalt, S. J., Maliakal, S. K., and Denslow, J. S. 2003. Changes in vegetation structure and composition along a tropical forest chronosequence: implications for wildlife. Forest Ecology and Management 182:139–151.

Didham, R. K., and Lawton, J. H. 1999. Edge structure determines the magnitude of changes in microclimate and vegetation structure in tropical forest fragments. Biotropica 31:17–30.

Duncan, R. S., and Chapman, C. 1999. Seed dispersal and potential forest succession in abandoned agriculture in tropical Africa. Ecological Applications 9:998–1008.

Duncan, R. S., and Duncan, V. E. 2000. Forest succession and distance from forest edge in an afro-tropical grassland. Biotropica 32:33–41.

FAO (Food and Agriculture Organization). 2001. The State of the World's Forests 2001. Rome, Italy: FAO.

Finegan, B. 1984. Forest succession. Nature 312:109–114.

Finegan, B. 1996. Pattern and process in neotropical secondary rain forests: the first 100 years of succession. Tree 11:119–124.

Finegan, B., and Delgado, D. 2000. Structural and floristic heterogeneity in a 30-year-old Costa Rican rain forest restored on pasture through natural secondary succession. Restoration Ecology 8(4):380–393.

Gonzales-Montagut, R. 1996. Establishment of three rainforest species along the riparian corridor-pasture gradient in Los Tuxtlas, Mexico. Dissertation, Harvard University, Cambridge, Massachusetts.

Gorchov, D. L., Cornejo, F., Ascorra, C., and Jaramillo, M. 1993. The role of seed dispersal in the natural regeneration of rain forest after strip-cutting in the Peruvian Amazon. Vegetatio 107/108:339–349.

Grau, H. R., Arturi, M. F., Brown, A. D., and Aceñolaza, P. G. 1997. Floristic and structural patterns along a chronosequence of secondary forest succession in Argentinean subtropical montane forests. Forest Ecology and Management 95:161–171.

Guariguata, M. R., and Ostertag, R. 2001. Neotropical secondary forest succession: changes in structural and functional characteristics. Forest Ecology and Management 148:185–206.

Guariguata, M. R., Chazdon, R. L., Denslow, J. S., Dupuy, J. M., and Anderson, L. 1997. Structure and floristics of secondary and old-growth forest stands in lowland Costa Rica. Plant Ecology 132:107–120.

Guevara, S., and Laborde, J. 1993. Monitoring seed dispersal at isolated standing trees in tropical pastures: consequences for local species availability. Vegetatio 107/108:319–338.

Guevara, S., Meave, J., Moreno-Casasola, P., and Laborde, J. 1992. Floristic composition and structure of vegetation under isolated trees in Neotropical pastures. Journal of Vegetation Science 3:655–664.

Guevara, S., Purata, S. E., and Van der Maarel, E. 1986. The role of remnant forest trees in tropical secondary succession. Vegetatio 66:77–84.

Harms. K. E., Wright, S. J., and Calderon, O. 2000. Pervasive density-dependent recruitment enhances seedling diversity in a tropical forest. Nature 404:493–495.

Healey, S. P., and Gara, R. I. 2003. The effect of teak (*Tectona grandis*) plantation on the establishment of native species in an abandoned pasture in Costa Rico. Forest Ecology and Management 176:497–507.

Holl, K. D. 1999. Factors limiting tropical forest regeneration in abandoned pasture: seed rain, seed germination, microclimate and soil. Biotropica 31:229–242.

Holl, K. D. 2002. Effect of shrubs on tree seedling establishment in an abandoned tropical pasture. Journal of Ecology 90:179–187.

Holl, K. D., Loik, M. E., Lin, H. V., and Samuels, I. A. 2000. Tropical montane forest restoration in Costa Rica: overcoming barriers to dispersal and establishment. Restoration Ecology 8:339–349.

Hooper, E., Condit, R., and Legendre, P. 2002. Responses of 20 native tree species to reforestation strategies for abandoned farmland in Panama. Ecological Applications 12:1626–1641.

Hooper, E. R., Legendre, P., and Condit, R. 2004. Factors affecting community composition of forest regeneration in deforested, abandoned and in Panama. Ecology 85(12):3313–3326.

Hooper, E. R., Legendre, P., and Condit, R. 2005. Barriers to forest regeneration of deforested and abandoned land in Panama. Journal of Applied Ecology 42:1165–1174.

Howe, H. F., and Smallwood, J. 1982. Ecology of seed dispersal. Annual Review of Ecology and Systematics 13:201–228.

Howorth, R. T., and Pendry, C. A. 2006. Post-cultivation secondary succession in a Venezuelan lower montane rain forest. Biodiversity and Conservation 15:693–715.

Hubbell S. P., Foster, R. B., O'Brien, S. T., Harms, K. E., Condit, R., Wechsler, B., Wright, S. J., de Lao, S. L. 1999. Light-gap disturbances, recruitment limitation, and tree diversity in a neotropical forest. Science 283:554–557.

Janzen, D. H. 1988. Management of habitat fragments in a tropical dry forest: growth. Annals of the Missouri Botanical Garden 75:105–116.

Kenoyer, L. A. 1929. General and successional ecology of the lower tropical rain-forest at Barro Colorado Island, Panama. Ecology 12:201–222.

Knight, D. A. 1975. A phytosociological analysis of species-rich tropical forest on Barro Colorado Island, Panama. Ecological Monographs 45:259–284.

Laska, M. S. 1997. Structure of understory shrub assemblages in adjacent secondary and old growth tropical wet forests, Costa Rica. Biotropica 29(1):29–37.

Legendre, P., and Anderson, M. J. 1999. Distance-based redundancy analysis: testing multispecies responses in multifactorial ecological experiments. Ecological Monographs 69:1–24.

Legendre, P., Galzin, R., and Harmelin-Vivien, M. L. 1997. Relating behavior to habitat: solutions to the fourth-corner problem. Ecology 78:547–562.

Lugo, A. E., and Helmer, E. 2004. Emerging forests on abandoned land: Puerto Rico's new forests. Forest Ecology and Management 190:145–161.

Maluf de Souza, F., and Batista, J. L. F. 2004. Restoration of seasonal semideciduous forests in Brazil: influence of age and restoration design on forest structure. Forest Ecology and Management 191:185–200.

McCook, L. J. 1994. Understanding ecological community succession: Causal models and theories, a review. Vegetatio 110:115–147.

Medellin, R. A., and Gaona, O. 1999. Seed dispersal by bats and birds in forest and disturbed habitats of Chiapas, Mexico. Biotropica 31:478–485.

Mesquita, R. C. G., Ickes, K., Ganade, G., and Williamson, G. B. 2001. Alternative successional pathways in the Amazon basin. Journal of Ecology 89:528–537.

Miller, P. M., and Kauffman, J. B. 1998. Seedling and sprout response to slash-and-burn agriculture in a tropical deciduous forest. Biotropica 30:538–546.

Myster, R. 2003a. Seed regeneration mechanisms over fine spatial scales on recovering coffee plantation and pasture in Puerto Rico. Plant Ecology 166:199–205.

Myster, R. 2003b. Effects of species, density, patch-type, and season on post-dispersal seed predation in a Puerto Rican pasture. Biotropica 35(4):542–546.

Myster, R. W. 2004. Regeneration filters in post-agricultural fields of Puerto Rico and Ecuador. Plant Ecology 172:199–209.

Myster, R. W., and Pickett, S. T. A. 1992. Effects of palatability and dispersal mode on spatial patterns of trees in oldfields. Bulletin of the Torrey Botanical Club 119(2):145–151.

Nepstad, D. 1989. Forest regrowth in abandoned pastures of eastern Amazonia: limitations to tree seedling survival and growth. Dissertation, Yale University, New Haven, Connecticut.

Nepstad, D., Uhl, C., and Cardosa da Silva, J. M. C. 1996. A comparative study of tree establishment in abandoned pasture and mature forest in eastern Amazonia. Oikos 76:25–39.

Nepstad, D., Uhl, C., and Serrao, E. A. 1990. Surmounting barriers to forest regeneration in abandoned pastures. In Alternatives to Deforestation: Steps Towards Sustainable Uses of Amazonian Forests, ed. A. Anderson, New York, NY: Columbia University Press.

Nepstad, D., Uhl, C., and Serrao, A. S. 1991. Recuperation of degraded Amazonian landscape: forest recovery and agricultural restoration. Ambio 20:248–255.

Otsamo, A., Adjers, G., Sasmito Hadi, T., Kuusipalo, J., Tuomela, K., and Vuokko, R. 1995. Effect of site preparation and initial fertilization on the establishment and growth of four plantation tree species used in reforestation of *Imperata cylindrica* (L.) Beuv. dominated grasslands. Forest Ecology and Management 73:271–277.

Pascarella, J. B., Aide, T. M., Serrano, M. I., and Zimmerman, J. K. 2000. Land use history and forest regeneration in the Cayey Mountains, Puerto Rico. Ecosystems 3:217–228.

Peña-Claros, M. 2003. Changes in structure and species composition during secondary forest succession in the Bolivian Amazon. Biotropica 35(4):450–461.

Peña-Claros, M., and H. De Boo. 2002. The effect of forest successional stage on seed removal of tropical rain forest tree species. Journal of Tropical Ecology 18:261–274.

Purata, S. E. 1986. Floristic and structural changes during old-field succession in the Mexican tropics in relation to site history and species availability. Journal of Tropical Ecology 2:257–276.

Reiners, W. A., Bouwman, A. F., Parsons, W. F. J., and Keller, M. 1994. Tropical rain forest conversion to pasture: Changes in vegetation and soil properties. Ecological Applications 4(2): 363–377.

Richards, P. W. 1964. The Tropical Rain Forest: An Ecological Study. Cambridge, UK: Cambridge University Press.

Rodrigues, R. R., Martins, S. V., and de Barros, L. C. 2004. Tropical rain forest regeneration in an area degraded by mining in Mato Grosso State, Brazil. Forest Ecology and Management 190:323–333.

Sampaio, E. V. S. B., Salcedo, I. H., and Kauffman, J. B. 1993. Effect of different fire severities on coppicing of caatinga vegetation in Serra Talhado, PE-Brazil. Biotropica 25:452–460.

Sheil, D. 1999. Developing tests of successional hypotheses with size-structured populations, and an assessment using long-term data from a Ugandan rain forest. Plant Ecology 140:117–127.

Slocum, M. G., and Horvitz, C. C. 2000. Seed arrival under different genera of trees in a neotropical pasture. Plant Ecology 149:51–62.

Thomlinson, J. R., Serrano, M. I., del M. Lopez, T., Aide, T. M., and Zimmerman, J. K. 1996. Land-use dynamics in a post-agricultural Puerto Rican landscape (1936–1988). Biotropica 28:525–536.

Toh, I., Gillespie, M., and Lamb, D. 1999. The role of isolated trees in facilitating tree seedlings recruitment at a degraded sub-tropical rainforest site. Restoration Ecology 7(3):288–297.

Uhl, C. 1987. Factors controlling succession following slash-and-burn agriculture in Amazonia. Journal of Ecology 75:377–407.

Uhl, C. 1998. Propensity of fire in the humid tropics. Conservation Biology 12:942–943.

Uhl, C., Buschbacher, R., and Serrao, E. A. S. 1988. Abandoned pastures in eastern Amazonia. I Patterns of plant succession. Journal of Ecology 76:663–681.

Uhl, C., Nepstad, D., Buschbacher, R., Clark, K., Kauffman, B., and Subler, S. 1990. Studies of ecosystem response to natural and anthropogenic disturbance provide guidelines for designing sustainable land-use systems in Amazonia. In Alternatives to Deforestation: Steps Toward Sustainable Use of the Amazon Rain Forest, ed. A. Anderson, New York, NY: Columbia University Press.

Vieira, I. G., Uhl, C., and Nepstad, D. 1994. The role of the shrub *Cordia multispicata* Cham. as a 'succession facilitator' in an abandoned pasture, Paragominas, Amazonia. Vegetatio 115:91–99.

Wijdeven, S. M. J., and Kuzee, M. E. 2000. Seed availability as a limiting factor in forest recovery processes in Costa Rica. Restoration Ecology 8:414–424.

Wills, C., Harms, K. E., Condit, R., King, D., Thompson, J., He, F., Muller-Landau, H. C., Ashton, P., Losos, E., Comita, L., Hubbell, S., LaFrankie, J., Bunyavejchewin, S., Dattaraja, H. S., Davies, S., Esufali, S., Foster, R., Gunatilleke, N., Gunatilleke, S., Hall, P., Itoh, A., John, R., Kiratiprayoon, S., Loo de Lao, S., Massa, M., Nath, C., Noor, M. N. S., Kassim, A. R., Sukumar, R., Suresh, H. S., Fang Sun, I., Tan, S., Yamakura, T., and Zimmerman, J. 2006. Nonrandom Processes Maintain Diversity in Tropical Forests. Science 311:527–531.

Willson, M. F., and Crome, F. H. J. 1989. Patterns of seed rain at the edge of a tropical Queensland rain forest. Journal of Tropical Ecology 5:301–308.

Yarranton, G. A., and Morrison, R. G. 1974. Spatial dynamics of a primary succession: nucleation. Journal of Ecology 62(2):417–428.

Yih, K., Boucher, D. H., Vandermeer, J. H., and Zamora, N. 1991. Recovery of the rain forest of southeastern Nicaragua after destruction by Hurricane Joan. Biotropica 23:106–113.

Zahawi, R. A., and Augspurger, C. K. 1999. Early plant succession in abandoned pastures in Ecuador. Biotropica 31:540–552.

Zimmerman, J. K., Pascarella, J. B., and Aide, T. M. 2000. Barriers to forest regeneration in an abandoned pasture in Puerto Rico. Restoration Ecology 8:350–360.

Chapter 7
Seed Dispersal by Cattle: Natural History and Applications to Neotropical Forest Restoration and Agroforestry

Clara Luz Miceli-Méndez, Bruce G. Ferguson and Neptalí Ramírez-Marcial

7.1 Introduction

Agricultural expansion in Latin America is intimately associated with tropical deforestation and biodiversity loss (e.g. Angelsen and Kaimowitz 2001). Although agricultural activities are frequently abandoned (Aide and Grau 2004), their effects on forest regeneration processes may be much more persistent (e.g. Ferguson et al. 2003). On land subjected to prolonged and/or heavy use, potential for forest recovery is limited by the dearth of tree seedlings, resprouts and seed banks (Uhl et al. 1990). At such sites, seed dispersal from off-site is the principal colonization route for woody plants. However, distance to seed sources, lack of seed disperser activity, seed predation, herbivory, poor soils, high light intensity, frequent fires and competition with herbaceous ground cover impede forest regeneration from seed rain (Holl 1998b, 1999, Meli 2003).

Among the most common of these barriers is the lack of animals capable of transporting seed between wooded and open areas (Holl 1999, Guariguata and Ostertag 2002, Meli 2003). Animals disperse as many as 66% of tree species in Neotropical forests, principally by swallowing and defecating their seed (Howe and Smallwood 1982, Wunderle Jr. 1997). The major dispersers are birds and bats, although primates, rodents and other mammals may play a complementary part. Many wild seed dispersers, particularly forest birds, do not cross open areas (Wunderle Jr. 1997). Frugivorous birds that do enter open areas generally limit their activities to patches of woody vegetation within the agricultural landscape, and their movements depend upon plant phenology, particularly fruiting period. The result is a scanty seed rain that is heterogeneous in both space and time (da Silva et al. 1996, Wunderle Jr. 1997).

Restoration ecologists have proposed techniques to overcome or bypass this barrier to tropical forest regeneration. Such techniques include the manual collection and planting of tree seeds and/or seedlings and the artificial establishment of structural diversity to attract dispersers to open areas (McClanahan and Wolfe 1993, Holl 1998a, Holl and Kappelle 1999, Parrotta and Knowles 1999). While these efforts have met with some success, they can be costly to implement at large scale (Lamb et al. 2005).

R. W. Myster, *Post-Agricultural Succession in the Neotropics.*

Another option for reestablishing tree cover over large areas arises from the observations of Janzen and Martin (1982), who argued that numerous Central American trees are adapted for dispersal by large herbivores. When much of the Pleistocene megafauna went extinct around 10,000 years ago, they maintain, numerous tree species were left without efficient dispersers. Since colonial times, seeds of many such trees have been dispersed by introduced livestock, particularly cattle and horses.

Under the right circumstances, cattle, a salient feature of many Neotropical landscapes, are indeed important seed dispersal agents. Where management permits, cattle cross easily between secondary forests, younger fallows, and pastures (Hernández and Benavides 1995), and may also be rotated through fields to graze on crop residues (Jiménez Ferrer et al. 2003). They eat the fruit of a wide variety of trees and shrubs (Moreno Casasola 1996), and their digestive processes are mild enough to leave many seeds intact (Razanamandranto et al. 2004, Pascacio Damián 2006). Furthermore, light grazing can diminish the competitive effect of grasses on woody colonizers (Posada et al. 2000) another important barrier to tree establishment (Holl et al. 2000, Posada et al. 2000).

Although many instances of seed dispersal by cattle (hereafter referred to as "bovinochory") are known, systematic studies of the relationship are lacking. We review ecological and agronomic literature to typify bovinochorous trees and shrubs and describe the methods used to study the phenomenon. We describe fruits, seeds and habitat of the species involved, quantify reports of bovinochory by plant family and summarize the authors' conclusions with regard to the impact of bovinochory. Next, taking the ecologically diverse state of Chiapas, Mexico as a case study, we analyze the management potential of bovinochory as a tool for reestablishing woody plant cover under both active cattle production and ecological restoration scenarios. We present preliminary interview and sampling data identifying woody bovinochores in Chiapas, then combine information from interviews and literature review to describe the distribution, physiological requirements, uses and aptitude as successional nuclei of these species. Finally, we propose a research agenda to support full realization of bovinochory's management potential.

7.2 Methods

7.2.1 Literature Review

Between September, 2004 and June, 2006, we consulted databases including Science Citation Index expanded (ISI web of science, 1951–2005), Google Scholar and the Livestock, Environment and Development Digital Library (lead.virtualcentre.org). Search terms included combinations of "seed dispersal," or "germination," and "livestock," "cattle," or "bovine," in both English and Spanish. We also searched for articles citing Janzen and Martin (1982). We carried out additional searches to develop descriptions of the characteristics of each species relevant to their dispersal, habitats, uses and potential function as successional nuclei.

7.2.2 Chiapas Field Studies

During 2005 and 2006 we conducted interviews on 60 ranches to identify bovinochorous, woody species in both the wet and dry lowlands of Chiapas. During both the wet and dry seasons, we surveyed dung pats on the same ranches for germinating seedlings. We also collected dung and sieved a portion for seeds while spreading the rest over soil-filled trays in the greenhouse to facilitate germination. This work is ongoing and the preliminary results are reported here as a case study of the management potential of bovinochory. Details of the methodology will be provided elsewhere (Miceli Méndez In prep.).

7.3 Results

7.3.1 Studies Identified and Their Methodologies

Our searches identified 38 articles that mention 24 bovinochorous, woody species (cited here by the scientific names utilized by the Missouri Botanical on mobot.mobot.org; Table 7.1). We classify these according to the methodology they employed to study bovinochory as experimental, descriptive and/or modeling. The experimental studies include germination trials of seed banks in pasture soil and dung (Gutiérrez and Armesto 1981, Janzen et al. 1985, Somarriba 1986, Peinetti et al. 1993, Campos and Ojeda 1997, Brown and Archer 1999, Doucette et al. 2001, Villagra et al. 2002, Kneuper et al. 2003, Razanamandranto et al. 2004) as well as trials of pregermination treatments (McCully 1951, Argaw et al. 1999, Rubio-Delgado et al. 2002). Other experimental trials compared the effectiveness of simulated bovine digestion with those of other scarification treatments (Janzen et al. 1985, Somarriba 1986). Studies described as observational/descriptive mainly documented foraging behaviour, presence of viable seed in dung and the distribution of seedlings. Their data come from both field observations and interviews. The modeling studies simulated the population dynamics of bovinochorous species under differing management scenarios (Radford et al. 2001, Tews et al. 2004).

7.3.2 Characteristics of Bovinochores

Sixteen of the 24 bovinochorous species identified belong to the Fabaceae family, two species are Rosaceae, and 6 other families, Boraginaceae, Caprifoliaceae, Myrtaceae, Rutaceae, Sterculiaceae, and Tiliaceae, are represented by a single species each (Fig. 7.1). Two of these families are in the same order (Malvales) but no other relationships are apparent at this taxonomic level. Of the Fabaceae, 14 species are in the Mimosoideae subfamily and two are Ceaesalpinoideae.

Table 7.1 Cattle-dispersed woody plants reported in the literature and further findings regarding bovinochory

Species	Type of study	Findings	Country	Study
Acacia caven (Mol.) Molina *Mimosoideae* Fabaceae	E	• Germination of seeds extracted from dung was 5 times that of seed extracted directly from pods.	Chile	(Gutiérrez and Armesto 1981)
Acacia dudgeoni Craib ex Hall *Mimosoideae* ***Acacia seyal*** Del. Mimosoideae ***Burkea africana*** *Hook. f.* *Caesalpinioideae* ***Prosopis africana*** (Guill., Perrott. & Rich.) Taub. *Mimosoideae* Fabaceae	E	• Due to the larger diameter of their digestive systems, cattle affect viability and germination of these four species less than sheep do. • The moist, acid environment of dung may accelerate germination. • Large herbivores may plan a major role in long-distance seed dispersal.	Burkina Faso	(Razanamandranto et al. 2004)
Acacia farnesiana (L.) Willd. *Mimosoideae* Fabaceae	D	• Colonization of pastures and other disturbed areas by these species results from bovinochory (a). • Cattle can disperse large amounts of viable seed (b)	USA México	a) (Parrotta 1992) b) (Pascacio Damián 2006)

Acacia nilotica L. Willd. ex Del. *Mimosoideae* Fabaceae	D M	• This is an invasive plant in Australian rangelands. • Its invasive power is a product of prolific seed output and efficient dispersal by cattle (80% of seeds defecated intact) • Only 1% of seed in sheep dung is intact. • Models predict patchy invasion related to microclimate and water sources for livestock.	Australia	(Radford et al. 2001)
Acacia pennatula Schldl. & Cham. Benth. *Mimosoideae* Fabaceae	D (a–c) E(d)	• The range of this species has increased due to human activity (d). • Cattle disperse its seeds (a–e). • Seedlings grow directly from cattle dung (b, c) • Germination is greatest in the shade (d). • The size distribution varies in response to canopy cover and disturbance intensity (d). • This species seems to facilitate colonization by other woody plants (d).	Mexico (a, d, e) Nicaragua (b) Costa Rica (c)	a) (Greenberg et al. 1997) b) (Nieto et al. 2001) c) (Ibrahim and Camargo 2001) d) (Rubio-Delgado 2001) e) (Pascacio Damián 2006)

(continued)

Table 7.1 (continued)

Species	Type of study	Findings	Country	Study
Acacia tortilis (Forsk.) *Mimosoideae* Fabaceae ***Dichrostachys cinerea*** *Mimosoideae* Fabaceae	D E	• A majority of seeds of both species, consumed in their pods by cattle, were still viable upon defecation (while seeds of *Acacia senegal*, *A. seyal*, and *Balanites aegyptiaca* were not). • Seed of all 5 species collected from the soil, under tree canopies, and from barns germinated in the laboratory. • The abundance and breadth of seed distribution of *A. tortilis* and *D. cinerea* is a result of bovinochory. • Seeds of both species were most numerous in barns.	Ethiopia	(Argaw et al. 1999)
Citrus limetta RissoRutaceae ***Cordia curassavica*** (Jacq.) Roem. & Schult. Boraginaceae	D	• Farmers report bovinochory. • Seedlings sprout from cattle dung.	Nicaragua	(BGF pers. obs.)
Dichrostachys cinerea (L.) Wight & Arn.Fabaceae (see also *Acacia tortilis*)	D (a, c) E (b)	• This treelet, introduced to Cuba from Madagascar, is considered an invasive pest although it has multiple domestic and industrial uses (a, c). • Cattle disperse this species (c,b) and are the principal cause of its invasion of Cuban pastures (a).	Cuba (c), Etiopía (a–b)	a) (Mabberley 1997) b) (Argaw et al. 1999) c) (Méndez Santos and Ramos Jalil 2004)

Enterolobium cyclocarpum (Jacq.) Griseb. Mimosoideae Fabaceae	D (a, c) E (b)	• Cattle and horses consume the seeds, favoring their germination from dung and soil (a, c). • *In vitro* simulations of bovine and equine digestion found that seeds defecated within a day of ingestion survive at rates of 11 and 9% respectively. Longer retention reduces viability (b).	Costa Rica (a–c)	a) (Janzen and Martin 1982) b) (Janzen et al. 1985) c) (Ibrahim and Camargo 2001)
Grewia flava Tiliaceae	M	• This is an invasive species in rangelands. • Cattle distribute the seed into open grassland. • Simulations predict that bovinochory promotes invasion.	South Africa	(Tews et al. 2004)
Guazuma ulmifolia Lam. Sterculiaceae	D (a–c)	• Cattle consume and disperse the seeds, favoring establishment on degraded lands (a, b). • Bovinochory favors germination (a). • This species was likely dispersed by extinct Pleistocene megafauna (a). • This species favors colonization by other woody plants (c).	Guatemala (b) Costa Rica (b, c)	a) (Janzen 1982b) b) (Ferguson 2001) c) (Ibrahim and Camargo 2001)

(continued)

Table 7.1 (continued)

Species	Type of study	Findings	Country	Study
Pithecellobium saman Benth *Mimosoideae* Fabaceae	D (a, b)	• The species is bovinochorous (a, b). • This is a multipurpose species suitable for silvopastoral systems (a).	Australia (a) Costa Rica (b)	a) (Durr 2001) b) (Ibrahim and Camargo 2001)
Prosopis caldenia Burkart *Mimosoideae* Fabaceae	D, E	• The fruit is valuable forage. • The gradual increase in this species' range is due to bovinochory. • The species can become a pest as a result of bovinochory. • Excreted seed exhibited lower viability and higher germinability than uningested seed.	Argentina	(Peinetti et al. 1993)
Proposis flexuosa DC *Mimosoideae* Fabaceae	E	• Bovinochory favors germination more than dispersal by other wild and domestic animals (a, b). • However, seeds obtained directly from the tree exhibit greater viability than those defecated by cattle (a). • Both cattle and Patagonian maras (*Dolichotis patagonum*) separate seeds from their endocarp, favoring germination (a, b).	Argentina	a) (Campos and Ojeda 1997) b) (Villagra et al. 2002)

Prosopis juliflora (Sw.) DC (synonym of ***Prosopis glandulosa*** Torr. var. Glandulosa) *Mimosoideae* Fabaceae	E (a, b) D (b, c)	• Cattle defecate viable seed (b, c). • However, seed consumed by sheep and goats are less viable (b). • Cattle may accelerate invasion of rangelands by this species (a). • Invasion is homogeneous and independent of ground cover (a).	USA (a, b) Ethiopia (c)	a) (Brown and Archer 1999) b) (Kneuper et al. 2003) c) (Shiferaw et al. 2004)
Psidium guajava L. Myrtaceae	D (a, b, d, e) E (c)	• Cattle disperse the seeds, favoring seedling establishment (d,e). • During peak fruiting, cattle may consume 49,500 seeds/day (a,b). • The number of seeds dispersed diminishes with increasing fruit size (a,b). • Simulated bovine digestion did not affect seed germination (c). • This species can become a pest in tropical rangelands as a result of bovinochory (d).	Costa Rica (a–d) Colombia (e)	a) (Somarriba 1985a) b) (Somarriba 1985b) c) (Somarriba 1986) d) (Somarriba 1995) e) (Esquivel Sheik and Calle Díaz 2002)

(continued)

Table 7.1 (continued)

Species	Type of study	Findings	Country	Study
Rosa bracteata Wendl. Rosaceae	D, E (a, b)	• Cattle defecate viable seed of this species (a, b). • Approximately 80% of seeds consumed were recovered between the second and third days following ingestion (a). • Eight % of defecated seeds were physically damaged (a). • Bovinochory favors germination (a). • Seeds that remain in dung maintain their viability (a). • The species is an invasive exotic in SE USA (a, b).	USA	a) (McCully 1951) b) (Douce et al.)
Rosa woodsii Lindl Rosaceae ***Symphoricarpos albus*** (L.) BlakeCaprifoliaceae	E	• Viability following excretion by cattle was 77.4 and 69.3% respectively. • Cattle could be managed to disseminate these species. • *R. woodsii* invades rangelands.	Canada	(Doucette et al. 2001)
Senna atomaria (L.) H.S.Irwin & BarnebyFabaceae		• Cattle can disperse large amounts of viable seed.		(Pascacio Damián 2006)

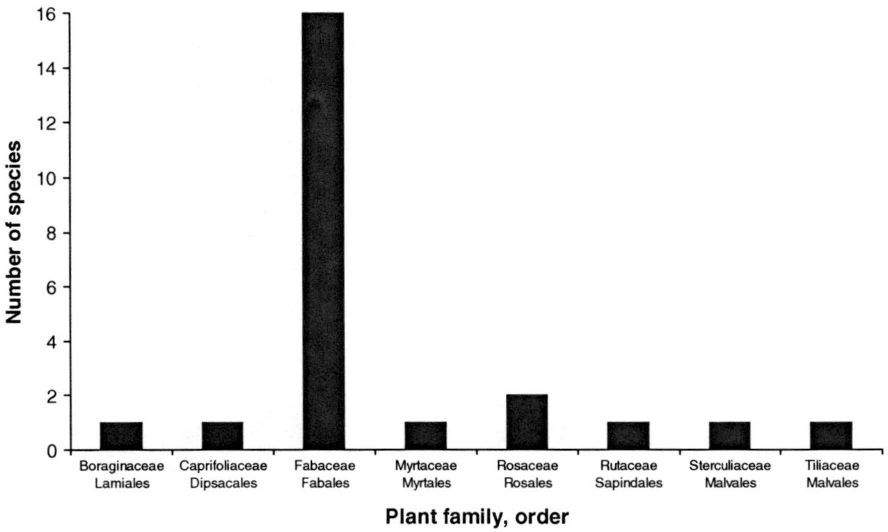

Fig. 7.1 Cattle-dispersed woody plants reported in the literature, by botanical family and order

The fruits of these species are variously dry and indehiscent (n=16), dry and dehiscent (n=1) or fleshy (n=7); drab (n=13), yellow (n=4), red-orange (n=4) or white (n=1); and may be fragrant or without odor. The mean fruit length for species for which data could be found is 10.3 ± 6.6 (standard deviation) cm. The typical bovinochorous fruit is a legume with dry, elongated, indehiscent, brown-black pods.

Seeds of bovinochorous are also highly variable. Their mean length is 7.0 ± 4.2 mm, with a range of 2–23 mm. Seed weight averages 83 ± 154 mg across species, ranging from 4 to 700 mg. Appendix 1 provides detailed descriptions of bovinochorous seeds and fruits.

7.3.3 Where does Bovinochory Occur?

The studies we identify report bovinochory of trees and shrubs in North, Central and South America, the Caribbean, Africa and Australia. They describe bovinochory in tropical and temperate savannas and, less frequently, in areas where the native vegetation is deciduous or semi-deciduous tropical dry forest, tropical rain forest or pine-oak forest (Table 7.1).

7.3.4 Effects of Bovinochory on Seed Fate

Cattle digestive processes often have minimal effect on seed viability and germination (Peinetti et al. 1993, Razanamandranto et al. 2004). However scarification in the bovine digestive tract may soften the seed coat, both diminishing the proportion of viable seed and accelerating germination of surviving seed (Janzen et al. 1985, Doucette et al. 2001). For several species (*Acacia dudgeoni, Acacia seyal, Rosa woodsii, Symphoricarpos albus, Burkea africana, Prosopis africana*), viability varies inversely with duration of retention in the digestive tract (Doucette et al. 2001, Razanamandranto et al. 2004). Small, smooth seeds tend to pass more rapidly and suffer less damage (Razanamandranto et al. 2004). Cattle digestion augments germination rate more than other physical and chemical treatments for some species, including *Enterolobium cyclocarpum, Psidium guajava, Rosa bracteata, Rosa woodsii*, and *Symphoricarpos albus* (McCully 1951, Janzen et al. 1985, Somarriba 1986, Doucette et al. 2001).

The molar mill may also play a role in scarification, as in the case of the *Acrocomia aculeata* (Jacq.) Lodd. ex Mart. In a series of scarification trials, this palm germinated only after being chewed and spit out by cattle (Orantes García 1999).

Cattle dung can be a propitious microsite for germination and establishment of some plants. Dung provides both an acid environment in which further scarification takes place, and abundant nutrients for seedling growth (Somarriba 1986, Razanamandranto et al. 2004). Species exhibiting increased establishment in dung microsites include *Acacia dudgeoni, Acacia seyal, Burkea africana, Prosopis africana, Prosopis flexuosa* (Villagra et al. 2002, Razanamandranto et al. 2004, Loth et al. 2005).

7.3.5 Bovinochory in Chiapas

The state of Chiapas, Mexico is an appealing context in which to explore the ecology and management of bovinochory for several reasons: (1) it affords a broad range of topographic and climatic conditions, resulting in a correspondingly broad variety of ecosystems (González-Espinosa et al. 2005); (2) these ecosystems host extraordinary biological diversity, including 9 of the 24 bovinochorous species reported in Table 7.1; (3) cattle ranching is of economic and social importance throughout the state, and is practiced under a variety of management scenarios. In this section, we report preliminary results of extensive surveys designed to identify further bovinochorous species in lowland Chiapas.

The ranchers surveyed identified a total of 17 woody species that germinate from cattle dung, including representatives of the Anacardiaceae, Burseraceae, Bignoniaceae, Cactaceae, Fabaceae, Malpighiaceae, Moraceae, Myrtaceae, Rutaceae, Sapotaceae, Sterculiaceae, Ulmaceae and Tiliaceae. Of these, we have thus far confirmed the presence of viable seed of 12 tree species and one shrub species in cattle dung. Eight of the trees have been previously reported as bovinochorous.

While our surveys focused on lowland Chiapas, our literature and database searches found that the 13 confirmed bovinochorous species are distributed across a broad range of ecosystems (Miranda 1998, Pennington and Sarukhán 1998). Most are not frost-resistant, but at least three species reach altitudes of 2,000 m or more.

All 13 grow in the dry lowlands and five are also present in the humid tropics. This suite of species also grows across a spectrum of soil types, from rich soils of volcanic origin to calcareous soils, to highly-weathered, acidic soils.

7.4 Discussion

Notably, most of the studies we identify in our literature review treat one or a few bovinochorous species, often as invasive plants that present a challenge to rangeland management (Brown and Acher 1989). Some of these cases present dramatic demonstrations of the influence of bovinochory on the composition and structure of plant communities. However, few studies (e.g. Argaw et al. 1999) attempt to identify the suite of bovinochorous plants in a given system and we found no systematic studies designed to identify bovinochorous species in the Neotropics. Nonetheless, our review and findings from Chiapas corroborate Janzen and Martin's (1982) generalizations regarding the importance of livestock as dispersal agents for neotropical trees and shrubs. We are able to draw some useful conclusions regarding characteristics of bovinochorous species, when and how bovinochory occurs, and how the phenomenon might be exploited for both enhancement of tropical rangelands and for catalyzing ecological restoration.

7.4.1 The Ecology of Bovinochory

As Janzen and Martin (1982) predicted, bovinochorous fruits are highly variable in size, color, morphology, odor, nutrient composition, seed size, etc., defying definition of a consistent bovinochorous dispersal syndrome (Howe 1985). Indeed, as Bruun and Poschlod (2006) note, cattle and other ungulates may be important vectors for many species without obvious dispersal adaptations. One generalization that does seem to hold true is that the seeds are hard and smooth, protected within a hard, smooth nut, or else very small (Janzen and Martin 1982).

While legumes with indehiscent pods are disproportionately favored by cattle, bovinochory manifests itself across numerous plant families (Fig. 7.1). Janzen and Martin (1982) mention species from 14 plant families in Pacific lowland, deciduous forest of Costa Rica that are dispersed by livestock and/or may have been dispersed by extinct megafauna. Eleven of these families (Anacardiaceae, Arecaceae, Bignoniaceae, Bromeliaceae, Fabaceae, Malpighiaceae, Moraceae, Myrtaceae, Sapotaceae, Sterculiaceae, Tiliaceae) along with three additional families (Burseraceae, Cactaceae, Rutaceae) are represented among the species we have thus far identified in Chiapas as bovinochorous.

Both the literature and our preliminary data suggest that bovinochory is more prevalent in dry forest areas than in the humid tropics. This observation is consistent with the hypothesis that bovinochory is a characteristic inherited from coevolution with large, extinct herbivores that maintained, savanna-like vegetation (Janzen and Martin 1982). Such plants must have been able to survive in the relatively dry microclimate of these open areas and could be expected to be most abundant in today's dry forests.

Bovinochory has been reasonably well documented in the range management context (e.g. McCully 1951, Brown and Acher 1989, Cox et al. 1993, Auman et al. 1998, Doucette et al. 2001, Bruun and Fritzboger 2002, Edwards and Younger 2006). However, its influence upon successional patterns in abandoned neotropical pastures seems to have been largely disregarded by ecologists since Janzen and Martin (1982), perhaps because of the more obvious association between cattle and deforestation (Hecht 1993).

Intriguingly, we and the ranchers we have interviewed have observed latency in seeds of cattle-dispersed taxa, including *Acacia*, spp. and *Guazuma ulmifolia*, species that are often important early-successional colonizers (e.g. Ferguson et al. 2003). Thus these species may germinate not from post-abandonment seed rain, but from a seed bank that accumulates in pastures during years of use. Draft animals and livestock grazing on crop stubble may also contribute to seed banks and successional patterns in croplands.

Furthermore, the structural diversity and resources provided by bovinochores that establish in active pasture doubtless also play a role in post-abandonment succession. In open vegetation, structural diversity is vital to maintaining seed dispersal by wild animals, and isolated trees offer key microsites for establishment of other woody species (Guevara et al. 1986, Vieira et al. 1994, da Silva et al. 1996, Ferguson 2001). In the Central Valleys of Chiapas, bovinochorous trees including *Enterolobium cyclocarpum*, *Pithecelobium dulce*, *Guazuma ulmifolia*, *Acacia pennatula,* and *Acacia farnesiana* are salient features of the landscape. These trees can offer important habitat for birds (e.g. Greenberg et al. 1997), contributing to successional dynamics.

7.4.2 Management Potential of Bovinochory in the Neotropics

Strategies that contribute simultaneously to rural livelihoods and biological conservation are urgently needed throughout the Neotropics (Sánchez et al. 2000, Angelsen and Kaimowitz 2001). Our findings with respect to the diversity of Neotropical bovinochorous trees and shrubs, the range of ecosystems in which they grow and the apparent importance of bovinochory in structuring vegetation in grazing land in the dry tropics all highlight the potential for management of this relationship. Establishment of trees via bovinochory holds particular promise for both silvopastoral systems and ecological restoration. This strategy could be used to plant trees at low cost by exploiting relationships and resources already familiar to rural people.

7.4.2.1 Silvopastoral Systems

Silvopastoral systems rely on trees, particularly multiple-use trees to improve the productivity of grazing land (Sánchez 1998). Our searches found multiple uses for all of the bovinochorous trees and shrubs we report from Chiapas. In addition to forage, these plants offer wood for diverse applications, extracts for tanning, shade for coffee or cattle, medicine, perfume, soil retention, wind breaks and live fences among other products and services (Miranda 1998).

Availability of quality forage in the dry season is a major determinant of the productivity of Neotropical rangelands (Botero and Russo 1998). Most bovinochorous species fruit during the dry season when grasses are less abundant and cattle seek alternative forage. Their fruits and seeds generally possess high concentrations of lipids, proteins and soluble carbohydrates that both attract and nourish livestock (Chargoy Zamora 1988, Razanamandranto et al. 2004). We have identified several ranchers in the Central Valleys who use trees dispersed by their cattle as key dry season forage. One of these ranchers relies on bovinochory to establish *Acacia pennatula*, thinning the trees as necessary to maintain a favorable balance between trees and grass. Two more ranchers use cattle-dispersed *Guazuma ulmifolia,* and one of them spread *G. ulmifolia* seed across his ranch intentionally by feeding its fruits to his cows. These ranchers report that within 3 months of germination, *G. ulmifolia* can withstand frequent browsing. In addition to high quality forage, *G. ulmifolia* provides firewood, wood, and medicine (CATIE 1991, Miranda 1998, Pennington and Sarukhán 1998). Other bovinochorous species that may represent considerable forage resources in Chiapas include *Acacia farnesiana, Enterolobium cyclocarpum, Manilkara achras, Nopalea dejecta, Parmentiera aculeata, Pithecellobium* spp.*, Prosopis juliflora, Psidium guajava, Senna atomaria y Senna spectabilis.*

7.4.2.2 Restoration

Where ecological restoration is the primary management objective, bovinochory may catalyze successional processes by increasing woody plant seed rain in open areas and diversifying microsites for establishment. Bovinochorous colonizers help re-establish the vegetation structure and food resources that attract wild seed vectors, which in turn will diversify the seed rain. Seeds deposited by wild dispersers will find improved microsites for germination and recruitment under bovinochorous plants (e.g. Ferguson 2001). Thus two of the most frequent barriers to forest regeneration in tropical pastures, lack of seed rain and of safe sites for establishment of woody plants (Holl et al. 2000), may be overcome by bovinochory.

Several bovinochores hold considerable potential as successional nuclei. All colonize open areas and most tolerate poor soils (CATIE 1991, Miranda 1998, Pennington and Sarukhán 1998). Many such early colonizers improve soil conditions or microclimate or attract seed-dispersing wildlife (Vieira et al. 1994, Meli 2003). Some, such as *A. pennatula*, *G. ulmifolia* and *P. dulce* are of particular importance for wildlife conservation (Greenberg et al. 1997, Leon-Cortes et al. 2004, B.G.F. pers ob.). However some bovinochorous species, including *Acacia* spp., *D. cinerea*, *Prosopis* sp. and *P. guajava* form dense thickets that could impede colonization by other species or restrict cattle movement and the productivity of rangelands (Peinetti et al. 1993, Somarriba 1995, Méndez Santos and Ramos Jalil 2004).

7.4.3 Research Priorities

Despite the success of some of the ranchers we interviewed at managing bovinochory, broad and efficient exploitation of this phenomenon will require deeper understanding of its ecology and natural history. Here we outline some pertinent areas of research.

The range of plant species susceptible to bovinochory has yet to be studied in most places. Interviews with ranchers, observations of germination in dung pats, controlled germination of seed from dung and simple experiments involving the addition of seeds or fruits to livestock diets may all quickly yield this kind of information.

Differences among livestock species and even breeds with respect to their capacity for seed dispersal should also be explored. We have focused on cattle as the most ubiquitous livestock in neotropical landscapes. However horses are also effective seed dispersers and they use different habitat and graze different species than do cattle (Cosyns et al. 2005). Sheep, while less likely than cattle to excrete whole seeds (Razanamandranto et al. 2004), are locally important in tropical highlands (Jiménez Ferrer et al. 2003), while tropical sheep are increasingly grazed in the lowlands (Nuncio-Ochoa et al. 2001).

The distribution and characteristics of various microsites arising from bovinochory influence seed and seedling fate. Patterns of dung deposition correspond to herbivores' habitat preferences and can be influenced by management (Wiegand et al. 1999, Cosyns et al. 2005). Dung pats can create short-lived gaps in dominant vegetation (Cosyns et al. 2006, Kohler et al. 2006), offer an acid environment that scarifies seed, and provide nutrients that favor seedling establishment (Razanamandranto et al. 2004). Dung beetles may act as secondary dispersers, diversifying the range of microsites encountered by bovinochores, but may also bury seed well below the soil surface where establishment is unlikely (Díaz Valdiviezo 2005). Microsite conditions arising from bovinochory and secondary dispersal are likely to favor some species in some environments, while filtering out others.

Changes in management of grazing lands bovinochory can alter the significance of bovinochory over a range of spatial scales. In northern Europe, for example, a reduction in the number of free-ranging livestock, increased fencing of pastures and less common long-distance seasonal movement of livestock have all contributed to a decline in seed dispersal in modern times (Bruun and Fritzboger 2002). Recent changes in Mexico's land reform law permitting division and sale of *ejido* land may similarly alter patterns bovinochory in that country. At the same time, enforcement of bans on burning seems to be favoring increased tree cover, particularly of bovinochores, in some ranching-dominated landscapes of the Central Valleys of Chiapas.

Grazing pressure likely interacts with bovinochory to determine germination and establishment rates. Intriguingly, because cattle generally prefer herbaceous forage, woody plant establishment can be greater at low livestock densities than in the absence of grazing where grasses are unchecked (Posada et al. 2000, Tobler et al. 2003). Thus cattle may simultaneously increase seed rain and lessen competition faced by establishing seedlings. However the form of the relationship between grazing pressure and woody plant establishment is little known as is the effect of timing of grazing in relation to germination. Further complicating this relationship, well-timed grazing may reduce fuel loads and thus the frequency and severity of fire during early succession (Fonseca and Beita In press).

For purposes of both restoration and silvopastoral systems, we need a better understanding of the interactions of bovinochorous shrubs and trees with other species. Which bovinochores permit or even favor grass growth beneath their crowns, and which grasses are most compatible with them (Hernández-Daumas 2000)? Conversely, which create shade levels or other microsite conditions that favor establishment of more woody plants (Vieira et al. 1994)? Which create habitat for wildlife species of interest (e.g. Greenberg et al. 1997), particularly for wild seed dispersers (Vieira et al. 1994)? Which species, under what conditions, are likely to form dense stands, and which maintain wider spacing that permits continued grazing and/or establishment of other species (Peinetti et al. 1993)? Does early dominance by bovinochores cause a "founder effect" that influences successional trajectory over the long term (Drake 1991)?

7.5 Conclusions

Cattle ranching is justifiably viewed as a major agent of neotropical deforestation (e.g. Hecht 1993). However under some conditions, seed dispersal and grazing by cattle can overcome major barriers to reestablishment of woody cover in open areas, catalyzing successional processes. Exploitation of this relationship holds promise as a cheap, efficient tool for both silvopastoral systems and ecological restoration. Further research will help realize the full potential of this strategy across a range of tropical ecosystems.

Acknowledgment We thank Mario Alberto López Miceli, Ivonne Anahí López Miceli, María Guadalupe Pascacio Damián, María Lesvia Dominguez Morales, Henry Castañeda Ocaña, Miguel Martínez Icó and Juan López Méndez for valuable support in the field, laboratory, greenhouse and herbarium. We are also grateful to Lorena Soto Pinto and René Pinto Ruíz for comments on an early draft. This research was supported by the Fondo sectorial CONACyT-SEMARNAP through the project "Uso sustentable de los Recursos Naturales en la Frontera sur de México", SEMARNAT-2002-C01-1109.

Appendix

Appendix 1 Characteristics of fruits and seeds of woody plants cited in the literature as cattle-dispersed

	Fruit			Seed				
Species and growth form	Type (Moreno 1984)	Color	Size Length × width × thickness (cm)	Shape	Mass (mg)	Size Length × width × thickness (mm)	Fruiting season	Source
Acacia caven (Mol.) Molina Shrub/Tree	Dry indehiscent, thick, oblong	Brown to black, shiny.	5 × 2	Compressed, smooth, ellispsoid	115.1	7.5	Dec.–April	(CIAT/FAO, Sánchez de Lorenzo-Cáceres, Piaggio 2003)
Acacia dudgeoni Craib ex Hall Tree/Shrub	Dry indehiscent, oblong, flattened, glabrous	Brown	5.5 × 2	NA	8.54	8.2 × 8.2 × 1.8	Dec.	(CIAT/FAO, Razanamandranto et al. 2004).
Acacia farnesiana (L.) Willd. Tree/Shrub	Dry indehiscent, sessile or on short stem, cylindrical-fusiform	Brown to black	5.5 × 1.3	Elliptical, reniform	83	6.6 × 4.7	Jan.–May	(PROCYMAF, Sánchez de Lorenzo-Cáceres, Standley et al. 1946–1975, Parrotta 1992, Vázquez-Yanes et al. 1999, Cordero and Boshier 2003, Pascacio Damián 2006)
Acacia nilotica L. Willd. ex Del. Tree	Dry indehiscent, woody	Brown to black	12.5 × 1.8	Ellipsoid, smooth, compressed	95.2	6.5 × 4.8	Mar.–June	(NewCROP, Sánchez de Lorenzo-Cáceres, Winrock International, Radford et al. 2001)

Acacia pennatula Schldl. & Cham. Benth. Tree	Dry indehiscent, woody, sessile, turgid	Brown	12.5 × 2.1 × 0.8	Globose, elliptical, hard, smooth	60.60	7 × 4.3 × 3.5	Nov.–Jan.	(Standley et al. 1946–1975, Missouri Botanical Garden, Martínez 1992, Miranda 1998, Rubio-Delgado 2001, Pascacio Damián 2006)
Acacia tortilis (Forsk.). Tree/Shrub	Dry indehiscent, glabrous or pubescent	Brown	10 × 0.7	NA	33.33	5.5	Oct.–Dec.	(FAO, Sánchez de Lorenzo-Cáceres, Winrock International, Duke 1983, Argaw et al. 1999, Loth et al. 2005)
Acacia seyal Del. Tree	Dry dehiscente, falcada, glabra	Brown	13.5 (length) × 0.7(thick-ness)	Elliptical, compressed, wrinkled, funiculate, elongate	3.83	6.6 × 4.2 × 3.1	April	(Sánchez de Lorenzo-Cáceres, Razanamandranto et al. 2004)
Burkea africana Hook. F.Tree	Dry indehiscent, pubescent, elliptical	NA	6 × 2.5	NA	10	8.1 × 6.1 × 2.5	Jan.	(CIAT/FAO, Razanamandranto et al. 2004)
Brosimum alicastrum *Sw*Tree	Globose with fleshy pericarp	Yellowish green, orangish red	2.3	Spherical with flattened tips, papery testa	1750	11.8 × 15 × 15	Mar.–May	(Pennington and Sarukhán 1998) Orantes et al. 2007.
Citrus limetta Risso. Tree	Fleshy, exocarp thin and smooth	Pale yellow	5.8 × 6.1	Ovoid	170	11.6 × 5.8 × 4.4	Apr.–Jun.	(Miranda 1998, Miceli Méndez In prep.)
Cordia curassavica Shrub	Fleshy	Red	3.7 × 4.2 × 5.9	Ovoid	NA	3.7 × 4.2 × 5.9	Apr.–Nov.	(Missouri Botanical Garden)

(continued)

Appendix 1 (continued)

Species and growth form	Fruit: Type (Moreno 1984)	Fruit: Color	Fruit: Size Length × width × thickness (cm)	Seed: Shape	Seed: Mass(mg)	Seed: Size Length × width × thickness (mm)	Fruiting season	Source
Dichrostachys cinerea (L.) Wight & Arn. Shrub/Tree	Dry indehiscent, glabrous, oblong	Brown	6 × 1.1	Biconvex, obovate, elliptical, subcircular	29.41	4 × 3	Aug.	(FAO Feed Resources Group, ISSG)
Enterolobium cyclocarpum (Jacq.) Griseb. Tree	Dry indehiscent, woody, curved	Brown	5.5 × 7.1 × 1.2	Ovoid, flattened and hard	700	23 × 15	Mar.–May	(Standley et al. 1946–1975, Janzen 1982a, Janzen et al. 1985, Miranda 1998, Pennington and Sarukhán 1998)
Grewia flava Shrub	Fleshy	NA	NA	NA	NA	NA	End of rainy season	(Tews et al. 2004)
Guazuma ulmifolia Lam. Tree	Dry indehiscent, woody, ellipsoid	Black with small protuberances, hard	2.2 × 1.8	Spheroid	6.1	1.8 (diameter)	Sep.– Apr.	(Missouri Botanical Garden, Janzen 1982b, Miranda 1998, Pennington and Sarukhán 1998, Villatoro 1998, Vázquez-Yanes et al. 1999)

Parmentiera aculeata (Kunth) Seemann/Tree	Fleshy, elongate with longitudinal furrows	Yellowish green	15 × 6.5	Thin, hard	NA	NA	Jan.–Dec.	(PROCYMAF, Miranda 1998, Pennington and Sarukhán 1998)
Pithecellobium saman Benth. Tree	Dry indehiscent, linear, compressed	Brown	17.5 × 1.2	Oblong	175.43	6.5	Feb.–May	(Standley et al. 1946–1975) (Sánchez de Lorenzo-Cáceres, Miranda 1998, Zamora et al. 2001)
Prosopis africana (Guill., Perrott. & Rich.) Taub. Tree	Dry indehiscent, cylindrical	Dark red	15 × 3	NA	15.71	8.7 × 5.8 × 4.0	Feb.–Mar.	(CIAT/FAO, Razanamandranto et al. 2004)
Prosopis caldenia Burkart. Tree	Dry indehiscent	Dark yellow with violet spots	15	Ovoid	24.09	6.1 × 3.7 × 2.2	Dec.–Mar.	(CIAT/FAO, Peinetti et al. 1993, Cosiansi et al. 2003)
Proposis flexuosa DC.Tree	Dry indehiscent, soft exocarp, tic mesocarp	Straw-colored with violet to black splotches	16.5 × 1.0	Ovoid	32.35	6.8 × 4.5 × 2.3	Dec.–Jan.	(CIAT/FAO, Campos and Ojeda 1997, Villagra et al. 2002, Cosiansi et al. 2003)
Prosopis juliflora (Sw.) DC Tree	Dry indehiscent, elongate, compressed, glabrous	Yellow to violet	18.5 × 1.3	Round to oval	36.36	7.5 × 5.3 × 3	Jan.–May	(CIAT/FAO, Martínez 1992, Moreno Casasola 1996, Miranda 1998, Shiferaw et al. 2004)

(continued)

Appendix 1 (continued)

	Fruit			Seed				
Species and growth form	Type (Moreno 1984)	Color	Size Length × width × thickness (cm)	Shape	Mass(mg)	Size Length × width × thickness (mm)	Fruiting season	Source
Psidium guajava L.Tree	Fleshy, globose to ovoid	Yellowish cream to pink, fragrant, sweet-sour	8 (diameter)	Round to triangular	NA	4	Jul.–Feb.	(Somarriba 1985b, 1986, Moreno Casasola 1996, Pennington and Sarukhán 1998)
Rosa bracteata Wendl. Shrub	Fleshy, globose	Orange to red	NA	NA	NA	NA	NA	(Douce, Moorhead and Bargeron)
Rosa woodsii Lindl Shrub	Fleshy, globose to elipsoid	Red	0.9 (diameter)	Oval, hard	13.6	4.5 × 2.5	Aug.	(Johnson and Hoagland, Doucette et al. 2001)
Senna atomaria (L.) H. S. Irwin & BarnebyTree	Dry indehiscent, linear-cylindrical	Dark brown	28.5 × 1	Hard, smooth	36.36	5.2 × 2.8	Feb.–Jun.	(Palma and Román 2000, Cordero and Boshier 2003, Pascacio Damián 2006)
Symphoricarpus albus (L.) Blake Shrub	Fleshy, globose	White	NA	NA	4.6	3 × 1.2	NA	(USDA, Doucette et al. 2001)

References

Aide, T. M., and Grau, H. R. 2004. Globalization, migration, and Latin American ecosystems. Science 305:1915–1916.

Angelsen, A., and Kaimowitz, D. Eds. 2001. Agricultural Technologies and Tropical Deforestation. Wallingford, UK: CABI.

Argaw, M., Teketay, D., and Olsson, M. 1999. Soil seed flora, germination and regeneration pattern of woody species in an Acacia woodland of the Rift Valley in Ethiopia. Journal of Arid Environments 43:411–435.

Auman, B. S., Call, C. A., and Wiedmeier, R. D. 1998. Crested wheatgrass establishment in livestock dung deposited on degraded rangeland vegetation types in the Intermountain West, USA. Arid Soil Research and Rehabilitation 12:317–333.

Botero, R., and Russo, R. O. 1998. Utilización de árboles y arbustos fijadoras de nitrógeno en sistemas sotenibles de producción animal en suelos ácidos tropicales. In Agroforestería para la producción animal en Latinoamérica (electronic conference), eds. M. Rosales M., and M. D. Sánchez, pp. 171–192. Rome: FAO. http://lead-es.virtualcentre.org/es/ele/conferencia1/Agrofor1.htm.

Brown, J. R., and Acher, S. 1989. Woody plant invasion of grasslands: establishment of honey mesquite (Prosopis glandulosa var. glandulosa) on sites differing in herbaceous biomass and grazing history. Oecologia 80:19–26.

Brown, J. R., and Archer, S. 1999. Shrub invasion of grassland: Recruitment is continuous and not regulated by herbaceous biomass or density. Ecology 80:2385–2396.

Bruun, H. H., and Fritzboger, B. 2002. The past impact of livestock husbandry on dispersal of plant seeds in the landscape of Denmark. Ambio 31:425–431.

Bruun, H. H., and Poschlod, P. 2006. Why are small seeds dispersed through animal guts: large numbers or seed size per se? Oikos 113:402–411.

Campos, C. M., and Ojeda, R. A. 1997. Dispersal and germination of Prosopis flexuosa (Fabaceae) seeds by desert mammals in Argentina. Journal of Arid Environments 35:707–714.

CATIE. 1991. Guácimo (Guazuma ulmifolia): Especie de árbol de uso múltiple en América Central. Turrialba, Costa Rica: Centro Agronómico Tropical de Investigación y Enseñanza.

Chargoy Zamora, C. I. 1988. Digestión Ruminal (Ecología y Evolución). Chapingo, Mexico: Universidad Autónoma Chapingo.

CIAT/FAO. Grassland species profiles.http://www.fao.org/ag/AGP/AGPC/doc/GBASE/. Accessed on February 2006.

Cordero, J., and Boshier, D. H. 2003. Árboles de Centroamérica: un Manual para Extensionistas. Oxford, UK, Costa Rica: Oxford Forestry Institute (OFI), Centro Agronómico Tropical de Investigación y Enseñanza (CATIE).

Cosiansi, J. F., Hayipanteli, S., Alvarez, V., Da Riva, D., and Milanesi, E. 2003. Efecto de las características del fruto de Prosopis caldenia "caldén" en la calidad y eficacia de trilla de las unidades con púas de flujo tangencial helicoidal. Agriscientia 20:45–52.

Cosyns, E., Bossuyt, B., Hoffmann, M., Vervaet, H., and Lens, L. 2006. Seedling establishment after endozoochory in disturbed and undisturbed grasslands. Basic And Applied Ecology 7:360–369.

Cosyns, E., Claerbout, S., Lamoot, I., and Hoffmann, M. 2005. Endozoochorous seed dispersal by cattle and horse in a spatially heterogeneous landscape. Plant Ecology 178:149–162.

Cox, J. R., Dealbaavila, A., Rice, R. W., and Cox, J. N. 1993. Biological and physical factors influencing Acacia constricta and Prosopis velutina establishment in the Sonoran Desert. Journal of Range Management 46:43–48.

da Silva, J. M. C., Uhl, C., and Murray, G. 1996. Plant succession, landscape management, and the ecology of frugivorous birds in abandoned Amazonian pastures. Conservation Biology 10:491–503.

Díaz Valdiviezo, J. L. 2005. Efecto de los Escarabajos Coprófagos (Coleoptera: Scarabaeidae) en el Destino de Semillas Presentes en Estiércol Vacuno. Licenciatura thesis. Universidad de Ciencias y Artes de Chiapas, Tuxtla Gutiérrez, Mexico.

Douce, G. K., Moorhead, D. J., and Bargeron, C. T. Invasive.org. www.invasive.org. Accessed on February 2006.
Doucette, K. M., Wittenberg, K. M., and McCaughey, W. P. 2001. Seed recovery and germination of reseeded species fed to cattle. Journal of Range Management 54:575–581.
Drake, J. A. 1991. Community-assembly mechanics and the structure of an experimental species ensemble. The American Naturalist 137:1–26.
Duke, J. A. 1983. Acacia tortilis (Forsk.). Hayne Handbook of Energy Crops.
Durr, P. A. 2001. The biology, ecology and agroforestry potential of the raintree, Samanea saman (Jacq.) Merr. Agroforestry Systems 51:223–237.
Edwards, A. R., and Younger, A. 2006. The dispersal of traditionally managed hay meadow plants via farmyard manure application. Seed Science Research 16:137–147.
Esquivel Sheik, M. J., and Calle Díaz, Z. 2002. Árboles aislados en potreros como catalizadores de la sucesión en la Cordillera Occidental Colombiana. Agroforestería en las Américas. CATIE 9(33–34):43–47.
FAO. Red Latinoamericana de Cooperación Técnica en Sistemas Agroforestales. http://www.fao.org/Regional/LAmerica/redes/sisag/arboles/.
FAO Feed Resources Group. Animal Feed Resources Information System. http://www.fao.org/AG/AGA/AGAP/FRG/AFRIS/default1.htm. Accessed on February 2006.
Ferguson, B. G. 2001. Seed dispersal by livestock accelerates succession in a degraded neotropical field. (Guatemala). Ecological Restoration 19:115–116.
Ferguson, B. G., Vandermeer, J., Morales, H., and Griffith, D. M. 2003. Post-agricultural succession in the Maya lowlands. Conservation Biology 17:818–828.
Fonseca, G. W., and Beita, A. M. In press. El bosque seco tropical en Costa Rica: caracterización ecológica y acciones para la restauración. In Restauración de Bosques en América Latina, ed. M. González-Espinosa, J. M. Rey-Benayas, and N. Ramírez-Marcial, México DF: Ediciones Mundi-Prensa y Fundación Internacional para la Restauración de Ecosistemas.
González-Espinosa, M., Ramírez-Marcial, N., Méndez-Dewar, G., Galindo-Jaimes, L., and Golicher, D. 2005. Riqueza de especies de árboles en Chiapas: variación espacial y dimensiones ambientales. In Diversidad Biológica en Chiapas, eds. M. González-Espinosa, N. Ramírez-Marcial, and L. Ruiz-Montoya, pp. 81–125. México, DF: Plaza y Valdés.
Greenberg, R., Bichier, P., and Sterling, J. 1997. Acacia, cattle and migratory birds in Southeastern México. Biological Conservation 80:235–247.
Guariguata, M. R., and Ostertag, R. 2002. Sucesión secundaria. In Ecología y Conservación de Bosques Neotropicales, ed. M. R. Guariguata and G. H. Kattan, pp. 591–623. Cartago, Costa Rica: Libro Universitario Regional.
Guevara, S., Purata, S. E., and Van der Maarel, E. 1986. The role of remnant forest trees in tropical secondary succession. Vegetatio 66:77–84.
Gutiérrez, J. R., and Armesto, J. J. 1981. El rol del ganado en la dispersión de las semillas de Acacia caven (Leguminosae). Ciencia e Investigación Agraria 8:3–8.
Hecht, S. 1993. The logic of livestock and deforestation in Amazonia. Bioscience 43:687–695.
Hernández-Daumas, S. 2000. Carbon and Nitrogen Balance in a Tree-Grass Inter-Cropping System in the Humid Tropics of Mexico. Doctoral thesis. University of Edinburgh, Edinburgh.
Hernández, S., and Benavides, J. 1995. Potencial forrajero de especies leñosas de los bosques secundarios de El Petén, Guatemala. Agroforestería en las Américas 2:15–22.
Holl, K. D. 1998a. Do bird perching structures elevate seed rain and seedling establishment in abandoned tropical pasture? Restoration Ecology 6:253–261.
Holl, K. D. 1998b. Effects of above- and below-ground competition of shrubs and grass on Calophyllum brasiliense (Camb.) seedling growth in abandoned tropical pasture. Forest Ecology and Management 109:187–195.
Holl, K. D. 1999. Factors limiting tropical rain forest regeneration in abandoned pasture: seed rain, seed germination, microclimate, and soil. Biotropica 31:229–242.
Holl, K. D., and Kappelle, M. 1999. Tropical forest recovery and restoration. Trends in Ecology and Evolution 14:378–379.
Holl, K. D., Loik, M. E., Lin, E. H. V., and Samuels, I. A. 2000. Tropical montane forest restoration in Costa Rica: overcoming barriers to dispersal and establishment. Restoration Ecology 8:339–349.

Howe, H. F., and Smallwood, J. 1982. Ecology of seed dispersal. Annual Review of Ecology and Systematics 13:201–228.
Howe, H. F. 1985. Gomphothere fruits: a Critique. American Naturalist 125:853–865.
Ibrahim, M., and Camargo, J. C. 2001. ¿Cómo aumentar la regeneración de árboles maderables en potreros? Agroforestería en las Américas. CATIE 8(32):35–41.
ICRAF. AgroForestryTree Database. http://www.worldagroforestrycentre.org/SEA/Products/AFDbases/AF/index.asp. Accessed on February 2006.
ISSG. Global Invasive Species Database. http://www.issg.org/database/welcome/. Accessed on February 2006.
Janzen, D. H. 1982a. Horse response to Enterolobium cyclocarpum (Leguminosae) fruit crop size in a Costa Rican deciduous forest pasture. Brenesia 19/20:209–219.
Janzen, D. H. 1982b. Natural history of guacimo fruits (Sterculiaceae, Guazuma umifolia) with respect to consumption by large mammals. American Journal of Botany 69:1240–1250.
Janzen, D. H., Demment, M. W., and Robertson, J. B. 1985. How fast and why do germinating guanacaste seeds (Enterolobium cyclocarpum) die inside cows and horses. Biotropica 17:322–325.
Janzen, D. H., and Martin, P. S. 1982. Neotropical anachronisms - the fruits the gomphotheres ate. Science 215:19–27.
Jiménez Ferrer, G., Soto Pinto, L., and Marquez Rosano, C. 2003. Ganadería en zona indígenas del sur de México: un reto para el desarrollo. In Sistemas Pecuarias Diversificados para el Alivio de la Pobreza Rural en América Latina (conferencia electrónica) FAO/CATIE. http://lead-es.virtualcentre.org/es/ele/conferencia4/programa.htm.
Johnson, F. L., and Hoagland, B. W. Catalog of the Woody Plants of Okalahoma. http://www.biosurvey.ou.edu/shrub/cover.htm. Accessed on February 2006.
Kneuper, C. L., Scott, C. B., and Pinchak, W. E. 2003. Consumption and dispersion of mesquite seeds by ruminants. Journal of Range Management 56:255–259.
Kohler, F., Gillet, F., Gobat, J. M., and Buttler, A. 2006. Effect of cattle activities on gap colonization in mountain pastures. Folia Geobotanica 41:289–304.
Lamb, D., Erskine, P. D., and Parrotta, J. A. 2005. Restoration of degraded tropical forest landscapes. Science 310:1628–1632.
Leon-Cortes, J., Perez-Espinoza, F., Marin, L., and Molina-Martinez, A. 2004. Complex habitat requirements and conservation needs of the only extant Baroniinae swallowtail butterfly. Animal Conservation 7:241–250.
Loth, P. E., de Boer, W. F., Heitkönig, I. M. A., and Prins, H. H. T. 2005. Germination strategy of the East African savanna tree Acacia tortilis. Journal of Tropical Ecology 21: 9509–517.
Mabberley, D. J. 1997. The Plant Book, 2nd edition. Cambridge: Reino Unido.
Martínez, C. B. 1992. Colecta de leguminosas silvestres con potencial forrajero. Licenciatura thesis. Universidad Autónoma de Chiapas, Villaflores, Chiapas, México.
McClanahan, T. R., and Wolfe, R. W. 1993. Accelerating forest succession in a fragmented landscape: the role of birds and perches. Conservation Biology 7:279–288.
McCully, W. G. 1951. Recovery and viability of Macartney rose seeds fed to cattle. Journal of Range Management 4:101–106.
Meli, P. 2003. Restauración ecológica de bosques tropicales. Veinte años de investigación académica. Interciencia 28:581–589.
Méndez Santos, I. E., and A. Ramos Jalil. 2004. El marabú: ¿plaga o recurso natural? http://www.cubasolar.cu/biblioteca/energia/Energia27/HTML/articulo03.htm.
Miceli Méndez, C. L. In prep. Efecto del Ganado Bovino y Ungulados Silvestres en la Dispersión de Semillas de Especies Leñosas en Agropaisajes. Doctoral thesis. El Colegio de la Frontera Sur, San Cristóbal de Las Casas, Chiapas.
Miranda, F. 1998. La Vegetación de Chiapas, 3rd edition. Tuxtla Gutiérrez, Mexico: Gobierno del Estado de Chiapas, CONECULTA.
Missouri Botanical Garden. W3TROPICOS. http://mobot.mobot.org/W3T/Search/vast.html. Accessed on February 2006.
Moreno Casasola, P. 1996. Vida y Obra de Granos y Semillas. México, D F: Fondo de Cultura Económica.

Moreno, N. P. 1984. Glosario Botánico. Xalapa, Veracruz, Mexico: Compañía Editorial Contienental, Instituto Nacional de Invesigación Sobre Recursos Naturales.

NewCROP. New Crops Resource Online Program. http://www.hort.purdue.edu/newcrop/. Accessed on February 2006.

Nieto, H., Somarriba, E., and M. Gómez. 2001. Contribución de Acacia pennatula (carbón) a la productividad agroforestal sostenible de la Reserva Natural Miraflor- Moropotente. Estelí, Nicaragua. Agroforestería en las Américas. CATIE 8(30):21–23.

Nuncio-Ochoa, G., Nahed-Toral, J., Díaz-Hernández, B., Escobedo-Amezcua, F., and Salvatierra-Izaba, B. 2001. Caracterización de los sistemas de producción ovina en el Estado de Tabasco. Agrociencia 35:469–477.

Orantes García, C. 1999. Importancia de la testa y el endocarpo en la latencia de Acrocomia mexicana Karwinsky ex martius (coyol). Licenciatura thesis. Universidad de Ciencias y Artes del Estado de Chiapas, Tuxtla Gutiérrez, Chiapas.

Orantes G., C., Méndez, C. L. M., Ramírez, E. R. G., Méndez, A. M. V., and Moreno, R. A. M. 2007. Generación de Metodologías para la Propagación Sexual en Vivero de Tres Especies Tropicales con Potencial de Uso Forestal, Forrajero y Alimenticio en Chiapas. UNICACH-Fundación Produce, Tuxtla Gutierrez, Mexico.

Palma, J. M., and Román, L. 2000. Frutos de especies arbøreas leguminosas y no leguminosas para la alimentación de rumiantes. In II Conferencia Electrønica sobre Agroforestería para la Producciøn Animal., eds. M. D. Sánchez and M. Rosales Méndez, Rome. FAO-CIPAV. http://www.virtualcentre.org/es/ele/conferencia2/vbconfe15.htm.

Parrotta, J. A. 1992. Acacia farnesiana (L.) Willd. Aroma, huisache, Leguminosae (Mimosoideae). Report SO-ITF-SM-49. New Orleans, USA: US Dept. of Agriculture Forest Service, Southern Forest Experiment Station.

Parrotta, J. A., and Knowles, O. H. 1999. Restoration of tropical moist forests on bauxite-mined lands in the Brazilian Amazon. Restoration Ecology 7:103–116.

Pascacio Damián, M. G. 2006. Influencia del paso por el tracto digestivo y defecación por ganado bovino sobre el destino de semillas de tres fabaceas de bosque tropical subcaducifolia, en el Ejido La Florida Municipio de Cintalapa, Chiapas. Licenciatura thesis. Universidad de Ciencias y Artes de Chiapas, Tuxtla Gutiérrez, Chiapas.

Peinetti, R., Pereyra, M., Kin, A., and Sosa, A. 1993. Effects of cattle ingestion on viability and germination rate of calden (Prosopis-caldenia) seeds. Journal of Range Management 46:483–486.

Pennington, T. D., and J. Sarukhán. 1998. Arboles tropicales de México, 2nd edition. México, DF: Universidad Nacional Autónoma de México/Fonda de Cultura Económica.

Piaggio, M. 2003. Plantas Nativas de Uruguay. http://micol.fcien.edu.uy/flora/uy_flora.htm.

Posada, J. M., Aide, T. M., and Cavelier, J. 2000. Cattle and weedy shrubs as restoration tools of tropical montane rainforest. Restoration Ecology 8:370–379.

PROCYMAF. Especies con Usos No Maderables en Bosques de Encino, Pino y Pino-Encino. http://www.semarnat.gob.mx/pfnm/indices.html. Accessed on February 2006.

Radford, I. J., Nicholas, D. M., Brown, J. R., and Kriticos, D. J. 2001. Paddock-scale patterns of seed production and dispersal in the invasive shrub Acacia nilotica (Mimosaceae) in northern Australian rangelands. Austral Ecology 26:338–348.

Razanamandranto, S., Tigabu, M., Neya, S., and Oden, P. 2004. Effects of gut treatment on recovery and germinability of bovine and ovine ingested seeds of four woody species from the Sudanian savanna in West Africa. Flora 199:389–397.

Rubio-Delgado, L. 2001. Distribución de Acacia pennatula (Schldl.& Cham.) Benth. en bosques perturbados del norte de Chiapas, México. MS thesis. ECOSUR, San Cristóbal de las Casas, Mexico.

Rubio-Delgado, L., Ramírez-Marcial, N., and Castellanos-Albores, J. 2002. Distribución y regeneración de Acacia pennatula (Schltdl. & Cham.) Benth. en bosques perturbados del norte de Chiapas, México. Brenesia 57–58:67–84.

Sánchez de Lorenzo-Cáceres, J. M. Árboles Ornamentales. http://www.arbolesornamentales.com/.

Sánchez, M. D. 1998. Sistemas agroforestales para intensificar de manera sostenible la producción animal en Latinoamérica tropical. In Agroforestería para la Producción Animal en Latinoamérica, ed. M. Rosales, E. Murgueitio, H. Osorio, M. D. Sánchez, and A. Speedy, pp. 1–12, Roma: FAO-CIPAV. http://lead.virtualcentre.org/es/ele/conferencia1/sanchez1.htm. Accessed on February 2006.

Sánchez, M. D., Rosales, M., and Murgueitio, E. 2000. Agroforestería pecuaria en América Latina. In Segunda Conferencia sobre Agroforestería para la Produccíon Animal en America Latina FAO-CIPAV.

Shiferaw, H., Teketay, D., Nemomissa, S., and Assefa, F. 2004. Some biological characteristics that foster the invasion of Prosopis juliflora (Sw.) DC at Middle Awash Rift Valley Area, northeastern Ethiopia. Journal of Arid Environments 58:135–154.

Somarriba, E. 1985a. Árboles de guayaba (Psidium guajava L.) en pastizales. I. Producción de fruta y potencial de dispersión de semillas. Turrialba 35(3):289–295.

Somarriba, E. 1985b. Arboles de guayaba (Psidium guajava L.) en pastizales. II. Consumo de fruta y dispersión de semillas. Turrialba 35:329–332.

Somarriba, E. 1986. Effects of livestock on seed germination of guava (Psidium guajava L.). Agroforestry Systems 4:233–238.

Somarriba, E. 1995. Guayaba en potreros: establecimiento de cercas vivas y recuperación de pasturas degradadas. Agroforestería en las Américas. CATIE. Año 2 No. 6:27–29.

Standley, P. C., Steyermark, J. A., Williams, L. O., and Gibson, D. N. 1946–1975. Flora of Guatemala (Fieldiana).

Tews, J., Schurr, F., and Jeltsch, F. 2004. Seed dispersal by cattle may cause shrub encroachment of Grewia lava on southern Kalahari rangelands. Applied Vegetation Science 7:89–102.

Tobler, M. W., Cochard, R., and Edwards, P. J. 2003. The impact of cattle ranching on large-scale vegetation patterns in a coastal savanna in Tanzania. Journal of Applied Ecology 40:430–444.

Uhl, C., Nepstad, D., Buschbacher, R., Clark, K., Kauffman, B., and Subler, S. 1990. Studies of ecosystem response to natural and anthropogenic disturbances provide guidelines for designing sustainable land-use systems in Amazonia. In Alternatives to Deforestation: Steps to Sustainable Use of the Amazon Rainforest, ed. A. Anderson, pp. 24–42. New York: Columbia University Press.

USDA. Plants Database. http://www.plants.usda.gov/. Accessed on February 2006.

Vázquez-Yanes, C., Batis Muñoz, A. I., Alcocer Silva, M. I., Gual Díaz, M., and Sánchez Dirzo, C. 1999. Árboles y arbustos potencialmente valiosos para la restauración ecológica y la reforestación. Report J084. México: CONABIO – Instituto de Ecología, UNAM.

Vieira, I. C. G., Uhl, C., and Nepstad, D. 1994. The role of the shrub Cordia multispicata Cham. as a 'succession facilitator' in an abandoned pasture, Paragominas, Amazônia. Vegetatio 115:91–99.

Villagra, P. E., Marone, L., and Cony, M. A. 2002. Mechanisms affecting the fate of Prosopis flexuosa (Fabaceae, Mimosoideae) seeds during early secondary dispersal in the Monte Desert, Argentina. Austral Ecology 27:416–421.

Villatoro, V. R. A. 1998. El Cuaulote (Guazuma ulmifolia Lam.) como Recurso Natural Antimicrobiano. Licenciatura thesis. Universidad de Ciencias y Artes de Chiapas, Tuxtla Gutiérrez, Mexico.

Wiegand, K., Jeltsch, F., and Ward, D. 1999. Analysis of the population dynamics of Acacia trees in the Negev desert, Israel with a spatially-explicit computer simulation model. Ecological Modelling 117:203–224.

Winrock International. http://www.winrock.org/. Accessed on February 2006.

Wunderle Jr., J. M. 1997. The role of animal seed dispersal in accelerating native forest regeneration on degraded tropical lands. Forest Ecology and Management 99:223–235.

Zamora, S., García, J., Bonilla, G., Aguilar, H., Harvey, C. A., and Ibrahim, M. 2001. ¿Cómo utilizar los frutos de guanacaste (Enterolobium cyclocarpum), guácimo (Guazuma ulmifolia), genízaro (Pithecellobium saman) y jícaro (Crescentia alata) en alimentación animal? Agroforestería en las Américas 8:45–49.

Chapter 8
The Roles of Disperser Behavior and Physical Habitat Structure in Regeneration of Post-Agricultural Fields

K. Greg Murray, Kathy Winnett-Murray, Jason Roberts, Katherine Horjus, William A. Haber, Willow Zuchowski, Mark Kuhlmann and Tammy M. Long-Robinson

8.1 Introduction

Since the 1980s, the plight of tropical forests and the factors that drive their conversion to human-dominated uses have received widespread attention in both the scientific and popular media. Worldwide, conversion to croplands and logging for timber and fuelwood have led the forces driving deforestation, but in Latin America creation of cattle pastures has been the major cause (Myers 1980). In Amazonia alone, more than 20 million hectares of forest had been converted to cattle pasture by the 1990s (Nepstad et al. 1996).

Despite their prevalence in many parts of the neotropics and their importance in local economies, productivity of cattle pastures often degrades rapidly through a combination of poor management, high pest pressures, and the loss of nutrients from what are often highly weathered soils (Serrão and Toledo 1990). The ultimate fate of most degraded pastures is abandonment, after which regeneration to forest via natural succession is possible. Forest regeneration on such abandoned pastures and other agricultural lands has received considerable attention (e.g., Uhl et al.1988 Nepstad et al. 1991, 1996, Buschbacher et al. 1992, Lugo 1992, Gerhardt 1993, Aide and Cavelier 1994, Brown and Lugo 1994, Kuusipalo et al. 1995, Sarmiento 1997a, Duncan and Chapman 1999, 2002 Holl 1999, Wijdeven and Kuzee 2000, Myster 2004b), much of it centered on the factors that limit the rate of regeneration and determine which forest species colonize. Soil degradation, competition with pasture grasses, seasonal drought, and herbivory on seeds and seedlings have been implicated as significant limiting factors in a number of studies (e.g., Uhl 1987, Holl et al. 2000, Zimmerman et al. 2000, Myster 2004a). Even more frequently, regeneration is thought to be limited by the availability of seeds dispersed into abandoned agricultural lands from the surrounding forest. Since most woody tropical species are adapted for dispersal by vertebrates (Howe and Smallwood 1982; Murray et al. 2000), many investigators have focused on the roles of seed dispersing animals and on those of remnant trees (i.e., those left standing when the pasture was created) as

 R. W. Myster, *Post-Agricultural Succession in the Neotropics.*

foci for the recruitment of forest plants (e.g., Guevara et al. 1986, 1992, Janzen 1988, Nepstad et al. 1990, Uhl et al 1990, Guevara and Laborde 1993, Vieira et al. 1994, Green 1995, Cardoso da Silva et al. 1996, Sarmiento 1997b, Wunderle 1997, Holl 1998, Harvey and Haber 1999, Otero-Arnaiz et al. 1999, Toh et al. 1999, Zahawi and Augspurger 1999, Harvey 2000, Harvey et al. 2000, 2007, Holl et al. 2000, Slocum 2001). Most studies have found that remnant trees or even artificial structures (e.g., Holl 1998) attract greater numbers of animal-dispersed seeds than adjacent sites without overhanging vegetation, presumably because fruit-eating animals are attracted to them. Similar effects of natural and artificial structures have been demonstrated at temperate latitudes as well (McDonnell and Stiles 1983, McClanahan and Wolfe 1993, Robinson and Handel 1993, 2000).

Given that secondary forests derived from former agricultural lands are likely to be increasingly important repositories of tropical biodiversity, understanding how recruitment foci operate will be a useful aid to understanding and directing the rate and trajectory of forest regeneration on such sites.

Despite the considerable number of recent studies on tropical pasture regeneration, almost none have followed the process over an extended period of time (but see Zahawi and Augspurger 1999, Aide et al. 2000, Finegan and Delgado 2000). In this chapter we address the role of physical habitat structure and seed dispersal in the regeneration process, specifically the role of remnant trees as recruitment foci. Using data from a montane site in northwestern Costa Rica, we first test the hypothesis that colonization of abandoned pastures is enhanced by the presence of remnant trees, and that the effect acts through influencing the behavior of fruit-eating animals rather than through shading effects. Second, we test the hypothesis that remnant trees that themselves produce fleshy fruits eaten by animals are more effective as recruitment foci for animal-dispersed plants than are trees that produce wind-dispersed seeds. Finally, we examine post-abandonment regeneration patterns over a period of 30 years to determine whether the effects of remnant trees on density and species composition in a regenerating forest persist over time.

8.2 Study Site and Methods

The Monteverde area (Provincia de Puntarenas) lies on a gently sloping plateau in the Cordillera de Tilarán in northwestern Costa Rica (10°18 ′N, 84° 48 ′W). Vegetation in the area is loosely termed tropical montane cloud forest, but includes life zones (Holdridge life zone classification system, Holdridge 1967) from tropical moist forest on the Pacific slope through lower montane rainforest along the continental divide, and down to tropical wet forest on the Caribbean slope. Climate and vegetation of the area are treated in detail by Clark et al. (2000) and Haber (2000), respectively.

Settled by Costa Rican farmers in the 1920s and 1930s, much of the area was substantially deforested for pasture and subsistence farming by the 1950s. Large-scale dairy farming began in 1951 after the arrival of North American settlers. Despite the continuing importance of dairy farming in the area, many pastures have been abandoned since the 1960s, and second-growth forests now occupy several areas that were active pastures as late as the 1950s and 1960s.

In order to examine the influence of remnant trees on patterns of pasture regeneration, we collected data on woody plant recruitment into two abandoned pastures on the Pacific slope, both of which were created by clearing of lower montane wet forest. Pasture A was 1.1 ha in area and was located at 1,525 meters elevation approximately 0.5 km west of the entrance to the Monteverde Cloud Forest Reserve (Fig. 8.1). It was surrounded on two sides by primary forest and on one side by approximately 40-year-old secondary forest. The remaining side was bordered by a dirt road and another pasture of similar age and structure. It was abandoned (cattle removed permanently) in 1984 after approximately 32 years of use. We censused woody recruits into the pasture in plots beneath remnant trees and in plots with no overhanging vegetation in 1987 and again in 1998 (hence, 3 years and 14 years post-abandonment).

Pasture B was located at approximately 1,500 meters elevation just within the Monteverde Cloud Forest Reserve, approximately 500 meters from pasture A. It was abandoned in approximately 1968 after approximately 20 years of use, and had a closed canopy dominated by the early-successional species *Conostegia oerstediana* (Melastomataceae) and *Hampea appendiculata* (Malvaceae) when it was censused in 1998 (hence, approximately 30 years post-abandonment). Including Pasture B in our design allowed us to determine whether the successional trajectories suggested by the first 14 years of regeneration are likely to persist over even longer periods of time. Although this design potentially confounds site effects with temporal ones, we feel that the proximity of the two pastures (0.5 km) and the similarity in species composition of forest surrounding them minimizes any site differences.

Census plots in pasture A centered on the boles of 77 remnant trees and 7 dead snags of canopy tree size, and on 35 sites without overhanging vegetation. Remnant trees and dead snags were those that had been left standing when the pasture was created, and the sites without overhanging vegetation ("open-sky" plots) were scattered systematically throughout the remainder of the pasture. In the 1987 census, sample plots were circular, with 2 m radii ($= 12.57\ m^2$). In 1998, we censused only one randomly chosen quadrant of each such plot ($= 3.14\ m^2$) due to higher overall recruit densities. Censuses consisted simply of counting and identifying all woody seedlings and saplings within the plot.

Pasture B was dominated by a closed canopy of even-aged early-successional species (as noted above) when we censused it in 1998, but we were able to locate and identify the trees that were remnants of the original forest at the time of abandonment as the only large late-successional species within the area. "Open sky" plots for this pasture were paired with each of the remnant trees sampled, and located 15 meters from the bole of the remnant tree in a randomly chosen compass direction. Plots were established and censused in this pasture in the same way as in pasture A in 1998.

Univariate statistical analyses reported here follow Sokal and Rohlf (1995) and were performed with SYSTAT 8.0.3 (Systat software, Inc., 1735 Technology Drive,

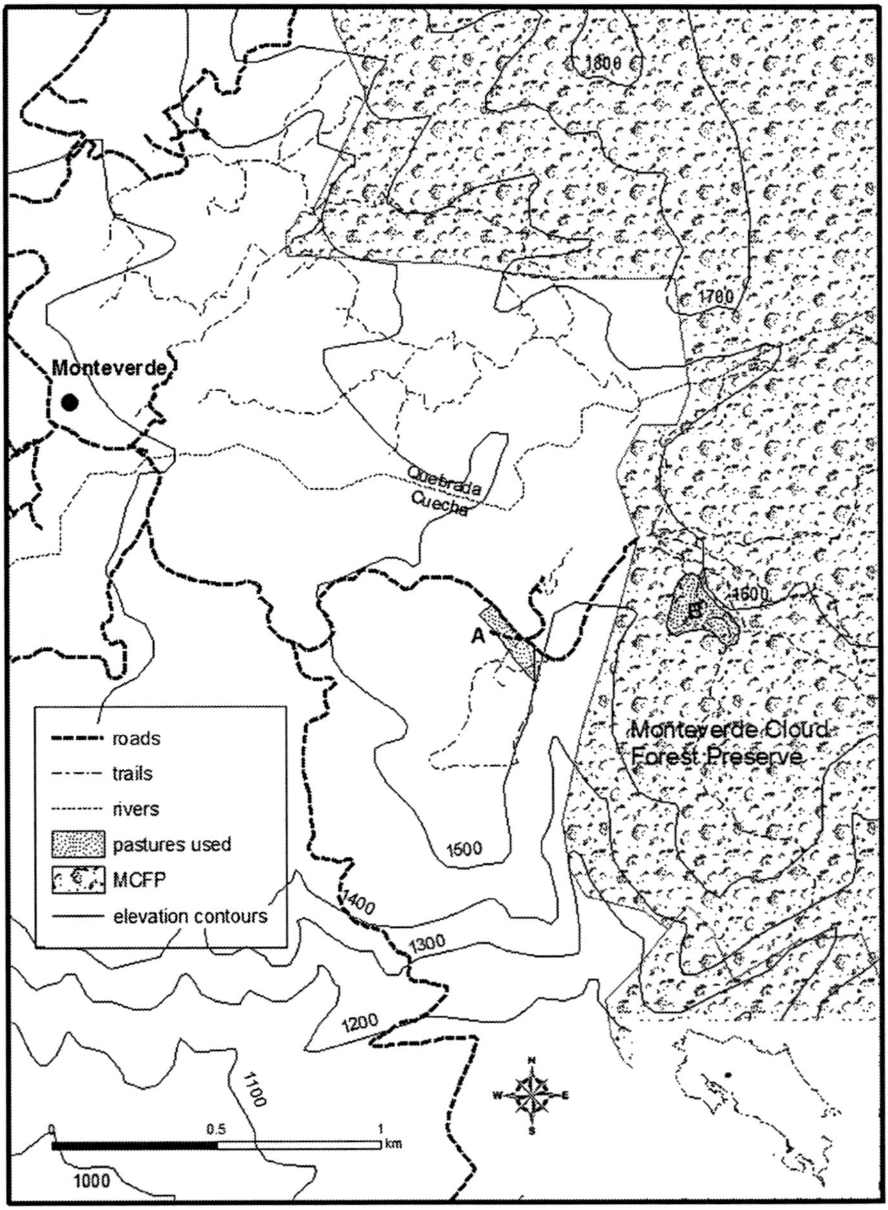

Fig. 8.1 Map of the Monteverde area, showing the locations of pastures A and B

Suite 430, San Jose, CA 95110). Ordination analyses (detrended correspondence analysis; Gauch 1982) were performed with DECORANA (Microcomputer Power, 111 Clover Lane, Ithaca, NY 14850 USA).

8.3 Results

8.3.1 Remnant Tree Species Present in Pastures

Of the 77 remnant trees in pasture A in 1987, 73 were identified at least to genus, and represented 26 species in 19 families (Table 8.1). Of these, 62 produce fruits adapted for dispersal by animals and 11 are adapted for dispersal by wind. Between the 1987 and 1998 censuses, some of the remnant trees in pasture A fell or were removed by the landowner. In addition, some of the original open sky sites were destroyed by the construction of pathways, such that we were able to include only 26 remnant tree and 16 open sky sites in the 1998 census.

We located and identified 11 remnant trees (all animal-dispersed) in pasture B, representing 7 species in 7 families (Table 8.1).

Table 8.1 Tree species present as pasture remnants. All remnants were sampled in pasture A, but in pasture B only those that were much larger than post-abandonment recruits were sampled. Names and dispersal modes follow Haber (2000); bd = bird, bt = bat, am = arboreal mammal, and tm = terrestrial mammal, wd = wind

Family	Species	Dispersal mode	# in pasture A	# in pasture B
Annonaceae	*Guatteria oliviformis*	bd	1	
Bombacaceae	*Quararibea costaricensis*	am	1	1
Boraginaceae	*Cordia cymosa*	bd		2
Celastraceae	*Maytenus reconditus*	bd	2	
Clusiaceae	*Chrysochlamys psychotriifolia*	bd	1	
Euphorbiaceae	*Sapium glandulosum*	bdx	5	2
	Hyeronima oblonga	bd	1	
Flacourtiaceae	*Xylosma oligandra*	bd	1	
	Hasseltia floribunda	bd		1
Lauraceae	*Beilschmiedia costaricensis*	bd	1	3
	Ocotea tonduzii	bd	3	
	Ocotea whitei	bd	6	
	Nectandra smithii	bd	4	
Malvaceae	*Hampea appendiculata*	bd		1
Meliaceae	*Guarea tonduzii*	bd	3	
	Cedrela tonduzii	wd		1
	Trichilia havanensis	bd	1	
Moraceae	*Ficus sp.*	bd/bt/am	2	
	Pseudolmedia mollis	bd	1	
Myrsinaceae	*Ardisia palmana*	bd	1	
Myrtaceae	*Eugenia sp.*	bd or bt	1	
Proteaceae	*Roupala glaberrima*	wd	11	
Rhizophoraceae	*Cassipourea elliptica*	bd	1	
Rubiaceae	*Chione sylvicola*	bd	2	
Rutaceae	*Stauranthus perforatus*	bd	8	
	Zanthoxylum sp.	bd	1	
Sabiaceae	*Meliosma idiopoda*	bd	4	
Sapotaceae	*Pouteria sp.*	am, tm	5	
Tiliaceae	*Mortoniodendron guatemalense*	bd	1	
Verbenaceae	*Citharexylum caudatum*	bd	2	
–	*Unidentified species / dead snags*	–	11	

8.3.2 Remnant Trees as Early Recruitment Foci

Remnant trees were exceedingly important as foci for early recruitment of woody plants. Three years after abandonment of pasture A, we found a total of 2,039 allospecific recruits (i.e., of species other than the overhanging remnant tree), representing 27 species and at least 19 families, within the 73 identified remnant tree plots (Fig. 8.2; Table 8.2). The 10 most common recruit taxa were *Eugenia guatemalensis* and *Myrcianthes fragrans* (we were unable to distinguish reliably between seedlings of these in 1987), *Conostegia oerstediana*, *Sapium glandulosum*, *Ocotea whitei*, *Hampea appendiculata*, *Viburnum costaricanum*, *Stauranthus perforatus*, *Ocotea tonduzii*, *Myrsine coriacea*, and *Citharexylum caudatum*. In contrast, we found only 34 woody recruits, representing 11 species in 8 families, in the 35 plots without overhanging vegetation (Fig. 8.2). The difference in mean density of recruits in the two plot types is highly significant (Mann-Whitney $U = 59.5$; $p < 0.0001$). The great majority of recruits were of species dispersed by birds or bats, although two species (*Persea americana* and *Pouteria fossicola*) dispersed primarily by arboreal mammals (e.g., monkeys, Kinkajous and Olingos) were also well-represented. Few recruits of wind-dispersed species were present, despite the fact that the single most common remnant tree species in pasture A (*Roupala glaberrima*) produces wind-dispersed seeds.

It might be argued that recruits would occur most commonly beneath remnant trees without any dispersal at all, since many seeds simply fall beneath parent plants. This source of bias was not important in our study, however: 12 of the 24 recruit species identified 3 years post-abandonment were not even present as remnant trees in the pasture, and therefore must have come from the surrounding forest. Moreover, data reported above do not include 344 recruits that were censused beneath remnant trees of the same species. Our approach therefore conservatively estimates the effects of remnant trees on recruitment patterns.

Clumping of recruits beneath remnant trees might also result simply from higher survival of seeds or seedlings there, e.g., due to favorable microclimate conditions. While we could not address that possibility directly, we did so indirectly

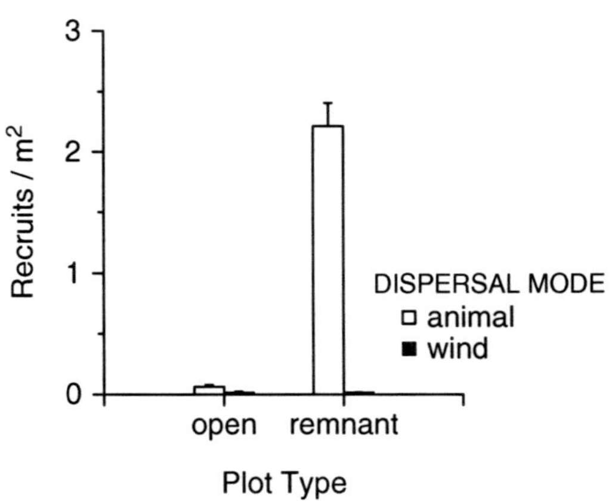

Fig. 8.2 Mean density of allospecific wind- and animal-dispersed recruits in 73 remnant tree and 35 "open sky" plots in pasture A, 3 years post-abandonment. Error bars represent one standard error of the mean

Table 8.2 Woody recruits in pasture A (3 and 14 years post-abandonment) and pasture B (30 years) in 1987 and 1998. Dispersal mode and successional status (early vs. late) is based on personal observation and Haber (2000)

Family	Species	Dispersal mode	Successional status	Pasture A, 3 years	Pasture A, 14 years	Pasture B, 30 years
Actinidiaceae	*Saurauia montana*	bd, am	early	x	x	
Apocynaceae	*Tabernaemontana longipes*	bd	late		x	
Aquifoliaceae	*Ilex lamprophylla*	bd	late		x	x
Araliaceae	*Schefflera rodrigueziana*	bd	early	x		
	Oreopanax oerstedianus	bd	late		x	
	Oreopanax xalapensis	bd	early			x
Asteraceae	*Neomirandea angularis*	wd	early		x	
	Koanophyllon pittieri	wd	late		x	
Capparidaceae	*Capparis discolor*	am	late		x	
Caprifoliaceae	*Viburnum costaricanum*	bd	early	x	x	
Celastraceae	*Maytenus reconditus*	bd	late	x		
	Perrottetia longistylis	bd	late			x
Cunoniaceae	*Weinmannia wercklei*	wd	late		x	
Elaeocarpaceae	*Sloanea brenesii*	bd	late			x
Euphorbiaceae	*Alchornea latifolia*	bd	early		x	
	Sapium glandulosum	bd	early	x	x	
Fabaceae	*Lonchocarpus sp.* (Haber 11102)	wd	late		x	
Flacourtiaceae	*Casearia tacanensis*	am	late		x	x
	Hasseltia floribunda	bd	late		x	x
	Xylosma oligandra	bd	late		x	
Lauraceae	*Beilschmiedia costaricensis*	bd	late	x		x
	Cinnamomum "paraplitrinerve"	bd	late		x	
	Ocotea meziana	bd	late		x	x
	Ocotea tonduzii	bd	late	x	x	x
	Ocotea whitei	bd	late	x	x	
	Nectandra smithii	bd	late		x	x
	Pleurothyrium palmanum	bd	late			x
	Persea americana	am	late	x		x
Malpighiaceae	Bunchosia veluticarpa	bd	late		x	

Malvaceae	*Hampea appendiculata*	bd	early	x	x	x
	Wercklea insignis	wd	early	x		
Melastomataceae	*Conostegia oerstediana*	bd, bt	early	x	x	x
Meliaceae	*Trichilia havanensis*	bd	late	x	x	
	Cedrela tonduzii	wd	late		x	
	Guarea rhopalocarpa	bd	late		x	
	Guarea tonduzii	bd	late			x
Moraceae	*Ficus hartwegii*	bd	late	x	x	
	Ficus pertusa	bd	late		x	
Myrsinaceae	*Ardisia palmana*	bd	late			x
	Ardisia compressa	bd	late		x	x
	Myrsine coriacea	bd	early	x	x	
Myrtaceae	*Eugenia guatemalensis* + *Myrcianthes fragrans**	bd or bt	late	x		
	Eugenia sp.	bd	late			x
	Calyptranthes pittieri	bd	late		x	
	Myrcia fallax	bd	late		x	
	Psidium guajava	tm, bt	early		x	
Oleaceae	*Forestiera cartaginense*	bd	late		x	
Proteaceae	*Roupala glaberrima*	wd	late	x	x	
Rhizophoraceae	*Cassipourea elliptica*	bd	late			x
Rosaceae	*Prunus skutchii*	bd	late	x	x	
	Prunus brachybotria	bd	late		x	x
Rubiaceae	*Guettarda poasana*	bd	early	x	x	
	Elaeagia auriculata	wd	late			x
	Chione sylvicola	bd	late			x
Rutaceae	*Stauranthus perforatus*	bd	late	x	x	
	Zanthoxylum sp.	bd			x	
Sabiaceae	*Meliosma idiopoda*	bd	late		x	
	Meliosma subcordata	am	late			
Sapindaceae	*Matayba oppositifolia*	bd	late		x	
Sapotaceae	*Pouteria fossicola*	am	late	x		

(continued)

Table 8.2 (continued)

Family	Species	Dispersal mode	Successional status	Pasture A, 3 years	Pasture A, 14 years	Pasture B, 30 years
	Pouteria sp.	am	late	x		
Symplocaceae	*Symplocos brenesii*	bd	late		x	
Thymelaeaceae	*Daphnopsis americana*	bd	early		x	x
Tiliaceae	*Heliocarpus americanus*	wd	early	x		
	Mortoniodendron guatemalense	bd	late			x
Verbenaceae	*Citharexylum caudatum*	bd	early	x	x	x
	unknown species	–		3	9	–
	total species			**27**	**54**	**25**
	total families			**19**	**29**	**17**

*We were unable to distinguish between seedlings of these species in 1987 by comparing numbers of wind-dispersed recruits in the two plot types. If seeds and seedlings simply survive better under the shade of remnant trees, we would expect the same distribution patterns among wind-dispersed and animal-dispersed species. Data on the few wind-dispersed recruits show clearly that this was not the case; wind-dispersed recruits were no more common beneath remnant trees than in plots without overhanging vegetation (Fig. 8.2; Mann-Whitney U = 1,270.5; $p > 0.9$), while animal-dispersed recruits were (Fig. 8.2; Mann-Whitney U = 55.5; $p < 0.0001$).

8.3.3 Effect of Distance to Nearest Possible Seed Source

To determine whether the number of recruits per plot might be a function of proximity to seed sources, we regressed the number of allospecific recruits of each species in each plot on two independent estimators of proximity to seed sources: distance to nearest conspecific remnant within the pasture, and distance to the nearest pasture/forest boundary. Distance to the nearest conspecific tree was a significant predictor of the number of recruits per plot for only two of the 7 taxa for which we had adequate sample sizes of conspecific remnants: *Ocotea whitei* and *Eugenia guatemalensis + Myrcianthes fragrans*. (Table 8.3).

Table 8.3 Results of multiple regression analyses of numbers of allospecific recruits on distance to nearest conspecific remnant and distance to pasture edge. Cells with dashes indicate species with insufficient numbers of conspecific remnants for analysis

Species	Distance to nearest conspecific remnant	Distance to pasture edge
Viburnum costaricanum	–	ns
Sapium glandulosum	ns	ns
Stauranthus perforatus	ns	ns
Beilschmeidia costaricensis	–	ns
Ocotea tonduzii	ns	ns
Ocotea whitei	***	ns
Persea americana	–	ns
Trichilia havanensis	–	ns
Hampea appendiculata	–	ns
Conostegia oerstediana	–	ns
Myrsine coriacea	–	ns
Eugenia guatemalensis + Myrcianthes fragrans	*	ns
Roupala glaberrima	ns	ns
Prunus sp.	–	ns
Pouteria exfoliata	ns	ns
Pouteria fossicola	–	*[a]
Heliocarpus americanus	–	ns
Citharexylum caudatum	–	ns

*$p < 0.05$ ***$p < 0.001$ ns $p > 0.05$

[a]regression coefficient significant, but in the opposite direction of that predicted.

The number of allospecific recruits was dependent upon distance to pasture edge for only one species, *Pouteria exfoliata*. In this case, however, the effect was in the direction opposite of that predicted: trees farther from the forest edge tended to have more *Pouteria* seedlings than those nearer to it.

8.3.4 Effect of Dispersal Mode of Remnant Tree

As a conservative test of the hypothesis that remnant trees are particularly effective as recruitment foci for species with similar fruit types, we compared the mean numbers of animal-dispersed recruits beneath animal-dispersed and wind-dispersed remnants. Surprisingly, the mean number of animal-dispersed recruits was actually somewhat higher beneath wind-dispersed remnants (Fig. 8.3), although the difference was not significant (Mann-Whitney U = 247; $p = 0.147$). In fact, even dead snags without foliage were as effective as animal-dispersed remnant trees as recruitment foci for animal-dispersed species (Fig. 8.3; Mann-Whitney U = 210; $p = 0.889$). Similarly, species richness of allospecific animal-dispersed recruits did not differ significantly among the two remnant tree types and dead snags (Fig. 8.3; Kruskal-Wallis one-way ANOVA statistic = 0.936, $p = 0.626$). Thus, remnant trees do not seem to attract animal-dispersed recruits only if they also serve as fruit sources for foraging animals. Rather, trees act as foci for seed input simply because they provide perch sites or sources of insect prey also eaten by frugivorous birds and mammals.

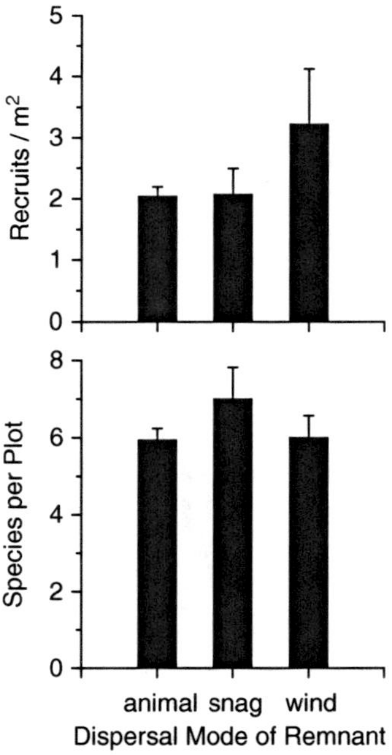

Fig. 8.3 Mean density and species richness of allospecific animal-dispersed recruits in plots beneath 11 wind-dispersed and 62 animal-dispersed remnant trees, and beneath 7 dead snags in pasture A, 3 years post-abandonment. Error bars represent one standard error of the mean

8.3.5 *Persistent Effects of Recruitment Foci on Regeneration*

Recruit density increased in both plot types between 3 and 14 years post-abandonment in pasture A, but remnant tree plots still had significantly higher stem densities than open sky plots (Fig. 8.4). By 30 years post-abandonment at pasture B, however, densities in both plot types had declined to levels below that typical of remnant plots at 3 years post-abandonment, presumably due to self-thinning. More importantly, recruit density in Pasture B (30 years post-abandoment) was similar in the two plot types.

Temporal trends in species richness of recruits paralleled those in density—species richness increased in both plot types from 3 to 14 years post-abandonment, with richness significantly higher in remnant tree plots. Richness then declined by 30 years, and was similar in both plot types.

8.3.6 *Effects of Remnant Trees on Species Composition of Recruits*

We investigated the long-term impact of remnant trees on regeneration of a species-rich assemblage of late-successional species in two ways. First, we compared density and species richness of early- vs. late-successional recruits in remnant tree and open sky plots as pastures aged. At 3 years post-abandonment,

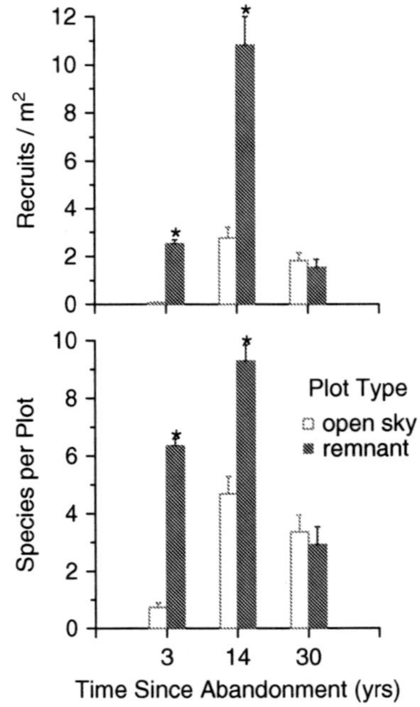

Fig. 8.4 Mean density and species richness of all woody recruits in remnant tree and "open sky" plots vs. time since pasture abandonment. Error bars represent one standard error of the mean, and asterisks denote significant differences at $p < 0.001$ (Mann-Whitney U test). Note that the 3 year post-abandonment data are based on 12.6 m^2 samples, while the 14 and 30 year data are based on 3.14 m^2 samples. Comparisons of species richness between 3 years post-abandonment and that for either the 14 or 30 year censuses must be made with caution

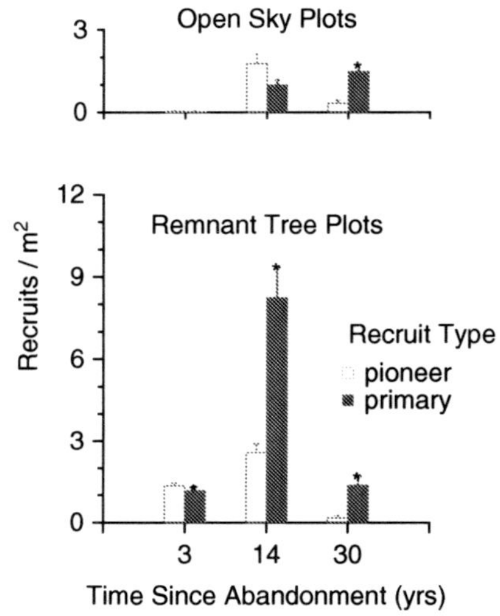

Fig. 8.5 Mean density of recruits of early- and late-successional species (see text) in both plot types vs. time since pasture abandonment. Error bars represent one standard error of the mean, and asterisks denote significant differences at $p < 0.001$ (Mann-Whitney U test)

early-successional species slightly outnumbered late-successional ones in remnant tree plots and approximately equaled them in open sky plots (Fig. 8.5). By 14 years, however, late-successional species far outnumbered early-successional ones in remnant tree plots. In open sky plots, early- and late-successional species were still similar in density. By 30 years post abandonment, early-successional species were significantly more dense in both plot types.

We also investigated the similarity in species composition among plot types and over time using detrended correspondence analysis (Gauch 1982). For purposes of our analysis, we included all remnant tree and open sky sites censused in the two pastures in 1987 and 1998. To retain the plots with no woody recruits in the analysis, we created a dummy species and included it as the only recruit in otherwise empty plots. In order to focus only on species composition rather than the differences in abundance discussed above, we normalized the overall density in each plot to 1.0; each species thus contributed a proportion to that overall density.

Species composition differed substantially between the two plot types at 3 years post-abandonment (Fig. 8.6A), largely because most open sky sites had few or no recruits. Open sky sites were more heterogeneous for the same reason. At 14 years post-abandonment, species composition was both more homogeneous within plot types and more similar between them (Fig. 8.6B), probably because by that time the original open sky sites had been colonized by recruits deposited by frugivores perched in new overhanging vegetation (i.e., from the original recruits that by this time were 3–5 meters tall) that didn't exist at the time of pasture abandonment. By 30 years, species composition was more heterogeneous in both plot types (Fig. 8.6C) probably because self-thinning reduced stem density and thus gave slight compositional differences a larger effect in the ordination. Direct comparisons between the 30-year data and that for 3 and 14 years must be done with caution, since the former was done in pasture B and the latter were done in pasture A half a kilometer away.

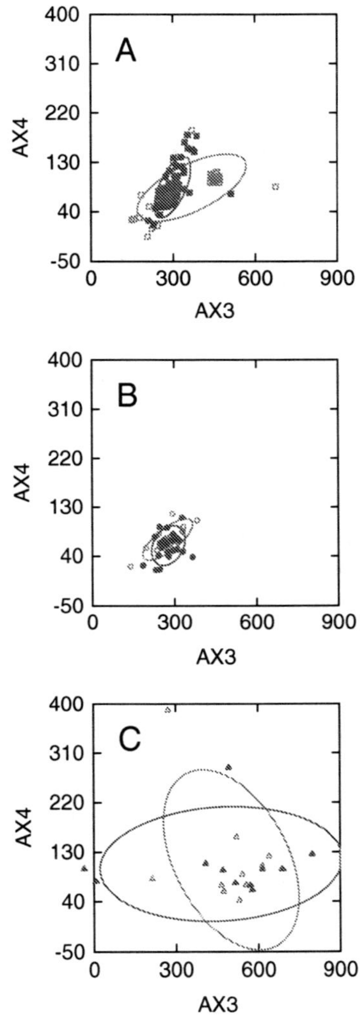

Fig. 8.6 Results of detrended correspondence analyses (DECORANA) on woody recruits in remnant tree (filled symbols) and "open sky" (open symbols) plots vs. time since pasture abandonment. (A) pasture A, 3 years post-abandonment. The tightly clustered open symbols represent open sky plots with no recruits. (B) pasture A, 14 years post-abandonment. (C) pasture B, 30 years post-abandonment. Confidence ellipses are centered on the bivariate means, and their sizes are such that 68.27% (1 standard deviation) of the data points are included within. Note that axes 3 and 4 are shown here for each ordination. Axes 1 and 2 separated plots more on the basis of the composition of a few idiosyncratic plots

8.4 Discussion

Data presented here show clearly that forest regeneration on abandoned pastures can occur very rapidly at Monteverde, and that the physical structure of the pasture plays a key role in determining the rate and successional trajectory of that regeneration. The fact that a species-rich assemblage of plants appears beneath remnant trees so soon after cattle are removed from a pasture suggests strongly that seeds are being deposited continuously even in active pastures, but that the cattle simply eat the resultant seedlings soon after germination. Studies of the seed rain in pastures verifies this pattern at Monteverde (Harvey 2000, R. Boyce and K. G. Murray, unpublished data) and elsewhere (e.g., Guevara and Laborde 1993, Vieira et al. 1994, Holl 1998).

8.4.1 Remnant Trees as Recruitment Foci

In Monteverde, seed dispersal by fruit-eating birds and mammals was largely responsible for the rapid colonization of pastures by woody plant species; of the 67 plant species identified as recruits in the study plots in our two pastures, all but 9 have seeds adapted for dispersal by animals.

Patterns of early recruitment were decidedly non-random with respect to physical habitat structure: significant recruitment of woody species occurred only beneath the crowns of remnant trees in pasture A. The reason for this pattern is obvious: fruit-eating animals forage in trees and shrubs, preferentially regurgitate or defecate while perched rather than in flight, and therefore deposit seeds beneath overhanging vegetation rather than in open pasture. Previous work at many other sites has also demonstrated this pattern (e.g., Debussche et al. 1982, McDonnell and Stiles 1983, Guevara et al. 1986, McDonnell 1986, McClanahan and Wolfe 1987, 1993, Bass 1990, Crome 1990, Uhl et al. 1991, Guevara 1992, Guevara and Laborde 1993, DeBussche and Isenmann 1994, Vieira et al. 1994, Toh et al. 1999, Zahawi and Augspurger 1999, Slocum 2001). This pattern also coincides with anecdotal observations of the behavior of fruit-eating birds (e.g., Cardoso da Silva et al. 1996, K. G. Murray, personal observation).

In contrast to Harvey (2000), we found no evidence that bat-dispersed species are more likely to colonize open pasture than are bird-dispersed ones: recruits of both *Conostegia oerstediana* and *Eugenia guatemalensis* + *Myrcianthes fragrans* were just as heavily concentrated beneath remnant trees as those of bird-dispersed species. Some authors have explained the deposition of bat-dispersed seeds in open pastures or gap interiors on the basis of bats' propensity to defecate in flight (e.g., Charles-Dominique 1986, Harvey 2000). The lack of such a pattern in our study may result from several factors. First, both bat-dispersed taxa are dispersed by birds as well as bats (Haber 2000). In fact, *C. oerstediana* is dispersed primarily by birds: only a small proportion of bat fecal samples collected by Dinerstein (1983) contained seeds of this species, while Wheelwright et al. (1984) found it to be one of the most commonly eaten species by frugivorous birds at Monteverde. Second, *E. guatemalensis* has large seeds that are never ingested by bats. Rather, they are discarded after the pulp is removed, which probably occurs at a feeding roost. Finally, bats may deposit the majority of the seeds they carry beneath feeding roosts despite the fact that they also defecate in flight, simply because they spend more time roosting than flying.

We found no evidence that the shade produced by remnant trees contributed to higher recruitment beneath them, as was suggested by Nepstad et al. (1991) for abandoned Amazonian pastures and by Holl et al. (2000) for a montane site in Costa Rica. This effect, which presumably results from amelioration of water stress during the dry season or inhibition of pasture grasses, is undoubtedly important at some locations—particularly at lower elevations where the dry season is more severe. Our study was conducted at an elevation where blowing mist usually prevents serious water stress even during the height of the dry season. Working at the same elevation in southern Costa Rica, Holl (1999) found no evidence of water stress among forest tree seedlings planted in abandoned pastures.

The particular effectiveness of animal-dispersed remnant trees as recruitment foci is widely cited (e.g., Janzen 1988), though few studies explicitly present data to support this conclusion. Groom (2000) found higher recruitment of woody plants beneath three species of animal-dispersed remnants than beneath one wind-dispersed species, and the pattern was especially pronounced for recruits in the Lauraceae, a family that depends disproportionately on large fruit-eating birds like Resplendant Quetzals and Three-wattled Bellbirds for dispersal. Similarly, Slocum and Horvitz (2001) found higher seed input (in terms of both density and species richness) beneath animal-dispersed remnants than beneath those with wind- or ballistically-dispersed seeds, but input beneath both types of remnants exceeded that at sites without overhanging vegetation. Censuses of established recruits showed similar patterns, but the differences between focal tree types were less pronounced (Slocum 2001). In contrast, Guevara et al. (1986) found that species richness was no higher beneath the five animal-dispersed remnants they censused (mean 11.2, s.d. 4.38) than beneath two wind-dispersed ones (mean 10.5, s.d. 3.54). They did not report data on the density of recruits. Similarly, Otero-Arnaiz et al. (1999) found no relationship between the species composition of recruits beneath pasture trees and the identity or dispersal mode of that tree. On abandoned farmlands in Queensland, Toh et al. (1999) reported that the presence of a fruit reward was less important than the structure of a remnant tree and its suitability as a perch in determining its effectiveness as a recruitment focus.

We too found no evidence that remnant trees that themselves produce fleshy fruits are more effective than other trees as recruitment foci. Instead, we found similar densities of animal-dispersed recruits beneath both animal-dispersed and wind-dispersed remnants, and even beneath dead snags. These findings and those reported above suggest that dispersers do not visit isolated pasture trees just for the fruit crops they contain, but perhaps as sources of cover (i.e., concealment from predators), as sites for courtship or territorial displays (e.g., singing perches), or as sources of other food, e.g. arthropods. Many of the bird species that frequent the pasture include substantial numbers of insects in their diets in addition to fruits, most notably many of the tanagers and thrushes. Cardoso da Silva et al. (1996) also found that tanagers in particular were the primary dispersers of seeds beneath isolated trees and shrubs in recently abandoned pastures in Brazil.

The finding that even artificial perches enhance recruitment (e.g., McDonnell and Stiles 1983, McClanahan and Wolfe 1993, Robinson and Handel 1993, 2000, Holl 1998) also suggests that the effectiveness of elevated structures as recruitment foci is largely due to their use as perches. Holl (1998) found that birds (and deposited seeds) were attracted by leafless tree branches placed upright in a southern Costa Rican pasture, but that they were no more effective when she added banana to them. And the greater seed input beneath animal-dispersed remnants in Slocum and Horvitz' (2000) study occurred even during times of the year when the remnant trees were not producing fruits. The importance of a remnant tree's own dispersal mode in determining its effectiveness as a recruitment focus thus remains ambiguous.

Virtually all remnant trees and other elevated structures facilitate secondary succession on post-agricultural lands by attracting animals, and sometimes (but not always) the effect is enhanced by the presence of fruits in the remnants themselves.

8.4.2 Distance to Seed Sources

We found little evidence that distance to nearest possible seed source had any effect on the density of most species of recruits in pasture A. Apparently the distances involved were small enough that most seed dispersers readily visited even the trees in the most central portions of the pasture. In contrast, Holl (1999) found almost no seed deposition beyond 5 meters from the edge of an abandoned montane pasture in southern Costa Rica, but this finding was more likely due to the fact that seed traps were placed in locations without overhanging vegetation. The only taxa in our study for which density was positively correlated with proximity to either pasture edge or nearest conspecific remnant (*Ocotea whitei* and *Eugenia guatemalensis* + *Myrcianthes fragrans*) both produce fruits with single large seeds that are dispersed by a few species of large frugivores (e.g., Resplendant Quetzals, Black Guans, Emerald Toucanets, and Three-wattled Bellbirds; Wheelwright et al. 1984). Such large seeds are voided rapidly by regurgitation (Wheelwright 1991), and are thus often not dispersed far from the parent plant (cf. Wenny and Levey 1998).

We do not suggest that the lack of a relationship between recruitment and distance to seed sources in this study is a general one, however. Pasture A is fairly typical of those in the Monteverde region in that it is relatively small and has forest in close proximity. Pastures are apparently small in much of Puerto Rico also, and like us Aide et al. (1996) found no evidence for a distance effect on pasture regeneration there. These conditions are not typical of pastures in much of South America, however, nor of those at lower elevations in much of Central America, and we would expect a much stronger effect of distance to pasture edge on recruitment beneath remnant trees in them. In large Amazonian pastures, for example, even tanagers characteristic of the forest/pastures boundary rarely venture more than 150 meters from the forest edge (Cardoso da Silva et al. 1996).

8.4.3 Successional Trends

Early-successional species were particularly well-represented among the early colonists of both remnant tree and open sky plots, perhaps because their small seed size permits dispersal by a greater number of bird species than do the larger seeds of late-successional species (e.g., Wheelwright 1985). Species with minute seeds (< 1mm, e.g., *Conostegia oerstediana*) in particular are often dispersed by small tanagers, which were abundant in the study area and readily forage within pastures at Monteverde (K. G. Murray, personal observation) as elsewhere (e.g., Cardoso da Silva et al. 1996). Tanagers fail to disperse seeds of species with larger seeds because they routinely separate seeds from the pulp before swallowing it (Levey 1986, Murray 1988). Most other studies have documented dominance by early-successional species in the early stages of post-agricultural regeneration as well (e.g., Aide et al. 1996, 2000, Myster 2004b).

By 14 years post-abandonment, remnant tree plots in our study were dominated by late-successional species, while open sky plots were still dominated by early-successional ones. While we did not address the reason for this trend directly, it probably derives from the combination of reduced germination of early-successional species beneath the increasingly shaded (by previous recruits) remnants and the simple accumulation of recruits from a larger pool of available primary species. By 30 years post-abandonment, both plot types were dominated by late-successional species to about the same degree.

Both remnant tree and open sky plots continued to accumulate species richness and total stem density for at least several years after pasture abandonment. Indeed, by 30 years no difference in either stem density or species richness was apparent between the two plot types. The reason for continued accumulation is obvious at least for remnant tree plots—fruit-eating animals continued to deposit seeds beneath perch sites. For open sky plots, however, the reason for eventual accumulation of recruits is less obvious, since they failed utterly to attract significant numbers of recruits soon after pasture abandonment. The probable answer is that rapid recruitment beneath the crowns of remnant trees leads to the formation of "vegetation islands" that expand in size and coalesce with one another and with regeneration proceeding into the pasture from the edge. At our study site, the density of remnant trees was high enough (and the pastures small enough) that most of the original open sky plots came to be beneath the crowns of trees recruited into the pasture by 14 years post-abandonment. As soon as these plots had overhanging vegetation, they began to attract recruits for the same reasons as the original remnant tree plots. By 30 years post-abandonment, the entire pasture had been covered for many years by the crowns of trees recruited after pasture abandonment. The elimination of the original "open sky" plots via the growth of vegetation islands, together with the self-thinning of recruits as light intensity near the ground decreased over time, led to the loss of any recognizable difference in either species richness or total stem density between the two plot types by 30 years post abandonment.

This relatively rapid convergence of density and species composition in remnant tree and open sky plots should not be taken as evidence that remnant trees have only transient effects on pasture regeneration. Indeed, the rapid convergence only occurred because the pastures we studied were relatively small and had large numbers of remnants (which were therefore close together). The most direct measure of the importance of remnant trees would require comparison of our data with that from a large pasture completely lacking remnant trees—a comparison we were unable to make. Working in Brazil, however, Chris Uhl and his colleagues have been able to study very large pastures (> 10 ha) in which any original remnant trees have been killed by fires. They estimate that without human intervention regeneration of species-rich forests on such pastures may take hundreds of years (Buschbacher et al. 1992).

Aide et al. (2000) also found that species richness (as well as density, basal area, and aboveground biomass) in abandoned pastures in Puerto Rico converged on that typical of old-growth forest by 40 years post-abandonment, but that the species composition of the secondary forest remained quite different from old growth. Clearly, we need more long-term studies of the trajectory of succession in post-agricultural landscapes to better understand whether, and how rapidly, regenerating communities converge on the species richness and composition of the communities that surround them.

8.4.4 Other Habitat Structural Elements

Other structural habitat features influence the rate and trajectory of succession on post-agricultural lands as well. Several investigators have documented facilitation of recruitment of woody forest species on rotting logs, (e.g., Peterson and Haines 2000, Slocum 2000). Such logs are often remnants of the original forest, and they may enhance recruitment by acting as perch sites for frugivores, because cattle can't trample seedlings that colonize them, because they are devoid of the pasture grasses that compete with tree seedlings, or because they have great water-holding capacity after they partially decompose.

Patches of ferns or shrubs also facilitate colonization by forest species, in some cases because they are avoided by cattle (e.g., Slocum 2000), and in others because they inhibit the herbaceous vegetation that competitively exclude woody forest species (e.g., Posada et al. 2000). As importantly, shrubs, like trees, provide elevated perch sites that attract foraging birds and mammals, and thus enhance seed deposition. Cardoso da Silva et al. (1996), for example, found that birds moving from forest into Amazonian pastures spent much of their time in shrubs, even when larger trees (e.g., *Cecropia* sp.) were present. It is possible, of course, that dense shrub patches in post-agricultural fields inhibit the growth rates of woody recruits that do establish via the shade they produce (e.g., Zahawi and Augspurger 1999). By enhancing seed deposition and inhibiting herbaceous vegetation, however, their role overall is more likely to be one of facilitation (e.g., Holl et al. 2000 and references cited therein).

8.4.5 Implications for Management

Secondary forests on previously abandoned tropical pastures are likely to become increasingly important repositories of tropical biodiversity and providers of other ecosystem services like erosion control and flood amelioration, even if they are ultimately converted again to some other land use type. Understanding the key roles played by frugivores and remnant trees can give those responsible for managing tropical landscapes an important incentive for conserving frugivore populations and seed sources, as well as allow them to manage such lands for maximum speed and species richness of recovery.

Results presented here and elsewhere suggest management practices that would speed succession to a species-rich forest on abandoned tropical pastures. Key among the factors that affect the rate and trajectory of regeneration is the physical structure of the pasture: all else being equal, small pastures with remnant trees are likely regenerate most rapidly. And while characteristics that would speed recovery after abandonment might best be addressed at the time of conversion from forest, farmers are more likely to have a greater incentive to make pastures large and to leave only enough trees to provide their cattle with a few sources of shade. After abandonment, however, some relatively simple manipulations might greatly accelerate seed input even in large barren pastures. The finding that remnant trees enhance recruitment primarily through a "perch effect" suggests that the rate of forest regeneration would be maximized by planting the most rapidly growing native trees locally available, regardless of the dispersal mode for which their own seeds are adapted. And since many seed dispersers will travel among forest patches through even narrow shelterbelts of trees (e.g., Harvey 2000), managers might speed recovery by connecting the areas bordering abandoned pastures with other forest patches via the construction of shelterbelts of rapidly growing species.

We acknowledge the danger of drawing conclusions about temporal trends from data on spatial ones (i.e., comparison of pasture B at 30 years post-abandonment with pasture A at 3 and 14 years post-abandonment), and of drawing conclusions about pasture regeneration from a study of just two pastures in the study described here. Because the pastures were only 0.5 km apart, however, and because the species composition of forest surrounding them is so similar we are confident that the patterns seen in the two pastures reflect a successional trajectory common to both. We also remain confident that our results contain important insights about the time scale on which the interaction between initial habitat structure and successional trajectory occurs, and that our results reflect processes that operate widely. Like several of the other studies cited above, which were conducted on four continents, this study demonstrates a profound effect of physical habitat structure on successional trajectory. Few studies of tropical pasture regeneration have followed the process for more than a few years post-abandonment, however, and we believe that future investigations will also find that some effects of initial habitat structure weaken with time (e.g., recovery of density and species richness concentrated beneath remnant trees), while others persist for much longer (e.g., species composition).

Finally, we acknowledge that drawing conclusions about secondary succession on post-agricultural lands in general from studies concentrated on abandoned pastures can lead to oversimplification. Like the work reported here, most of the literature on regeneration of former agricultural lands concerns abandoned cattle pastures rather than croplands. That bias is partly geographical; most studies are from the Neotropics, where conversion to cattle pasture is the single greatest driver of deforestation (Myers 1980). It may also reflect greater permanence of other types of agricultural land use. Different agricultural uses result in profound differences in soil characteristics, disperser, herbivore and pathogen communities, and physical habitat structure at the time of abandonment, however (e.g., Myster 2004b), and given that all of these factors have been shown to influence the rate and trajectory of regeneration we should expect details of the process to reflect the unique history and characteristics of each site.

It would be naïve to suggest that the pristine tropical forests remaining today can be protected permanently from conversion to agriculture, and there is widespread agreement that maintaining or restoring biodiversity over wide areas will require regeneration of lands already deforested. Over the long term, sustainable use of many tropical lands will involve repeated cycles of conversion to agriculture and regeneration, and learning to manage regeneration should remain a high priority.

Acknowledgment We are especially grateful to Alan Pounds and Marlene Brenes for allowing us to collect data in pasture A. The Tropical Science Center, and especially the former director of the Monteverde Cloud Forest Preserve, Robert Carlson, greatly facilitated the project by providing laboratory space and giving us access to pasture B. Nalini Nadkarni and Rodrigo Solano kindly shared previous data on the identities and locations of remnant trees at the pasture B site. The work was funded by Faculty Development Grants from Hope College and by NSF grants to Hope College (BBS-8712566 and DBI-9322220) and to K. G. Murray (DEB-9208229). Christopher Lepczyk, Dan Wenny, and Chris J. Peterson provided useful comments on an earlier version of the manuscript. Finally, KGM thanks Chris Uhl for many stimulating discussions about pasture regeneration, and for the opportunity to see it in Amazonia.

References

Aide, T. M., and Cavelier, J. 1994. Barriers to lowland tropical forest restoration in the Sierra Nevada de Santa Marta, Colombia. Restoration Ecology 2:219–229.

Aide, T. M., Zimmerman, J. K., Pascarella, J. B., Rivera, L., and Marcano-Vega, H. 2000. Forest regeneration in a chronosequence of tropical abandoned pastures: implications for restoration ecology. Restoration Ecology 8:328–338.

Aide, T. M., Zimmerman, J. K., Rosario, M., and Marcano, H. 1996. Forest recovery in abandoned cattle pastures along an elevational gradient in northeastern Puerto Rico. Biotropica 28: 537–548.

Bass, D. A. 1990. Pied currawongs and seed dispersal. Corella 14:24–27.

Brown, S., and Lugo, A. E. 1994. Rehabilitation of tropical lands: a key to sustaining development. Restoration Ecology 2: 97–111.

Buschbacher, R., Uhl, C., and Serrão, E. A. S. 1992. Reforestation of degraded Amazon pasture lands. In Ecosystem Rehabilitation: Ecosystem Analysis and Synthesis, Vol. 2, ed. M. K. Wali, pp. 257–274. SPB Academic Publishing, The Hague, The Netherlands.

Cardoso da Silva, J. M., Uhl, C., and Murray, G. 1996. Plant succession, landscape management, and the ecology of frugivorous birds in abandoned Amazonian pastures. Conservation Biology 10:491–503.

Charles-Dominique, P. 1986. Inter-relations between frugivorous vertebrates and early-successional plants: Cecropia, birds and bats in French Guyana. In Frugivores and Seed Dispersal, ed. A. Estrada and T. H. Fleming, pp. 119–135. Dordrecht, The Netherlands: W. Junk.

Clark, K. L., Lawton, R. O., and Butler, P. R. 2000. The physical environment. In Monteverde: Ecology and conservation of a tropical cloud forest, ed. N. M. Nadkarni and N. T. Wheelwright, pp. 15–38. New York, NY: Oxford University Press.

Crome, F. H. J. 1990. Vertebrates and successions. In Australian Tropical Rainforests: Science-Value-Meaning, ed. L. J. Webb and J. Kikkawa, pp. 53–64. Canberra: CSIRO.

Debussche, M., and Isenmann, P. 1994. Bird-dispersed seed rain and seedling establishment in patchy Mediterranean vegetation. Oikos 69:414–426.

Debussche, M., Escarré, J., and Lepart, J. 1982. Ornithochory and plant succession in Mediterranean abandoned orchards. Vegetatio 48:255–266.

Dinerstein, E. 1983. Reproductive ecology of fruit bats and seasonality of fruit production in a Costa Rican cloud forest. Unpublished Ph.D. dissertation, University of Washington, Seattle, Washington, USA.

Duncan, R. S., and Chapman, C. A. 1999. Seed dispersal and potential forest succession in abandoned agriculture in tropical Africa. Ecological Applications 9:998–1008.

Duncan, R. S., and Chapman, C. A. 2002. Limitations of animal seed dispersal for enhancing forest succession on degraded lands. In Seed Dispersal and Frugivory: Ecology, Evolution, and Conservation, ed. D. J. Levey, W. R. Silva, and M. Galetti, pp. 437–450. Wallingford, Oxon, UK: CAB International.

Finegan, B., and Delgado, D. 2000. Structural and floristic heterogeneity in a 30-year-old Costa Rican rain forest restored on pasture through natural secondary succession. Restoration Ecology 8:380–393.

Gauch Jr., H. G. 1982. Multivariate Analysis in Community Ecology. Cambridge, England: Cambridge University Press.

Gerhardt, K. 1993. Tree seedling development in tropical dry abandoned pasture and secondary forest in Costa Rica. Journal of Vegetation Science 4:95–102.

Green, R. J. 1995. Using frugivores for regeneration: A survey of knowledge and problems in Australia. In Integrating People and Wildlife Fore a Sustainable Future, ed. J. A. Bissonette and P. R. Krausman, pp. 651–657. Bethesda, MD: The Wildlife Society.

Groom, M. 2000. Patterns in the regeneration of lauraceous trees in abandoned pastures. In Monteverde: Ecology and Conservation of a Tropical Cloud Forest, ed. N. M. Nadkarni and N. T. Wheelwright, pp. 442–446. New York, NY: Oxford University Press.

Guevara, S., and Laborde, J. 1993. Monitoring seed dispersal at isolated standing trees in tropical pastures: consequences for local species availability. Vegetatio 107/108:319–338.

Guevara, S., Meave, J., Moreno-Casasola, P., and Laborde, J. 1992. Floristic composition and structure of vegetation under isolated standing trees in neotropical pastures. Journal of Vegetation Science 3:655–664.

Guevara, S., Purata, S. E., and Van der Maarel, E. 1986. The role of remnant trees in tropical secondary succession. Vegetatio 66:77–84.

Haber, W. A. 2000. Vascular plants of Monteverde. In Monteverde: Ecology and Conservation of a Tropical Cloud Forest, ed. N. M. Nadkarni and N. T. Wheelwright, pp. 39–94. New York, NY: Oxford University Press.

Harvey, C. A. 2000. Windbreaks enhance seed dispersal into agricultural landscapes in Monteverde, Costa Rica. Ecological Applications 10:155–173.

Harvey, C. A., and Haber, W. A. 1999. Remnant trees and the conservation of biodiversity in Costa Rican pastures. Agroforestry Systems 44:37–68.

Harvey, C. A., Guindon, C. F., Haber, W. A., DeRosier, D. H., and Murray, K. G. 2000. The importance of forest patches, isolated trees and agricultural windbreaks for local and regional biodiversity: the case of Monteverde, Costa Rica. XXI IUFRO World Congress, International Union of Forestry Research Organizations, Subplenary sessions Vol. 1, pp. 787–798.

Harvey, C. A., Guindon, C. F., Haber, W.A., DeRosier Hamilton, D., and Murray, K. G. 2007. Importancia de fragmentos de bosque, árboles dispersos y cortinas rompevientos para la conservación de biodiversidad en el agropaisaje de Monteverde, Costa Rica. In Evaluación y Conservación de la Biodiversidad en Agropaisajes de Mesoamérica, ed. C. A. Harvey and J. C. Sáenz. San Jose, Costa Rica: Editorial EUNA, (in press)

Holdridge, L. 1967. Life Zone Ecology. San Jose, Costa Rica: Tropical Science Center.

Holl, K. D. 1998. Do bird perching structures elevate seed rain and seedling establishment in abandoned tropical pasture? Restoration Ecology 6:253–261.

Holl, K. D. 1999. Factors limiting tropical rain forest regeneration in abandoned pasture: seed rain, seed germination, microclimate, and soil. Biotropica 31:229–242.

Holl, K. D., Loik, M. E., Lin, E. H. V., and Samuels, I. A. 2000. Tropical montane forest restoration in Costa Rica: overcoming barriers to dispersal and establishment. Restoration Ecology 8: 339–349.

Howe, H. F., and Smallwood, J. 1982. Ecology of seed dispersal. Annual Review of Ecology and Systematics 13:201–228.

Janzen, D. H. 1988. Management of habitat fragments in a tropical dry forest: growth. Annals of the Missouri Botanical Garden 75:105–116.

Kuusipalo, J., Ådjers, G., Jafarsidik, Y., Otsamo, A., Toumela, K., and Vuokko, R. 1995. Restoration of natural vegetation in degraded *Imperata cylindrica* grassland: understory development in forest plantations. Journal of Vegetation Science 6:205–210.

Levey, D. J. 1986. Methods of seed processing by birds and seed deposition patterns. In Frugivores and Seed Dispersal, eds A. Estrada and T. H. Fleming, pp. 147–158. Dordrecht, The Netherlands: Dr. W. Junk Publishers.

Lugo, A. E. 1992. Tree plantations for rehabilitating damaged forest lands in the tropics. In Ecosystem Rehabilitation: Ecosystem Analysis and Synthesis, Vol. 2, ed. M. K. Wali, pp. 247–255. The Hague, The Netherlands.

McClanahan, T. R., and Wolfe, R. W. 1987. Dispersal of ornithochorous seeds from forest edges in central Florida. Vegetatio 71:107–112.

McClanahan, T. R., and Wolfe, R. W. 1993. Accelerating forest succession in a fragmented landscape: the role of birds and perches. Conservation Biology 7:279–288.

McDonnell, M. J. 1986. Old field vegetation height and the dispersal pattern of bird-disseminated woody plants. Bulletin of the Torrey Botanical Club 113:6–11.

McDonnell, M., and Stiles, E. W. 1983. The structural complexity of old field vegetation and the recruitment of bird-dispersed plant species. Oecologia 56:109–116.

Murray, K. G. 1988. Avian seed dispersal of three neotropical gap-dependent plants. Ecological Monographs 58:271–298.

Murray, K. G., Kinsman, S., and Bronstein, J. 2000. Plant-animal interactions. In Monteverde: Ecology and Conservation of a Tropical Cloud Forest, ed. N. M. Nadkarni and N. T. Wheelwright, pp. 245–302. New York, NY: Oxford University Press.

Myers, N. 1980. Conversion of Tropical Moist Forests. Washington DC: National Research Council.

Myster, R. W. 2004a. Regeneration filters in post-agricultural fields of Puerto Rico and Ecuador. Plant Ecology 172:199–209.

Myster, R. W. 2004b. Post-agricultural invasion, establishment and growth of Neotropical trees. The Botanical Review 70:381–402.

Nepstad, D., Uhl, C., and Serrão, E. A. 1990. Surmounting barriers to forest regeneration in abandoned, highly degraded pastures: a case study from Paragominas, Pará, Brazil. In Alternatives to Deforestation: Steps Toward Sustainable Use of the Amazon Rain Forest, ed. A. B. Anderson, pp. 215–229. New York: Columbia University Press.

Nepstad, D. C., Uhl, C., and Serrão, E.A.S. 1991. Recuperation of a degraded Amazonian landscape: forest recovery and agricultural restoration. Ambio 20(6):248–255.

Nepstad, D. C., Uhl, C., Pereira, C.A., and Cardoso da Silva, J. M. 1996. A comparative study of tree establishment in abandoned pasture and mature forest of eastern Amazonia. Oikos 76: 25–39.

Nielsen, K., and DeRosier, D. 2000. Windbreaks as corridors for birds. In Monteverde: Ecology and Conservation of a Tropical Cloud Forest, eds. N. M. Nadkarni and N. T. Wheelwright, pp. 448–450. New York, NY: Oxford University Press.

Otero-Arnaiz, A., Castillo, S., Meave, J., and Ibarra-Manríquez, G. 1999. Isolated pasture trees and the vegetation under their canopies in the Chiapas coastal plain, México. Biotropica 31: 243–254.

Peterson, C. J., and Haines, B. L. 2000. Early successional patterns and potential facilitation of woody plant colonization by rotting logs in premontane Costa Rican pastures. Restoration Ecology 8:361–369.

Posada, J. M., Aide, T. M., and Cavelier, J. 2000. Cattle and weedy shrubs as restoration tools of tropical montane rainforest. Restoration Ecology 8:370–379.

Robinson, G. R., and Handel, S. N. 1993. Forest restoration on a closed landfill: rapid addition of new species by bird dispersal. Conservation Biology 7:271–278.

Robinson, G. R., and Handel, S. N. 2000. Directing spatial patterns of recruitment during an experimental urban woodland reclamation. Ecological Applications 10:174–188.
Sarmiento, F. O. 1997a. Arrested succession in pastures hinders regeneration of Tropandean forests and shreds mountain landscapes. Environmental Conservation 24:14–23.
Sarmiento, F. O. 1997b. Landscape regeneration by seeds and successional pathways to restore fragile Tropandean slopelands. Mountain Research and Development 17:239–252.
Serrão, E. A., and Toledo. J. M. 1990. The search for sustainability in Amazonian Pastures. In Alternatives to Deforestation: Steps Toward Sustainable Use of the Amazon Rain Forest, ed. A. B. Anderson, pp. 195–214. New York, NY: Columbia University Press.
Slocum, M. G. 2000. Logs and fern patches as recruitment sites in a tropical pasture. Restoration Ecology 8:408–413.
Slocum, M. G. 2001. How tree species differ as recruitment foci in a tropical pasture. Ecology 82:2547–2559.
Slocum, M. G., and Horvitz, C. C. 2000. Seed arrival under different genera of trees in a neotropical pasture. Plant Ecology 149:51–62.
Sokal, R. R., and Rohlf, F. J. 1995. Biometry, 3rd edition. New York, NY: W. H. Freeman and Co.
Toh, I., Gillespie, M., and Lamb, D. 1999. The role of isolated trees in facilitating tree seedling recruitment at a degraded sub-tropical rainforest site. Restoration Ecology 7:288–297.
Uhl, C. 1987. Factors controlling succession following slash-and-burn agriculture in Amazonia. Journal of Ecology 75:377–407.
Uhl, C., Buschbacher, R., and Serrão, E. A. S. 1988. Abandoned pastures in eastern Amazonia. I. Patterns of plant succession. Journal of Ecology 76:663–681.
Uhl, C., Nepstad, D., Buschbacher, R., Clark, K., Kauffman, B., and Subler, S. 1990. Studies of ecosystem response to natural and anthropogenic disturbances provide guidelines for designing sustainable land-use systems in Amazonia. In Alternatives to Deforestation: Steps Toward Sustainable Use of the Amazon Rain Forest, ed. A. B. Anderson, pp. 24–42. New York, NY: Columbia University Press.
Uhl, C., Nepstad, D., Vieira, I. C., and Silva, J. M. C. 1981. Restauração da floresta em pastagens degradadas. Ciência Hoje 13:22–31.
Vieira, I. C. G., Uhl, C., and Nepstad, D. 1994. The role of the shrub *Cordia multispicata* Cham. as a 'succession facilitator' in an abandoned pasture, Paragominas, Amazonia. Vegetatio 115: 91–99.
Wenny, D. G., and Levey, D. J. 1998. Directed seed dispersal by bellbirds in a tropical cloud forest. Proceedings of the National Academy of Sciences 95:6204–6207.
Wheelwright, N. T. 1985. Fruit size, gape width, and the diets of fruit-eating birds. Ecology 66:808–818.
Wheelwright, N. T. 1991. How long do fruit-eating birds stay in the plants where they feed? Biotropica 23:29–40.
Wheelwright, N. T., Haber, W. A., Murray, K. G., and Guindon, C. 1984. Tropical fruit-eating birds and their food plants: a survey of a Costa Rican lower montane forest. Biotropica 16:173–192.
Wijdeven, S. M. J., and Kuzee, M. E. 2000. Seed availability as a limiting factor in forest recovery processes in Costa Rica. Restoration Ecology 8:414–424.
Wunderle, J. M. 1997. The role of animal seed dispersal in accelerating native forest regeneration on degraded tropical lands. Forest Ecology and Management 99:223–235.
Zahawi, R. A., and Augspurger, C. K. 1999. Early plant succession in abandoned pasturesin Ecuador. Biotropica 31:540–552.
Zimmerman, J. K., Pascarella, J. B., and Aide, T. M. 2000. Barriers to forest regeneartion in an abandoned pasture in Puerto Rico. Restoration Ecology 8:350–360.

Chapter 9
Neotropic Post-Dispersal Seed Predation

Randall W. Myster

9.1 Introduction

The dynamics of seeds and seedlings determine much of plant community development after disturbance (Grubb 1977). For old field succession, whether in the temperate zone or in the Neotropics, such regeneration comes overwhelmingly from outside the field (Purata 1986, Uhl, Buschbacher and Serrao 1988, Myster 1993, Medellin and Gaona 1999) because few species sprout from remnant trees or leftover rootstocks (although some shrubs can recruit asexually: *Miconia* spp. in Neotropical fields [Myster 2004b] and *Rhus* spp. in temperate fields [Myster 1993]). Since most species depend on dispersed seeds for their recruitment, dispersal and the fate of seeds once they have landed in fields are of primary concern.

As previous chapters have shown, large numbers of seeds disperse into Neotropic pastures and old fields which, when combined with the low seedling emergence from Neotropic seed banks (Vazquez-Yanes and Orozco-Segovia 1993), points to a critical mechanism that removes large amounts of seeds from old fields after dispersal. Numerous studies in both temperate and tropical America have shown that seed predators can eat up to 100% of seeds presented to them in old field experiments (reviewed in Myster 1993, 2004b), implying that seed predation is one of those key mechanisms.

In temperate old fields, small mammals account for most of those losses (Gill and Marks 1991, Myster and Pickett 1993). As a field ages and becomes more complex structurally and as more mammal refuges become available, there occurs an increase in predation levels. In general, predators in old fields prefer woody patches to herbaceous patches (Webb and Willson 1985) and avoid open areas (Burton 1989). Also because of woody cover predators choose seeds that are closer to the forest, which becomes less true over time as fields develop patches of tall vegetation (Gill and Marks 1991). There are other sources of variation that affect the behavior of seed predators. For example, additions of litter have been shown to decrease predation (Myster and Pickett 1993) while spring rain have been shown to increase it (DeSteven 1991). Furthermore, studies have generally shown a strong ranking of species-specific predator preferences, which may correlate more with tree taxa (i.e., conifers are the least desired) than with common seed sizes (DeSteven 1991) which change as a field ages (Myster and Pickett 1993).

 R. W. Myster, *Post-Agricultural Succession in the Neotropics.*

The investigation of seed predation in Neotropic old fields has built on the results from temperate old fields to further examine how the interactions among various factors affect rates of seed removal, with the aim of more precisely quantifying the seed predation niche for Neotropic old field plant species. Thus, the study of predator preferences in the Neotropics has been a priority, but also litter interactions, density of seed, age of field, distance to forest edge, field-type, patch-type, and season.

9.2 Results from Past Experiments in Puerto Rican and Ecuadorian Old Fields

A number of seed predation experiments were performed in (1) pastures, Banana plantations, Sugarcane plantations, and a Coffee plantation in the Luquillo Experimental Forest of Puerto Rico (Myster 2003a, 2003b, 2004a) and in (2) pastures, Banana plantations, and Sugarcane plantations in the Maquipucuna Reserve in Ecuador (Myster 2004a, 2007: see Chapter 1 for a description of these study sites). At both locations, seeds were put out in 9cm diameter plastic Petri dishes (except for the Coffee plantation where a plastic grid was used). After 2 weeks, seed loss was scored in each dish (or grid cube) and assumed to be due to predation, not wind or rain, since plastic seed mimics of similar size, shape, and weight were not lost.

The analysis of data from the Puerto Rican pastures indicated that most seeds were lost to predators with significant differences in species (*Syzgium jambos* had 50% losses, *Solanum tovum* had 65% losses, *Miconia prasina* had 76% losses, and *Piper aduncum* had 79% losses), in density (78% losses at 5 seeds per dish, 69% losses at 10 seeds per dish, 51% losses at 20 seeds per dish), and in patch type (grass patches had 69% losses, *Miconia* spp. patches had 65% losses). In addition, there were significant interaction effects (1) between species and density for *P. aduncum* and *S. tovum* (88% losses at low density vs. 85% losses at medium density vs. 56% losses at high density, 83% losses at low density vs. 70% losses at medium density vs. 27% losses at high density, respectively) and (2) between species and patch-type for *M. prasina* (93% losses in grass patches vs. 55% losses in *Miconia* spp. patches). Another experiment in the same pastures and in a Puerto Rican Coffee plantation showed large species differences (*Guarea guidonia* had 34% losses, *Gonzalugunia spicata* had 55% losses, *Miconia racemosa* had 66% losses, and *Casearia arborea* had 72% losses), with predators taking seeds at spatial scales of 2–11cm. Puerto Rican Banana plantations and Sugarcane plantations also demonstrated significant differences in species (*Gonzalagunia spicata* had 80% losses, *Palicourea riparia* had 63% losses, *Piper aduncum* had 74% losses, and *Miconia pracina* had 86% losses).

Experiments in Ecuadorian pastures, Banana plantations, and Sugarcane plantations yielded similar results: most seeds were lost to predators while there were significant differences in species (*Cecropia monostachya* had 76% losses,

Solanum ovlifolium had 82% losses, *Piper aduncum* had 57% losses, and *Cestrom megalothyllum* had 80% losses). There were certain interactions between species and predator taxa. *C. monostachya* had 86% of its predation from insects rather than mammals, *S. ovlifolium* had 48%, *Piper aduncum* had 56% with most of the predation occurring in the forest, and *C. megalothyllum* had 40%. More seeds were eaten in the pasture (97% losses) than in Sugarcane (76% losses) and Banana fields (43% losses).

9.3 Discussion

The results suggest that those aspects of plant niches that deal with seed predation are very important in determining plant-plant replacements during old field succession. Therefore, seed predation is critical in creating invasion and establishment patterns. Studies on seed predation that were conducted in different Neotropic areas (Puerto Rico, Ecuador) and that involved different species, field-types, densities, and patch-types all report significant differences in the way predators eat seeds of different species. Due to this dominant species effect, other sources of variation in the working of seed predation may best be seen as the results of interaction with the significant species effects which are always present. In addition to the species effect, there has been found a general negative correspondence between rates of seed predation and seed size (as in Uhl 1987, Nepstad et al. 1996, alternatively see Holl and Lulow 1997). Other Neotropic studies also indicated that seed losses were large and dominated by species differences (Table 9.1).

The effect of woody patches on seed predation is variable. Whereas in the present series of studies woody patches decreased predation levels (patch-type effects in

Table 9.1 Other seed predation studies of Neotropic old fields and pastures. All studies were done in lowland wet rainforest, with the exception of Sarmiento which was done in montane wet rainforest at the Maquipucuna Reserve. For dry forest seed predation results see Chapter 11.

% lost	time period	field	variation	country	reference
10–100%	19 days	pasture	species field age	Colombia	Aide and Cavelier 1994
up to 98%	2 months	field	species field age	Mexico	Hammond 1995
7–100%	24 days	pasture	species patch-type year	Costa Rica	Holl 2002
20–100%	30 days	pasture	species forest distance patch-type	Costa Rica	Holl and Lulow 1997
15–100%	200 days	pasture	species	Brazil	Nepstad et al. 1996
4–100%	28 weeks	field	species	Peru	Notman and Gorchov 2001
90%	2 weeks	pasture	species	Ecuador	Sarmiento 1997
35–100%	4 months	field	species field age	Brazil	Uhl 1987
5–100%	6–8 months	pasture	species	Brazil	Vieira and Scariot 2006

Holl and Lulow 1997) due to increased predation on mammals by birds in those patches, other studies showed more predation in woody patches and in the forest (Nepstad et al. 1996, Aide and Cavelier 1994, Holl 2002) perhaps due to the greater litter accumulation there. As in temperate fields, the large seed losses caused by predation and the highly specific preferences of predators (Holl and Lulow 1997) suggest that the issues of variation in predator density and distribution in old fields and in predator foraging behavior, in particular how these aspects of their biology and ecology are affected by the availability and structure of plant cover, should be an essential component of research into seed predation.

Holl and Lulow (1997), Norman and Gorchov (2001), and Vieira and Scariot (2006) used enclosures to implicate small mammals as seed predators on the Neotropic mainland. The presented case studies implicated both small mammals and insects (mainly ants) in Ecuador, but only insects (again ants) in Puerto Rico, as the major seed predator taxa. Nepstad et al. (1993) reported similar results that implicated both ants and small mammals as seed predators while suggesting that seeds larger than 0.4g would be taken by mammals and smaller than 0.4g by insects. Researchers have shown both a decrease (Uhl 1987, Hammond 1995) and an increase (Aide and Cavilier 1994) in predation ever successional time, and effects of forest edge on predation are also not conclusive (Holl and Ludlow 1997).

A comparison of the results from Neotemperate and Neotropical studies brings forward the following points. Old fields in both places show high predation rates, up to 100% losses for some species, and more predation both under shrubs and at low density (Myster and Pickett 1993). Seasonal effects, however, are more pronounced in temperate studies (reviewed in Myster 1993, 2003b). Reduced rates of predation on seeds under litter and distance away from the forest (Gill and Marks 1991) have not yet been examined in Neotropical old fields.

All studies point to seed predation as a critical mechanism that determines old field invasion patterns. Results show that, first, predators prefer some species over others. This predilection gives the less preferred species an advantage in that they can establish themselves both before and more extensively than the preferred, more eaten species. Second, predation may be reduced under "nurse" trees, especially at the higher seed densities that are naturally found under shrubs and trees (satiation hypothesis: Janzen 1971, Myster 2006), which allows regeneration to radiate from these trees. Finally, patch-types with less predation may be able to facilitate more regeneration while field-types with more predation, such as pastures, may lag behind developmentally for years.

References

Aide, T. M., and Cavelier, J. 1994. Barriers to lowland tropical forest restoration in the Sierra Nevada de Santa Marta, Colombia. Restoration Ecology 2:219–229.

Burton, P. J. 1989. Constraints to tree invasion on a nutrient rich site during old-field succession. Ph.D dissertation. University of Illinois, Urbana.

DeSteven, D. 1991. Experiments on mechanisms of tree establishment in old field succession: seedling emergence. Ecology 72:1066–1075.

Gill, D. S., and Marks, P. L. 1991. Tree and shrub seedling colonization of old fields in central New York. Ecological Monographs 61:183–205.

Grubb, P. J. 1977. The maintenance of species-richness in plant communities: the important of the regeneration niche. Biological review of the Cambridge Philosophical Society 52:107–145.

Hammond, D. S. 1995. Post-dispersal seed and seedling mortally of tropical dry forest trees after shifting agriculture, Chiapas, Mexiso. Journal of Tropical Ecology 11:295–313.

Holl, K. D. 2002. Effect of shrubs on tree seedling establishment in an abandoned tropical pasture. Journal of Ecology 90:179–187.

Holl, K. D., and Lulow, M. E. 1997. Effects of species, habitat and distance from edge on post-dispersal seed predation in a tropical rainforest. Biotropica 29:459–468.

Janzen, D. H. 1971. Seed predation by animals. Annual Review of Ecology and Systematics 2:465–492.

Medellin, R. A., and Gaona, O. 1999. Seed dispersal by bats and birds in forest and disturbed habitats of Chiapas, Mexico. Biotropica 31:478–485.

Myster, R. W. 1993. Tree invasion and establishment in old fields at Hutcheson Memorial Forest. The Botanical Review 59:251–272.

Myster, R. W. 2003a. Seed regeneration mechanisms over fine spatial scales on recovering Coffee plantation and pasture in Puerto Rico. Plant Ecology 166:199–205.

Myster, R. W. 2003b. Effects of species, density, patch-type and season on post-dispersal seed predation in a Puerto Rican pasture. Biotropica 35:542–546.

Myster, R. W. 2004a. Regeneration filters in post-agricultural fields of Puerto Rico and Ecuador. Plant Ecology 172:199–209.

Myster, R. W. 2004b. Post-agricultural invasion, establishment, and growth of Neotropical trees. The Botanical Review 70:381–402.

Myster, R. W. 2006. Shrub vs. grass patch effects on the seed rain and the seed bank of a five-year pasture in Puerto Rico. Ecotropica 12:12–19.

Myster, R. W. 2007. Early successional pattern and process after Banana, Sugarcane and Pasture in Ecuador. The New Zealand Journal of Botany 45:35–44.

Myster, R. W., and Pickett, S. T. A. 1993. Effects of litter, distance, density and vegetation patch type in postdispersal tree seed predation in old fields. Oikos 66:381–388.

Nepstad, D. C., Uhl, C., Pereira, C. A., and Cardoso-Da Silva, J. M. 1996. A comparative study of tree establishment in abandoned pasture and mature forest of eastern Amazonia. Oikos 76:25–39.

Notman, E., and Gorchov, D. L. 2001. Variation in post-dispersal seed predation in mature Peruvian low-land tropical forest and fallow agricultural sites. Biotropica 33:621–636.

Purata, S. E. 1986. Floristic and structural changes during old field succession in the Mexican tropics in relation to site history and species availability. Journal of Tropical Ecology 2:257–276.

Sarmiento, F. O. 1997. Landscape regeneration by seeds and successional pathways to restore fragile Tropandean slopes. Mountain Research and Development 17:239–252.

Uhl, C. 1987. Factors controlling succession following slash-and-burn agriculture in Amazonia. Journal of Ecology 75:377–407.

Uhl, C., Buschbacher, R., and Serrao, E. A. S. 1988. Abandoned pastures in eastern Amazonia. I. Patterns of plant succession. Journal of Ecology 73:663–681.

Vazquez-Yanes, C., and Orozco-Segovia, A. 1993. Patterns of seed longevity and germination in the tropical rainforest. Annual Review of Ecology and Systematics 24:69–87.

Vieira, D. L. M., and Scariot, A. 2006. Effects of logging, liana tangles and pasture on seed fate of dry forest tree species in Central Brazil. Forest Ecology and Management 230:197–205.

Webb, S. L., and Willson, M. F. 1985. Spatial heterogeneity in post-dispersal predation on *Prunus* and *Uvula* seeds. Oecologia 57:150–153.

Chapter 10
Facilitation Versus Competition in Neotropical Old-Fields: A Case Study After *Pinus taeda* Cultivation in Brazil

Gislene Ganade, Lessandra Zanini and Ingo Hübel

10.1 Facilitation and Competition in Neotropical Old-Fields

Plantations of exotic tree monocultures for wood production are extremely widespread in the Neotropics (Richardson 1998). In Brazil, *Pinus* trees introduced from Europe and the US have been planted and commercialized for more than half a century. Most of these monoculture plantations have short clear-cut cycles that range from 7 to 10 years. Shortly after monocultures are removed, open fields are re-invaded by *Pinus* seedlings that regenerate from the seed bank or arrive in the site by dispersion from nearby parent trees (Higgins and Richardson 1998, Nathan et al. 2002). After clear-cut, a new tree monoculture can be naturally re-established in the site over a period of 10–15 years (Chilvers and Burdon 1983, Rouget et al. 2001). Therefore, old field succession after *Pinus* plantation suffers from the inevitable threat of aggressive exotic tree invasion (Kulmatiski 2006). The present chapter will discuss how the early successional vegetation established in these old-field sites could influence either the likelihood of *Pinus* invasion or the chances of native tree species to establish in those sites.

Two mechanisms of plant- plant interactions are considered to be crucial to define the structure and successional dynamics of old field communities: competition and facilitation (Callaway and Walker 1997). Competition is the mechanism by which plants are suppressed by neighbors due to resource depletion while facilitation is the mechanism by which plants are benefited by the presence of neighbor vegetation due to improvements of microclimatic and soil conditions (Callaway and Walker 1997). In old fields areas competition seems to be a much more frequent process influencing succession (Loik and Holl 2001, Myster 2004) while facilitation seems to occur more frequently in cases in which disturbance creates harsh conditions that are ameliorated by neighbors (Wilson 1999). Additionally, as old field succession proceeds, facilitation is suggested to become rarer while competition starts to dominate the plant interaction scenario (Wilson 1999).

The "stress gradient hypothesis" states that facilitation is more expressive in harsh environments where plants can experience a great deal of stress, e.g. deserts or alpine areas, while competition would be more important in low stress environments (Bertness and Callaway 1994). This hypothesis has been confirmed by works based

R. W. Myster, *Post-Agricultural Succession in the Neotropics.*

on meta-analysis which revisited studies performed in arid and semi arid environments (Gomez-Aparicio et al. 2004, Lortie and Callaway 2006). Nevertheless, productive rainforest areas would be expected by this hypothesis to hold competition as the most important process driving the interplay of species interactions during succession. Experimental studies, however, have shown that facilitation may be much more frequent than expected in neotropical old-fields (Vieira et al. 1994, Ganade and Brown 2002, Zanini and Ganade 2005, Zanini et al. 2006).

Switches from facilitation to competition in a given study site are quite frequently registered in the literature (Callaway and Walker 1997, Rousset and Lepart 2000, Alados et al. 2006). The occurrence of these switches may depend on levels of resource availability and / or environmental stress (Callaway et al. 2002). However, the degree in which a plant suffers from stress does not depend exclusively on resource conditions. Different life stages of a single plant may respond to environmental stress in distinct ways. For example, old fields can be extremely harsh environments for seedlings but not for adult trees. In old fields of Brazilian Amazonia seedlings of rain forest species rely on their plant neighbors to improve microclimatic conditions for germination, nevertheless, saplings have their growth reduced by light competition with neighbor vegetation (Ganade and Brown 2002, Ganade 2007). This same pattern has been registered in old-fields of temperate and sub tropical areas (De Steven 1991a, 1991b, Gill and Marks 1991, Hastwell and Facelli 2003). It is possible, therefore, that the facilitation process may be quite common in Neotropical old-fields during seedling establishment phase but may become rarer as seedlings grow through sapling and adult stages.

We exemplify this argument by a graphic model which shows how the outcome of facilitation and competition can change depending on the interplay between plant life stage and environmental stress (Fig. 10.1). For all life stages, facilitation is more pronounced when environmental stress increases. However, switches from competition to facilitation along a stress gradient differ for the distinct plant life stages. Seedlings experience facilitation by neighboring vegetation at a lower level

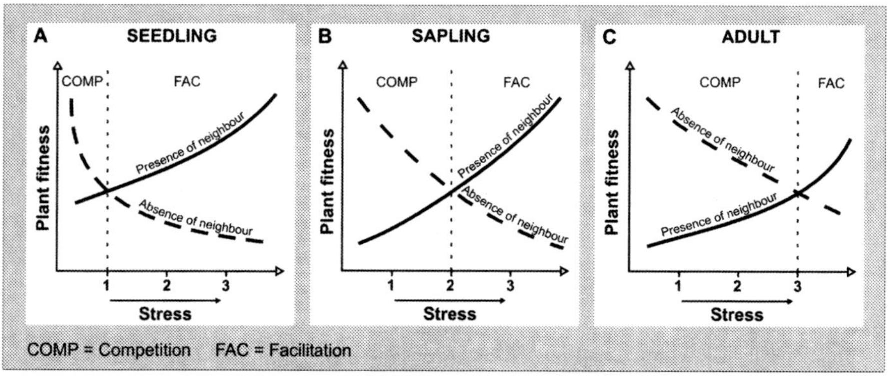

Fig. 10.1 Graphical model showing the outcome of facilitation and competition correlated with the interplay between plant life stages and environmental stress

of environmental stress (Fig. 10.1a) than saplings and adults trees (Fig. 10.1b, c). In the next section, we show the results of different experiments conducted in a Neotropical old field developed after the removal of a *Pinus taeda* plantation. These results confirm that neighbor vegetation quite frequently improves seedling performance in Neotropical old fields.

10.2 Old Field Succession After Pinus Plantation: A Study Case in Southern Brazil

We performed various field experiments in an old field area located at the National Forest of São Francisco de Paula (29°23‘S; 50°23‘W), RS, southern Brazil (Fig. 10.2). The reserve covers 1,600 ha and is mostly dominated by native Araucaria forest interspaced with plantations of exotic tree species of the genera *Pinus* and *Eucalyptus* as well as plantations of *Araucaria angustifolia* (Araucariaceae) the brazillian pine. The Araucaria forest is characterized by the canopy prevalence of *A. angustifolia.* The study site encompasses a 2 ha area that has been formally occupied by a *Pinus taeda* plantation for 30 years before clear-cut. In January 2001 the pine trees were cut at soil level and removed from the site for wood commercialization. After tree removal, the area was characterized by a mosaic of native early successional grasses and herbs as well as a small number of shrubs. Tree saplings of various species, including *A. angustifolia* and *P. taeda* in which

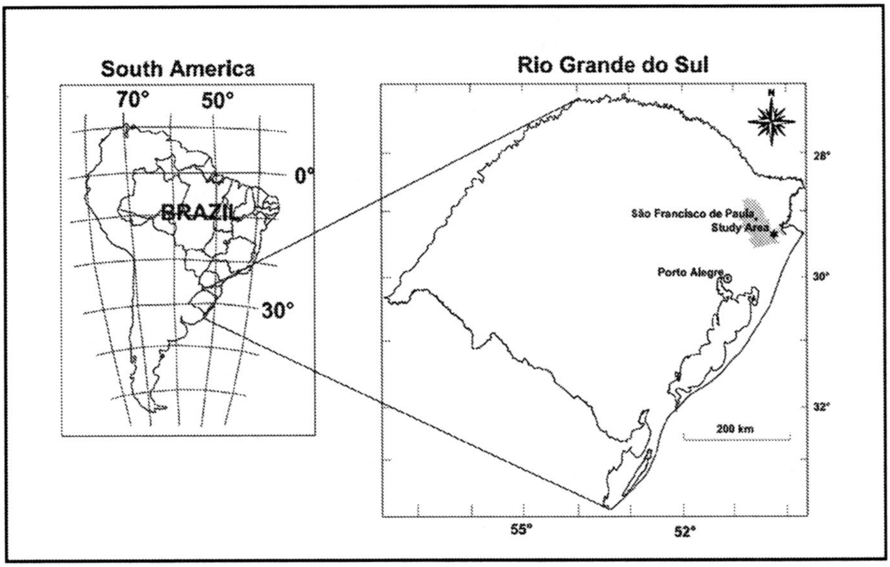

Fig. 10.2 Map showing the location of the study area in Rio Grande do Sul state, southern Brazil

the height could reach up to approximately 0.5 m were also present at the site. The edge of the nearest native forest was located at about 50 m from the study area.

10.3 Methods

We present here results of four experimental studies on seedling establishment that were carried out during the first years of old field succession. The studies aimed to test the effect of neighbor vegetation on establishment and performance of colonizing woody seedlings. The presence and absence of vegetation was assigned in 3×3m plots arranged in a randomized block design using 10 replicated blocks, 20 plots in total. Blocks were located at least 20 m apart. Pairs of plots within each block were randomly assigned to have vegetation removed or left intact. The vegetation removal treatment comprised the removal of all above ground plants, surface litter, and all non woody plants that became established throughout the experiment.

Experiment 1: To test the influence of neighbor vegetation on exotic *Pinus* invasion, we conducted an experiment where seeds of *P. taeda* were introduced in old-field plots where early successional vegetation was removed or left intact. A group of 20 seeds, 5 cm apart, were placed in each treatment and the number of seedlings established was monitored over an 8 months period.

Experiment 2: To investigate how spontaneous establishment of the woody species community would be affected by the presence of neighbour vegetation, we monitored the number and richness of woody seedlings naturally established in plots were vegetation was removed or left intact. Established seedlings were monitored within 1 m^2subplots placed inside each 3×3m plot. Seedlings were marked counted and morphotyped at monthly intervals over a 1 year period.

Experiment 3: To investigate which process competition or facilitation with neighbour vegetation could influence the survival and growth of a late successional tree we transplanted seedlings of *Inga virescens* (Mimosaceae) in all 3×3 m plots. Five seedlings averaged 15 cm in height were transplanted in each plot, at least 50 cm apart. The number of seedlings that survived and their heights were registered monthly during 16 months. Seedling growth was calculated by subtracting the registered seedling height (cm) by its height measured at the beginning of the experiment.

Experiment 4: To investigate how availability of soil nutrients could influence neighbour effects (competition or facilitation) seedlings of *Araucaria angustifolia* (Araucariaceae) were transplanted in the following four treatment combinations: (1) presence of neighbour vegetation without nutrients (control), (2) presence of neighbour vegetation with nutrients, (3) absence of neighbour vegetation without nutrients and (4) absence of neighbour vegetation with nutrients. Treatments were assigned in 3×3 m plots, in a randomized block design using 10 replicated blocks, 40 plots in total. Five seedlings averaged 30 cm in height were transplanted at least 50 cm apart in each plot. Seedling height was registered monthly during 22 months. Seedling growth was again estimated by the difference between a given seedling height and its initial height.

10.4 Results

Old-field invasion by the exotic species *Pinus taeda* was facilitated by neighbour vegetation (Fig. 10.3). Seedling establishment in the first 2 months did not differ from vegetation treatments, however, by the third month the number of seedlings established was significantly enhanced in plots where vegetation was left intact (significant interaction: time x vegetation treatment, $F_{7,63} = 5.4$; $P < 0.001$, repeated measures ANOVA).

The establishment of other woody species was also improved by neighbour vegetation. The presence of vegetation increased abundance ($F_{1,38} = 11.4$; $P < 0.01$, Fig. 10.4) and richness ($F_{1,38} = 8.9$; $P < 0.01$, Fig. 10.5) of native woody seedlings compared to plots where vegetation was removed.

Survival of *I. virescens* seedlings was improved by the presence of early successional vegetation. By the end of the experiment, seedling survival was 6 fold higher in intact vegetation plots in relation to plots where vegetation was removed (significant interaction: time × vegetation treatment $F_{9,162}= 7.7$; $P < 0.001$, repeated measures ANOVA, Fig. 10.6). Growth of *I. virescens* seedlings was also enhanced by the presence of neighbour vegetation. Seedling growth was similar between vegetation treatments in the first 6 months of the experiment, but, by its end, seedling growth increased 10 fold in plots where vegetation was left intact (significant interaction: time x vegetation treatment, $F_{15,135}= 12.9$; $P < 0.001$, repeated measures ANOVA, Fig. 10.7).

Nutrient availability was able to modulate the influence of early successional vegetation on growth of *A. angustifolia* seedlings. A significant third order interaction between time, vegetation and nutrient treatments confirmed this results ($F_{21,105} = 2.295$; $P < 0.01$). Without nutrient application, seedling growth was higher when vegetation was removed, indicating competition between colonizing seedlings and neighbour vegetation (Fig. 10.8a). Nevertheless, when nutrients were applied, seedling growth was higher when vegetation was left intact (Fig. 10.8b), indicating that once competition for nutrients is overcome due to increases in

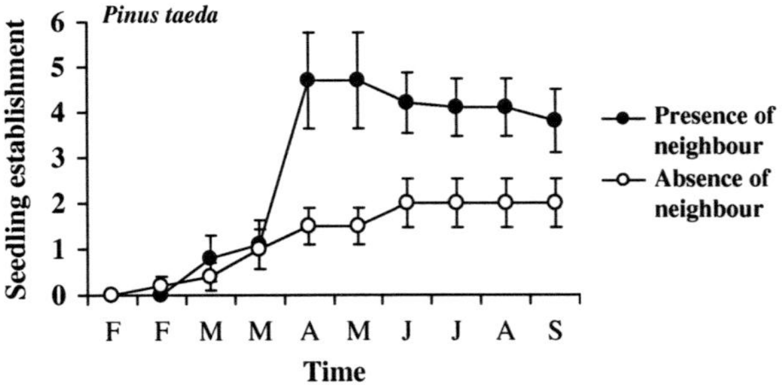

Fig. 10.3 Mean number of *Pinus taeda* seedlings established in the presence (vegetation intact, black symbols) or absence (vegetation removed, white symbols) of neighbour vegetation over a 8 months period. The study was undertaken in a neotropical old field created after a *Pinus* plantation clear-cut. Error bars represent ± 1 standard error

Fig. 10.4 Number of native woody seedlings per m^2 established at the presence (vegetation intact) and absence (vegetation removed) of neighbour vegetation. The study was undertaken in a neotropical old field created after a *Pinus* plantation clear-cut. Error bars represent ± 1 standard error

Fig. 10.5 Richness of native woody seedling morphotypes per m^2 established at the presence (vegetation intact) and absence (vegetation removed) of neighbour vegetation. The study was undertaken in a neotropical old field created after a *Pinus* plantation clear-cut. Error bars represent ± 1 standard error

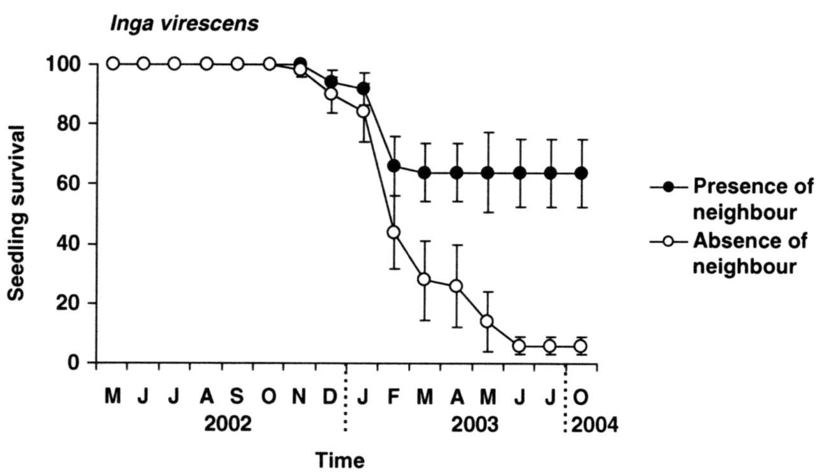

Fig. 10.6 Mean percentage of survival of *Inga virescens* seedlings transplanted in the presence (vegetation intact, black symbols) or absence (vegetation removed, white symbols) of neighbour vegetation over a 16 months period. The study was undertaken in a neotropical old field created after a *Pinus* plantation clear-cut. Error bars represent ± 1 standard error

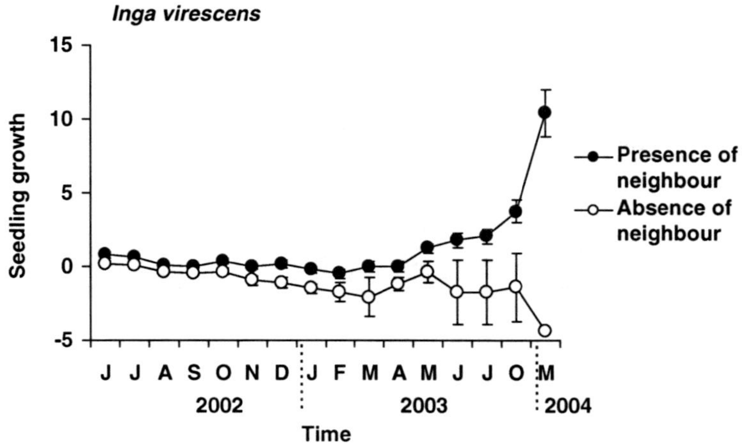

Fig. 10.7 Mean seedling growth (cm) of *Inga virescens* seedlings transplanted in the presence (vegetation intact, black symbols) or absence (vegetation removed, white symbols) of neighbour vegetation over a 16 months period. The study was undertaken in a neotropical old field created after a *Pinus* plantation clear-cut. Error bars represent ± 1 standard error

nutrient available, the presence of vegetation may improve seedling growth in relation to plots where vegetation is absent.

10.5 Discussion

Facilitation was the most common process occurring in the old field studied, because early successional herbaceous vegetation, in most cases, improved establishment, survival and growth of colonizing woody species. Facilitation might have occurred because vegetation had positive effects on local microclimate, protecting establishing seedlings from low winter temperatures, summer drought and even seed predators (Zanini and Ganade 2005, Zanini et al. 2006). In this study, however, facilitation occurred during the growth phase which is a rather rare phenomenon. Improvements of local microclimatic conditions provided by the pioneer vegetation could be responsible for the facilitation process registered for *I. virescens*. In this case, vegetation provided protection against consecutive frosts during winter months, which directly avoided leaf loss (Zanini et al. 2006).

Facilitation of woody seedling establishment by the herbaceous community has been reported in old fields of temperate areas (De Steven 1991a, 1991b, Gill and Marks 1991). However, various studies in Neotropical old fields report that competition with grasses is one of the main factors that restrain seedling establishment, slowing down forest regeneration (Nepstad et al. 1991, Aide and Cavelier 1994). This pattern may be due to the pasture fields studied being dominated by highly aggressive exotic grasses. These grasses have fast growth rates and use a great deal of water and soil resources that would otherwise be available for seedlings. Additionally its thick root and litter layer may restrict the availability of microsites

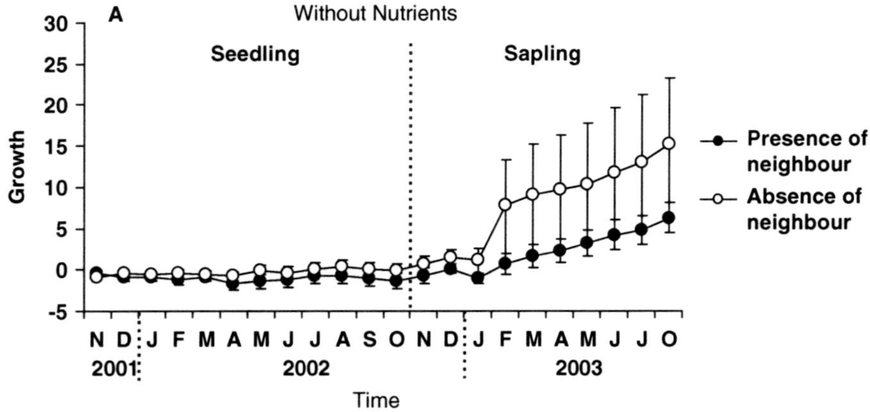

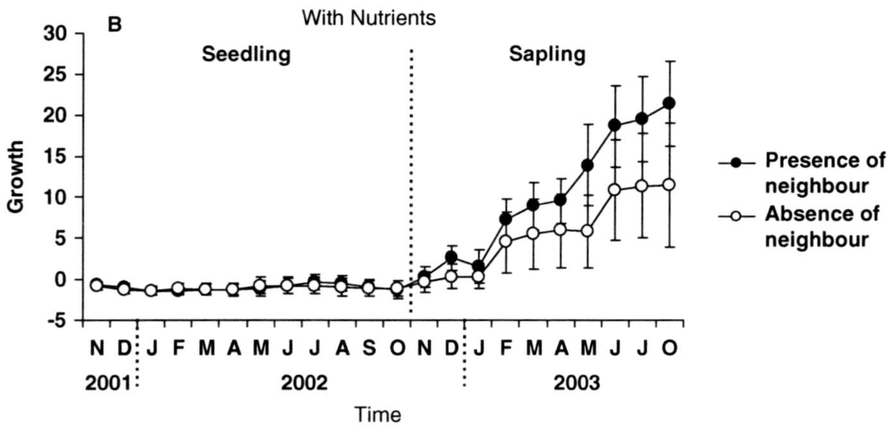

Fig. 10.8 Mean seedling growth (cm) of *Araucaria angustifolia* seedlings transplanted in the presence (vegetation intact, black symbols) or absence (vegetation removed, white symbols) of neighbour vegetation over a 22 months period. Graph A represents plots without nutrients (soil fertilizer not added) and graph B represents plots with nutrients (soil fertilizer added). The first year of measurement represent seedling phase and the second year represents the sapling phase. Error bars represent ± 1 standard error

for seedling establishment. In our study, we do not know which particular species improved seedling performance (Myster and Pickett 1992). We, nevertheless, know that the average effect of grasses and herbs on woody seedlings was positive (diffuse facilitation). This pattern may have occurred because the herbaceous vegetation of this study was composed by less aggressive species which are native grasses and herbs endemic from highland fields. Experimental works performed in this area have confirmed that these native species do not show strong competitive responses such as fast biomass accumulation after nutrient addition (Zanini and Ganade 2005).

As explained by the graph model proposed in Fig. 10.1, seedlings were indeed more likely to have their performance improved by neighbor vegetation. This is so because tree seedlings of mid and late successional species have evolved to establish under the shade of a tree canopy. Thus, although resources such as light may be more available during early old-field succession, the benefits of light may be outweighed by the stress of environmental exposure. Therefore, the seedling stage may be in greater need of a neighbor to improve its microclimatic conditions for survival than other more advanced developmental stages such as saplings and adults. In this study, the only case in which competition was revealed was related to the growth of *A. angustifolia* saplings. Our results showed that competition did not occur when transplants were still small seedlings but competition started to occur when transplants achieved a larger size during sapling phase (Fig. 10.8a). However, when fertilizer was applied competition switched to facilitation (Fig. 10.8b). The results indicate that nutrient competition by neighbour vegetation starts to occur only when seedlings achieve a larger size. However, when nutrients are added to the soil and competition is overcome, the presence of vegetation still improves seedling growth.

Invasion of the exotic pine has also been improved by the presence of early successional vegetation. Indeed, pine invasion in the studied old-field was quite aggressive over time. When naturally established pine were counted 5 years after clear-cut we found 1,224 pine individuals established at the site, almost half of them were already in sub-adult stage. By that time the site has also been dominated by shrubs and small early successional trees. The pairwise effects of these shrubs and trees on colonization of late successional species are still unknown, however, ongoing studies have shown that the spatial distribution of *Pinus* species is negatively correlated with the spatial distribution of early successional trees of the genus *Vernonia* (Asteraceae). Future experimental studies may look at possible inhibition patterns.

10.6 Concluding remarks

This chapter provides evidences that mechanisms of facilitation may occur quite frequently in neotropical old-fields. Most of the reasons underlying this pattern relates to seedlings being more sensitive to environmental stress than other life stages such us saplings and adult trees (Fig. 10.1). Moreover, early-successional vegetation not only improves the survival of native tree seedlings but it also improves the survival of undesirable exotic *Pinus* seedlings that reinvade old field areas after the *Pinus* plantation is clear-cut. These results have important consequences for programs of restoration and management of such areas. Removal of early successional vegetation may be applied in situations where desirable species have already established and achieved sapling phase. Further investigations on species by species interactions are needed to provide specific information on species that might inhibit exotic *Pinus* growth while not affecting native seedlings performance (Myster and Pickett 1992).

References

Aide, T. M., and Cavelier, J. 1994. Barriers to lowland tropical forest restoration in the Sierra Nevada de Santa Marta, Colombia. Restoration Ecology 2:219–229.

Alados, C. L., Gotor, P., Ballester, P., Navas, D., Escos, J. M., Navarro, T., and Cabezudo, B. 2006. Association between competition and facilitation processes and vegetation spatial patterns in alpha steppes. Biological Journal of the Linnean Society 87:103–113.

Bertness, M. D., and Callaway, R. M. 1994. Positive interactions in communities. Trends in Ecology and Evolution 9:191–193.

Callaway, R. M., and Walker, L. R. 1997. Competition and facilitation: a synthetic approach to interactions in plant communities. Ecology 78:1958–1965.

Callaway, R. M., Brooker, R. W., Choler, P., Kikvidze, Z., Lortie, C. J., Michalet, R., Paolini, L., Pugnaire, F. I., Newingham, B., Aschehoug, E. T., Armas, C., Kikodze, D., and Cook, B. J. 2002. Positive interactions among alpine plants increase with stress. Nature 417:844–848.

Chilvers, G. A., and Burdon, J. J. 1983. Further studies on a native Australian eucalypt forest invaded by exotic pines. Oecologia 59:239–245.

De Steven, D. 1991a. Experiments on mechanism of tree establishment in old-field succession: seedling emergence. Ecology 72:1066–1075.

De Steven, D. 1991b. Experiments on mechanism of tree establishment in old-field succession: seedling survival and growth. Ecology 72:1076–1088.

Ganade, G. 2007. Processes affecting succession in old fields of Brazilian Amazonia. In Old-Fields: Dynamics and Restoration of Abandoned Farmland, ed. V. A. Cramer and R. J. Hobbs. New York: Island Press.

Ganade, G., and Brown, V. K. 2002. Succession in old pastures of central Amazônia: role of soil fertility and plant litter. Ecology 83:743–754.

Gill, D. S., and Marks, P. L. 1991. Tree and shrub seedling colonization of old-fields in central New York. Ecological Monographs 61:183–205.

Gomez-Aparicio, L., Zamora, R., Gomez, J. M., Hodar, J. A., Castro, J., and Baraza, E. 2004. Applying plant facilitation to forest restoration: a meta-analysis of the use of shrubs as nurse plants. Ecological Applications 14:1128–1138.

Hastwell, G. T., and Facelli, J. M. 2003. Differing effects of shade-induced facilitation on growth and survival during the establishment of a chenopod shrub. Journal of Ecology 91:941–950.

Higgins, S. I., and Richardson, D. M. 1998. Pine invasions in the southern hemisphere: modeling interactions between organism, environment and disturbance. Plant Ecology 135:79–93.

Kulmatiski, A. 2006. Exotic plants establish persistent communities. Plant Ecology 187:261–275.

Loik, M. E., and Holl, K. D. 2001. Photosynthetic responses of tree seedlings in grass and under shrubs in early-successional tropical old fields, Costa Rica. Oecologia 127:40–50.

Lortie, C. J., and Callaway, R. M. 2006. Re-analysis of meta-analysis: support for the stress-gradient hypothesis. Journal of Ecology 93:7–16.

Myster, R. W. 2004. Post-agricultural invasion, establishment, and growth of neotropical trees. The Botanical Review 70:381–402.

Myster, R. W., and Pickett S. T. A. 1992. Dynamics of associations between plants in ten old fields during 31 years of succession. Journal of Ecology 80:291–302.

Nathan, R., Katul, G. G., Horn, H. S., Thomas, S. M., Oren, R. Avissar, R., Pacala, S. W., and Levin, S. A. 2002. Mechanisms of long-distance dispersal of seeds by wind. Nature 418:409–413.

Nepstad, D., Uhl, C., and Serrão, E. A. S. 1991. Recuperation of a degraded Amazonian landscape: forest recovery and agricultural restoration. Ambio 20:248–255.

Richardson D. 1998. Forestry trees as invasive aliens. Conservation Biology 12:18–26.

Rouget, M., Richardson, D. M., Milton, S. J., and Polakow, D. 2001. Predicting invasion dynamics of four alien Pinus species in a highly fragmented semi-arid shrubland in South Africa. Plant Ecology 152:79–92.

Rousset, O., and Lepart, J. 2000. Positive and negative interactions at different life stages of a colonizing species (*Quercus humilis*). Journal of Ecology 88:401–412.

Vieira, I. C. G., Uhl, C., and Nepstad, D. 1994. The role of the shrub *Cordia multispicata* Cham. as a succession 'facilitator' in an abandoned pasture, Paragominas, Amazonia. Vegetatio 115: 91–99.

Wilson, S. D. 1999. Plant interactions during secondary succession. In Ecosystems of Disturbed Ground, ed. L. R. Walker, pp. 231–254. Amsterdam: Elsevier.

Zanini, L., and Ganade, G. 2005. Restoration of *Araucaria* Forest: The role of perches, pioneer vegetation and soil fertility. Restoration Ecology 13:507–514.

Zanini, L., Ganade, G., and Hübel, I. 2006. Facilitation and competition influence succession in a subtropical old field. Plant Ecology 185:179–190.

Chapter 11
Recruitment of Dry Forest Tree Species in Central Brazil Pastures

Aldicir Scariot, Daniel L. M. Vieira, Alexandre B. Sampaio, Ernestino Guarino and Anderson Sevilha

11.1 Introduction

Despite its importance as a wealthy source of unique biodiversity (Janzen 1988) and that it covers 42% of the tropical vegetation worldwide (Murphy and Lugo 1995), the dry forest is the most endangered terrestrial ecosystem and one of the least protected (Scariot and Sevilha 2005). In Central America for example, less than 1% of the original 500,000 km^2 is preserved in conservation units (Janzen 1988, Sanchez-Azofeifa et al. 2003). In Brazil, the seasonal deciduous forests—a type of dry forest—covers 27,367,815 ha (3.21%) of the territory, from which 1,072,946 ha (3.9%) are protected by conservation units. Among conserved areas of deciduous dry forest, only 117,980 ha (0.43%) are in conservation units of restricted use as national parks and biological reserves. Most of the conservation units (402,456 ha or 1.47%) allow limited economic use of the land and natural resources. There is no information of the conservation unit assigned in official database for 552,509 ha (2.02%) of dry forests (Sevilha et al. 2004).

Dry forests occur in many soil classes. In Brazil they occur in at least 13 soil classes of the Brazilian Soil Classification (Scariot and Sevilha 2005), but deciduous dry forests occur predominantly in rich soils. There are many valuable trees used for timber. These forests are the main targets of farmers, who log the trees and sell the timber to acquire financial resources to establish farms. In Central Brazil, where some of the most important remnants of dry forests of the country occur (Fig. 11.1), most of the area originally occupied by dry forests has been replaced by pasture composed by exotic grass in extensive farmlands on rich soils.

Recommendations have been made for the creation of new conservation units, but given the high degree of landscape fragmentation any new conservation unit will contain significant proportion of its area covered by pasture and forest fragments with different levels of disturbance caused by logging, cattle grazing, fire and invasive species. Thus, in this scenario it is fundamental to understand the mechanisms of tree regeneration in pasture fields to address vegetation recovery and use this information to facilitate forest regeneration.

There are four to five times more studies on the natural regeneration of rainforests than on dry forests, and even less on restoration (Vieira and Scariot 2006a). Taking into account the threat level, the lack of conservation units, the lack of information

R. W. Myster, *Post-Agricultural Succession in the Neotropics.*

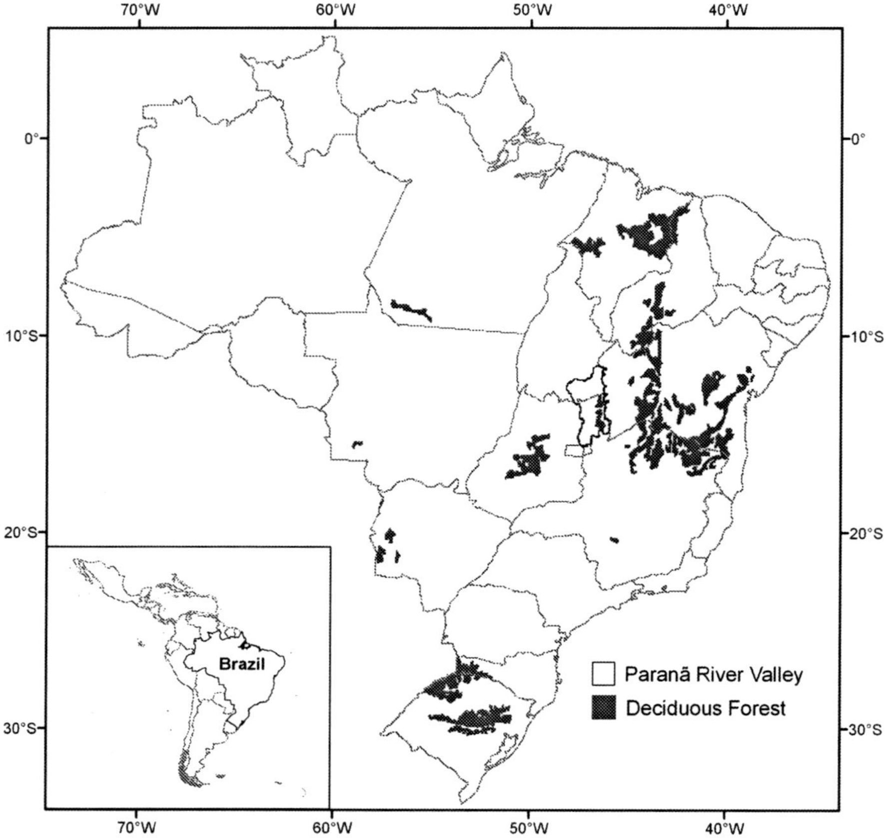

Fig. 11.1 Map of Brazil showing the distribution of the deciduous forests (shaded areas) and the location of the Paranã River Valley in Central Brazil

on regeneration mechanisms, and understanding how regeneration can be enhanced are crucial concerns for the conservation, restoration and management of the dry forests.

11.2 The Case of Paranã River Valley in Central Brazil

The Paranã River valley has an area of 59,403 km^2 in the Cerrado biome. It is embedded in a transition zone between the wet Amazon region and the semi-arid Caatinga region. It has a Tropical climate with two well defined seasons (AW of Köppen) varying to Cwa (Altitude Tropical Climate) (IBGE 1995).

This region presents a mosaic of vegetation types, and has some of the last remnants of seasonally deciduous forests in Brazil. These occur from 400–600 m above sea level, with annual precipitation of 1,236 mm ± 255 (SD, data from 1969 to 1994). 89% of the rains fall between October and March. The mean annual temperature is 23°C, fairly constant throughout the year. The landscape is flat and consists of limestone geology, with occasional limestone outcrops containing a moderately distinct tree flora (Silva and Scariot 2003). Nitosoil is the main soil class upon flat to lightly rugged (Krejci et al. 1982). Rocky limestone outcrops are common in the region.

This basin is between two well known vascular plant distribution patterns in Brazil. One of them is formed by forest species with different levels of leaf shedding that depend mainly on the presence of fertile soils within the Cerrado domain and occurs mainly in the northeast-southeast arch, connecting the Caatinga to the Chaco border. The other pattern is formed by species from the Amazon and Atlantic rainforests, crossing the Cerrado biome through riparian forests (Oliveira-Filho and Ratter 1995).

On the flat lands, the forest canopy is 17–23m height with a basal area of 23–28m^2/ha (Scariot and Sevilha 2005). The authors found 128 tree species in 90 genera and 41 families. Most species are commonly found in large Brazilian biomes such as Cerrado, Caatinga, Amazonia and Atlantic Forest (Scariot and Sevilha 2005). Cattle farming, logging and incipient agriculture are the main economic activities of the region. Wood has been harvested mainly for corral and fence building, for the production of woodchip, and for civil construction. The main exploited species are *Astronium fraxinifolium* Schott, *Myracrodruon urundeuva* Fr. All., *Schinopsis brasiliensis* Engl. and *Tabebuia impetiginosa* (Mart. ex DC.) Standl., which have been commercialized locally, regionally and also in the Southern Brazilian markets (IBGE 1995).

This region has been acknowledged for biodiversity conservation efforts by Brazilian researchers and the Government (Brasil 2002). However, despite its importance, the region is dominated by cattle farms and forest fragments rarely exceeding 250 ha (Andahur 2001). There are three conservation units, but only the smallest one, with less than 2,000 ha, is fully implemented.

11.3 Seed Germination in Pastures

Seed dispersal is the main barrier for plant recruitment on abandoned pastures (Holl 1999, Holl et al. 2000, Wijdeven and Kuzee 2000), but it can be overcome either by direct seeding (Engel and Parrotta 2001, Camargo et al. 2002, Doust et al. 2006) or by installation of bird perching (Melo et al. 1997, Holl 1998, Duncan and Chapman 1999, Melo et al. 2000). However, after sowing germination can also be highly limited by (1) seed predation (Nepstad et al. 1996, Holl and Lulow 1997, Jones et al. 2003, Myster 2003, 2004a,b); (2) competition with exotic grasses (Desteven 1991, Holl 1999, Holl et al. 2000) and (3) water shortage (Lieberman and Li 1992, Ray and Brown 1995), and (4) pathogens (Myster 1997, 2004a). The relative importance of each factor varies widely among species and studied areas (Holl 1999).

Although studies are scarce, seed predation has been found to have high impact on seed germination of dry forest species in pastures in Central Brazil (Vieira and Scariot 2006b), similarly to findings from other tropical regions (Nepstad et al. 1996,

Camargo et al. 2002, Myster 2004a,b). The intensity of seed predation is highly dependent on the seed species (Osunkoya 1994, Holl and Lullow 1997). Medium sized seeds (0.2–4 g) are usually more susceptible to rodents, since they are easier to find and they are easier to manipulate than larger seeds (Price and Jenkins 1986, Nepstad et al. 1990, Osunkoya 1994, but see Hammond 1995, Holl and Lulow 1997 for different results). Less rodent abundance in pastures may result in less medium-size seed predation than in forests (Jones et al. 2003, Meiners and LoGiudice 2003). Small seeds are eaten mainly by insects, such as ants (Nepstad et al. 1990, 1996, Osunkoya 1994), resulting in higher predation of small seeds in pastures (Nepstad et al. 1990 1996, Duncan and Duncan 2000, Jones et al. 2003). From 6 tree species studied in active pastures of Central Brazil dry forests, the small seeded *Tabebuia impetiginosa* (0.11g) and *Astronium fraxinifolium* Schott ex Spreng. (0.02g) were highly preyed or removed, mostly by ants (*Tabebuia*: 50% and *Astronium*: 28%), when compared to seeds dispersed in forest remnants. In recently abandoned pastures, small seeds are preyed and removed at rates varying from 46–73% under grass cover, although one species, *Guazuma ulmifolia* experiences only 1% of seed loss (Guarino 2004). The large seeded (>7.5g) *Cavanillesia arborea* (Willdenow) K. Schum. and *Swartzia multijuga* Vogel, had over 90% predation by cattle in pastures (Vieira and Scariot 2006b).

Harsh climatic conditions during the dry season and dry spells in the wet season can be critical for seed survival in dry forest regions (Gerhardt 1994, Ray and Brown 1995, McLaren and McDonald 2003). Desiccation killed ca. 75% of the thin-coated and high water-containing seeds of *Swartzia multijuga* and *Eugenia dysenterica* DC. in the pasture (Vieira and Scariot 2006b). A small amount of shade in the forests, however, was enough to prevent seed desiccation. Small seeds with low water content and small mass can survive the harsh environmental conditions of pastures (Vieira and Scariot 2006b). In the wet season, grasses maintain soil moisture to the same levels found under deciduous dry forests canopy (22–23%) thus contributing to a significant increase in seed germination in 6 out of 12 tree species in Central Brazil pastures (Guarino 2004). Three species (*Cedrela fissilis*, *Amburana cearensis* and *Sterculia striata*) showed increases in seed germination by 2 to 10 fold under grass cover (Fig. 11.2).

Pathogen attack was also a significant source of seed mortality, mostly for *Aspidosperma pyrifolium*. *A. pyrifolium* and *Tabebuia impetiginosa* experienced high pathogen attack in greenhouse seedlings, mainly in most shaded conditions (Vieira 2006). Even though this is well known for rain forest species (e.g. Augspurguer 1983, Dalling et al. 1998), seed mortality by pathogens has been recently reported for dry forest species as well (Grogan and Galvão 2006, Vieira and Scariot 2006b). Despite the lack of information, it has been reported that *Fusarium* and *Colletotrichum* fungi are the second most important seed mortality agents in abandoned pastures and early succession areas in Puerto Rico and Ecuador (Myster 1997, 2004a). However, fungi from *Phoma* and *Phyllosticta* genera and *Colletotrichum gloeosporioides* although abundant on *Inga* and *Cecropia schreberiana* respectively, resulted in modest seedling leaf loss (Myster 2002).

The use of direct seeding in pastures must consider that adequate species choice and seeding time are fundamental for restoration success. In the Paranã River Valley

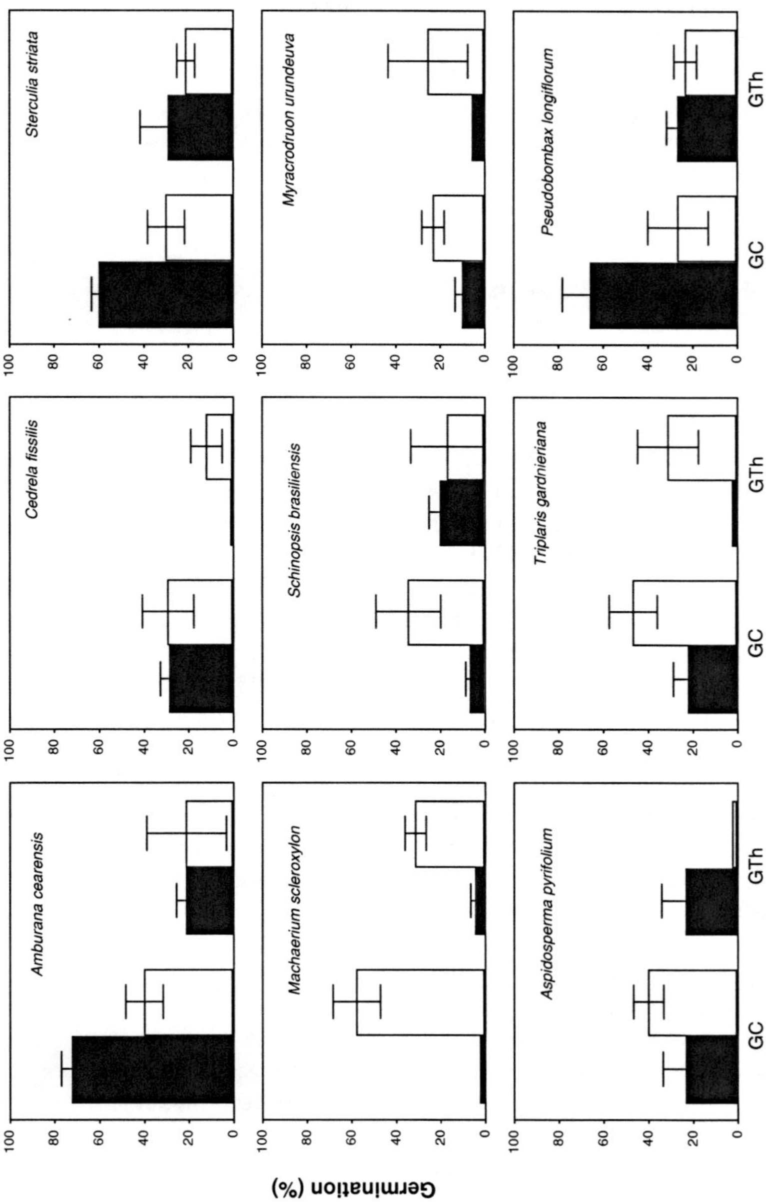

Fig. 11.2 Seed germination (mean (1 standard error) of nine dry forest tree species in abandoned pastures in the Paranã River Valley, Central Brazil. Shaded bars indicate buried seeds and opens bars unburied seeds (Grass Cover (GC), Grass thinning (GTh))

dry forests, priority must be given to species with seeds with small mass and low water content that can tolerate the harsh environmental conditions found in early successional stages and are less affected by predation and removal.

The positive effect of grass cover on seed germination must be further explored to better understand seedling establishment and growth. Burying seeds instead of disposing them on the soil surface, seems to be a good strategy to decrease mortality by predation (Woods and Elliott 2004) and desiccation (Negreros-Castillo et al. 2003) in agricultural lands previously covered by dry forests and degraded soils (Doust et al. 2006). However, Guarino (2004) found contradictory results, highlighting that this issue deserves further investigation.

11.4 Seedling Establishment

Several factors have been indicated to restrict the establishment of forest tree seedlings in pastures. These factors can act differently in germination and seedling establishment processes (Schupp 1995). For instance, some species can germinate under grass cover but will not develop due to grass competition (Zimmerman et al. 2000, Holl 2002). Herbivory by wild (Nepstad et al. 1996, Holl and Quiros-Nietzen 1999, Sweeney et al. 2002) or domestic animals (Posada et al. 2000) have been considered to affect seedling establishment in pastures. Other factors include limited water and nutrient availability (Aide and Cavelier 1994, Nepstad et al. 1996), competition with exotic or invasive plants (Holl 2002, Sweeney et al. 2002, Hau and Corlett 2003), high light availability, high air and soil temperatures (Nepstad et al. 1996, Loik and Holl 1999, Alvarez-Aquino et al. 2004) and soil compaction (Nepstad et al. 1996). Among these, grass competition is considered to be the most limiting, especially in tropical rain forest regions (Holl et al. 2000). Compared to forest environments, the establishment of seedlings in pastures is considerably reduced (Aide and Cavelier 1994, Nepstad et al. 1996, Alvarez-Aquino et al. 2004).

In two abandoned pastures originally covered by seasonal deciduous forests in Central Brazil, tree seedling survival was low but apparently was not related to grass competition. Despite the low survival, thinning the grass cover of *Andropogon gayanus* Kunth. did not interfered on the survival of 6 out of 7 species of planted tree seedlings after 14 months (Fig. 11.3). Some grass species which develop in tussocks, such as *A. gayanus*, leave bare soil patches, covered by a canopy of grass leaves, in between the grass tussocks. In the water deficient ecosystem of dry forests, grass canopy improves the microclimate allowing germination and establishment of tree seedlings in between the grass tussocks (Aide and Cavelier 1994, Guarino 2004). Pasture temperature is higher in plots where the grass cover is cleared (Gerhardt 1996, Zimmerman et al. 2000). Therefore the cooler and moister environment created by the grass cover seems to balance the negative effects of grass competition (Gerhardt 1996).

The benefits of grass cover might be even higher than the possible negative effects of grass competition. One of the 7 planted species in abandoned pastures in Central Brazil, *Talisia esculenta* (A. St.-Hil.) Radlk., had lower survival for grass thinning

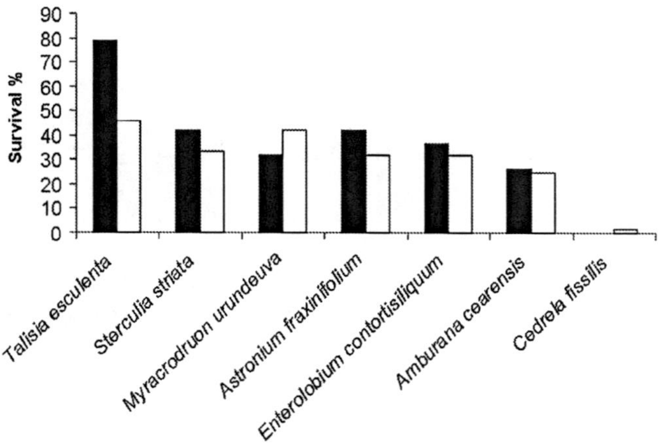

Fig. 11.3 Percentage survival after 14 months of planting of tree seedlings in two pastures in areas originally covered by Dry Forests (data from both pastures aggregated). Dark bars indicate no grass thinning (N = 19 seedlings per species) and open bars indicate grass thinning (N = 57 seedlings per species; grass thinning once, twice or four times). The only species that showed significant effect of grass thinning was *Talisia esculenta* (Pearson Chi-square = 6.4, df = 1, $p = 0.01$)

treatment (Fig. 11.3). *Talisia esculenta* seems to survive less in water stressed conditions of bare soil during the dry season and seems to be able to survive under low light levels created by the grass cover. The soil tends to be moister under the grass canopy (Hooper et al. 2002), allowing higher survival and growth of some tree species than in areas with less grass cover (Aide and Cavelier 1994). In dry forest regions water might be the most limiting resource (Gerhardt 1996, Cabin et al. 2002, Marod et al. 2002) and the grass cover in pastures might ameliorate the water deficit for tree regeneration (Aide and Cavelier 1994, Gerhardt 1996). The removal of dominating shrubs in a deforested seasonal moist evergreen forest in Uganda affected tree regeneration in a negative way, indicating possible facilitation interactions between trees and shrubs (Duncan and Chapman 2003). This was similar to the interactions for tree seedlings survival and grass cover found in our studies in Central Brazil. The interactions among grasses and tree regeneration in harsh environments such as pastures in dry forests tend to be more a facilitation relationship rather than competition (Callaway and Walker 1997).

Despite of the apparent positive effect of grass cover on tree seedling survival, seedling growth can be reduced by grass competition (Holl 1998, Hooper et al. 2002, Schaller et al. 2003). A study carried out by Sampaio et al. 2007 in the same region in Central Brazil dry forests that cleared out the grass cover leaving only the naturally established tree regeneration found a higher average increase in height compared to control plots, indicating competition between natural tree regeneration (mostly root-suckers) and grasses. The growth of dry forest tree seedlings planted in greenhouses under different light conditions (10, 40 and 72% of full sun photon flux density) seems to be higher in the most elevated light levels for the majority of tested species, however seedling survival did not follow a consistent pattern according to the light gradient (Vieira 2006). Grass root system also inhibits tree root development resulting in low growth due to grass competition (Putz and Canham 1992, Schaller et al. 2003). In areas with pronounced dry season, grass root competition may have decreased seedling survival and growth. However, clearing out the grass cover (i.e. mowing treatment) can transform pastures in even harsher environments, as detrimental to seedling establishment as grass competition (Hooper et al. 2002).

Light seems to promote seedling growth, and grass cover might out compete tree seedlings creating shadow, inhibiting tree root development and therefore reducing growth. In contrast, low light levels as those created by grass cover may not promote tree seedling survival and some species regeneration may even be facilitated by grass cover. This may indicate water deficit reduction of bare soil of dry forest regions. Therefore, light and water interacting with grass cover seem to be promoting tree seedling establishment at the studied pasture sites in the dry forest regions. These results suggest that seedlings can be introduced in pastures among grass tussocks for pasture restoration (as in *A. gayanus* pastures in Central Brazil, Sampaio et al. 2007). However, grass cover should be cleared just around the seedlings to improve tree seedling development (Vieira and Scariot 2006a). This will avoid excessive exposure to the plants under harsh conditions.

11.5 Resprouting Ability

Root-suckers from trees, shrubs and lianas are found in recently established pastures in tropical forest areas (Uhl et al. 1988, de Rouw 1993, Nepstad et al. 1996). The density of these shoots can be high enough to allow forest regeneration if not eliminated. Actually, shoot elimination by clipping or by use of selective herbicides and root collection after plowing, are common management practices in Central Brazil dry forests region. These practices respond to a large proportion of annual money and labor expenditures in pasture management. Even though farmers battle forest plants exhaustively, many dry forest woody plants persist resprouting many years after pasture establishment (Vieira et al. 2006c, Sampaio et al. in press).

The real contribution of root-suckers relative to seedling establishment for forest regeneration in pastures along tropical forest areas is neglected. To some degree this is due to the laborious work need to detect resprouting which involves digging the superficial soil. The other reason is simply because root-suckers are depleted by sequential cutting, fire, and intensive tractor use (Uhl et al. 1988, de Rouw 1993, Sampaio et al. 1993). Much more attention has been given to evaluate limiting factors and how to improve seed dispersal and survival and seedling survival and growth (reviewed in Holl 2002, Myster 2004b).

Plowing the soil in the middle of the dry season and seeding grass at the beginning of the rainy season is a common practice among farmers of Central Brazil to restore grass cover in pastures. Plowing completely eliminates plant cover, allowing plants to start resprouting soon. Vieira et al. (2006c) investigated the regeneration from root-suckers after plowing in a 10 year-old pasture, a 25 year-old pasture and an early successional forest recently clear cut in Central Brazil dry forests. They found that species richness and density were extremely high compared to other disturbed tropical sites with comparable history of land use (e.g. studies by Uhl et al. 1988, Nepstad et al. 1996, Zahawi and Augspurger 1999, Holl et al. 2000). They found 14–182 individuals per 100 m^2 of 30–42 species (6, 5–17, 6 species per 100 m^2) among the three sites, representing 80% of the species richness found in intact forest remnants (Vieira et al. 2006c). In another investigation on tree regeneration in active pastures in dry forests of Central Brazil, stem density, regardless of the origin, did not varied with pasture age (from ≤ 6 up to 40 years old, n=25), while richness decreased significantly with age (Sampaio et al. in press). Despite the long time since forest removal, pastures still maintained a significant number of dry forest species. The 40 year-old pastures still had 12 dry forest tree species regenerating in relatively high densities (19–58 stems/100 m^2, 2nd and 3rd quartiles). Species composition was quite similar among pasture ages and pastures and forest remnants (Vieira et al. 2006c, Sampaio et al. in press), indicating the potential of these areas for dry forest return.

These optimistic results corroborate others findings that dry forest species are strong resprouters and contribute to forest regeneration after disturbance. In other tropical dry forests, studies show a high frequency of resprouting after disturbance (Kennard et al. 2002, McLaren and McDonald 2003). The reasons why resprouting is a particularly important recovery mechanism in dry forests are not certain. Some probable reasons for this trait are (1) seeds have a lower probability of establishment, so the survivorship by resprouting was favored in this forest ecosystem (Ewel 1980); (2) trunk bases are less prone to decay, having more chance to resprout (Ewel 1980); and (3) dry forest plants are adapted for above ground mortality or drought, so they usually resprout (Sampaio et al. 1993, Bond and Midgley 2001).

Although some species are strong resprouters others are not and resprouting ability will depend on the frequency, intensity and duration of the disturbance (Sampaio et al. 1993, Kammesheidt 1999, McLaren and McDonald 2003). Information available from actual literature does not allow one to predict what species are able to resprout in a forest community (Bond and Midgley 2001, Vesk and Westoby 2004). Despite this limitation on the current knowledge it is essential to understand the traits or identities of strong resprouters to be able to (1) help to understand present and future community composition (Kennard 2002, Saha and Howe 2003); (2) use root and branch cuttings of these species as nurse trees in restoration plans; and (3) invest in other restoration mechanisms for weak or non-sprouting species in early successional forests dominated by resprouters.

In dry forests of Central Brazil, species show high density in forest remnants, but that does not happen in ploughed sites (see above in this section; Vieira et al. 2006c). Species with low-density wood such as *Cavanillesia arborea* and *Chorisia pubiflora* (A. St.-Hil.) G. Dawson. *Spondias mombin* L. and *Pseudobombax tomentosum* (C. Martius and Zuccarini) Robyns, and soft wood species were found only in forest fragments and early successional sites. Species with low-density wood could lose their resprouting ability faster and sometimes did not resprout at all because of decay. In a chronosequence of pasture implementation, most species appear to be able to continue resprouting for up to 11–16 years after disturbance (Sampaio et al. in press). However, in 40 year-old pastures *Eugenia dysenterica* dominated the community, which was poorer in tree species than younger pastures. *E. dysenterica* occurs in Central Brazil dry forests, but occurs in high frequency in the surrounding savanna vegetation. Species from the Brazilian savannas tend to resprout more than cogeneric species typical from riparian forests in the Cerrado biome, where root systems is not as developed as well as in the dry soils of savannas (Hoffmann et al. 2003). Therefore, savanna species like *E. dysenterica* that develop deep root system to survive in dry soils of savannas and dry forests might be able to survive longer and even increase its density via vegetative reproduction in pastures. The composition of the secondary vegetation developed in abandoned pastures will be dominated by strong resprouters, probably hard wood and savanna-related species.

Dry forests of Central Brazil show a high degree of resilience after pasture implementation due to root resprouting. Thus, planting seedlings to enhance early succession in abandoned pastures may not be desirable, particularly if standard planting techniques are used. Soil and vegetation disturbances, such as plowing or mechanical hole digging, reduce the density of naturally regenerating trees and consequently slow recovery (Sampaio et al. 2007). Non-resprouting species can be efficiently introduced by cautiously planting nursery-grown seedlings, minimizing damage to natural regeneration. Considering that resprouting ability is positively related to branch cutting success (Itoh et al. 2002), using branch cuttings to improve dry forest recovery should be tested. The rooting ability of branch cuttings was tested, with relative success, for tropical rain forests (Itoh et al. 2002, Zahawi 2005), but we have not found published studies for dry forests. Root cuttings taken from fallen trees from deforested areas may also be tested to enhance dry forest restoration, since planting them is equivalent to root-suckering after plowing.

References

Aide, T. M., and Cavelier, J. 1994. Barriers to lowland tropical forest restoration in the Sierra Nevada de Santa Marta, Colombia. Restoration Ecology 2(4):219–229.

Alvarez-Aquino, C., Williams-Linera, G., and Newton, A. C. 2004. Experimental native tree seedling establishment for the restoration of a Mexican cloud forest. Restoration Ecology 12(3):412–418.

Andahur, J. P. 2001. Florestas e questões de gestão ambiental na bacia do Rio Paranã. M.Sc. Thesis, Departamento de Engenharia Florestal, Universidade de Brasília, Brasília, DF, Brazil.

Augspurguer, C. K. 1983. Seed dispersal of the tropical tree, Platypodium elegans, and the escape of its seedlings from fungal pathogens. Journal of Ecology 71(3):759–771.

Bond, W. J., and Midgley, J. J. 2001. Ecology of sprouting in woody plants: the persistence niche. Trends in Ecology and Evolution 16(1):45–51.

Brasil. 2002. Biodiversidade Brasileira. Avaliação e identificação de áreas prioritárias para conservação, utilização sustentável e repartição de benefícios da biodiversidade brasileira. Ministério do Meio Ambiente, Brasília.

Cabin, R. J., Weller, S. G., Lorence, D. H., Cordell, S., and Hadway, L. J. 2002. Effects of microsite, water, weeding, and direct seeding on the regeneration of native and alien species within a Hawaiian dry forest preserve. Biological Conservation 104(2):181–190.

Callaway, R. M., and Walker, R. 1997. Competition and facilitation: a synthetic approach to interactions in plant communities. Ecology 78(7):1958–1965.

Camargo, J. L. C., Ferraz, I. D. K., and Imakawa, A. M. 2002. Rehabilitation of degraded areas of central Amazonia using direct sowing of forest tree seeds. Restoration Ecology 10(4):636–644.

Dalling, J., Swaine, M., and Garwood, N. 1998. Dispersal patterns and seed bank dynamics of pioneer trees in moist tropical forest. Ecology 79(2):564–578.

de Rouw, A. 1993. Regeneration by sprouting in slash and burn rice cultivation, Taï Rain Forest, Côte d'Ivoire. Journal of Tropical Ecology 9(4):387–408.

Desteven, D. 1991. Experiments on mechanisms of tree establishment in old-field succession - seedling emergence. Ecology 72(3):1066–1075.

Doust, S. J., Erskine, P. D., and Lamb, D. 2006. Direct seeding to restore rainforest species: Microsite effects on the early establishment and growth of rainforest tree seedlings on degraded land in the wet tropics of Australia. Forest Ecology and Management 234(1–3):333–343.

Duncan, R. S., and Chapman, C. A. 1999. Seed dispersal and potential forest succession in abandoned agriculture in tropical Africa. Ecological Applications 9(3):998–1008.

Duncan, R. S., and Chapman, C. A. 2003. Tree-shrub interactions during early secondary forest succession in Uganda. Restoration Ecology 11(2):198–207.

Duncan, R. S., and Duncan, V. E. 2000. Forest succession and distance from forest edge in an Afro-tropical grassland. Biotropica 32(1):33–41.

Engel, V. L., and Parrotta, J. A. 2001. An evaluation of direct seeding for reforestation of degraded lands in central Sao Paulo state, Brazil. Forest Ecology and Management 152(1–3): 169–181.

Ewel, J. 1980. Tropical succession: manifold routes to maturity. Biotropica 12(2):2–7.

Gerhardt, K. 1994. Seedling development of four tree species in secondary tropical dry forest in Guanacaste, Costa Rica. Comprehensive Summaries of Uppsala Dissertations from the Faculty of Science and Technology 39. Stockholm, Sweden, Uppsala.

Gerhardt, K. 1996. Effects of root competition and canopy openness on survival and growth of tree seedlings in a tropical seasonal dry forest. Forest Ecology and Management 82(1):33–48.

Grogan, J., and Galvão, J. 2006. Factors limiting post-logging seedling regeneration by big-leaf mahogany (*Swietenia macrophylla*) in southeastern Amazonia, Brazil, and implications for sustainable management. Biotropica 38(2):219–228.

Guarino, E. S. G. 2004. Germinação de sementes e estabelecimento de plântulas de árvores em florestas estacionais deciduais e pastagens abandonadas. M.Sc. Thesis. Departamento de Ecologia, Universidade de Brasília, Brasília, DF, Brazil.

Hammond, D. S. 1995. Postdispersal seed and seedling mortality of tropical dry forest trees after shifting agriculture, Chiapas, Mexico. Journal of Tropical Ecology 11:295–313.

Hau, B. C. H., and Corlett, R. T. 2003. Factors affecting the early survival and growth of native tree seedlings planted on a degraded hillside grassland in Hong Kong, China. Restoration Ecology 11:483–488.

Hoffmann, W. A., Orthen, B., and do Nascimento, P. K. V. 2003. Comparative fire ecology of tropical savanna and forest trees. Functional Ecology 17:720–726.

Holl, K. D. 1998. Do bird perching structures elevate seed rain and seedling establishment in abandoned tropical pasture? Restoration Ecology 6(3):253–261.

Holl, K. D. 1999. Factors limiting tropical rain forest regeneration in abandoned pasture: Seed rain, seed germination, microclimate, and soil. Biotropica 31(2):229–242.

Holl, K. D. 2002. Tropical moist forest. In Handbook of Ecological Restoration, Vol. II, ed. M. Perrow and A. J. Davy, pp. 539–558 Cambridge: Cambridge University Press.

Holl, K. D., and Lulow, M. E. 1997. Effects of species, habitat, and distance from edge on post-dispersal seed predation in a tropical rainforest. Biotropica 29(4):459–468.

Holl, K. D., and Quiros-Nietzen, E. 1999. The effect of rabbit herbivory on reforestation of abandoned pasture in southern Costa Rica. Biological Conservation 87:391–395.

Holl, K. D., Loik, M. E., Lin, E. H. V., and Samuels, I. A. 2000. Tropical montane forest restoration in Costa Rica: Overcoming barriers to dispersal and establishment. Restoration Ecology 8:339–349.

Hooper, E., Condit, R., and Legendre, P. 2002. Responses of 20 native tree species to reforestation strategies for abandoned farmland in Panama. Ecological Applications 12:1626–1641.

IBGE, 1995. Zoneamento geoambiental e agroecológico do Estado de Goiás: região nordeste. Instituto Brasileiro de Geografia e Estatística/Divisão de geociências do Centro-Oeste, Rio de Janeiro, Brasil, p. 178.

Itoh, A., Yamakura, T., Kanzaki, M., Ohkubo, T., Palmiotto, P. A., LaFrankie, J. V., Kendawang, J. J., and Lee, H. S. 2002. Rooting ability of cuttings relates to phylogeny, habitat preference and growth characteristics of tropical rainforest trees. Forest Ecology and Management 168:275–287.

Janzen, D.H., 1988. Guanacaste National Park: tropical ecological and biocultural restoration. In Rehabilitating Damaged Ecosystems, Vol. II, ed. J. J. Cairns, pp. 143–192. Boca Raton, FL: CRC Press.

Jones, F. A., Peterson, C. J., and Haines, B. L. 2003. Seed predation in neotropical pre-montane pastures: Site, distance, and species effects. Biotropica 35(2):219–225.

Kammesheidt, L. 1999. Forest recovery by root suckers and above-ground sprouts after slash-and-burn agriculture, fire and logging in Paraguay and Venezuela. Journal of Tropical Ecology 15:143–157.

Kennard, D. K. 2002. Secondary forest succession in a tropical dry forest: patterns of development across a 50-year chronosequence in lowland Bolivia. Journal of Tropical Ecology 18:53–66.

Kennard, D. K., Gould, K., Putz, F. E., Fredericksen, T. S., and Morales, F. 2002. Effect of disturbance intensity on regeneration mechanisms in a tropical dry forest. Forest Ecology and Management 162:197–208.

Krejci, L. C., Fortunato, F. F., and Côrrea, P. R. S. 1982. Pedologia – Levantamento exploratório de solos. In Projeto RADAMBRASIL Folha SD. 23 Brasília; geologia, geomorfologia, vegetação e uso potencial da terra. Ministério de Minas e Energia, Secretária-Geral, Vol. 29, ed. H. H. Moreira, p. 660. Brasil: Rio de Janeiro..

Lieberman, D., and Li, M. G. 1992. Seedling recruitment patterns in a tropical dry forest in Ghana. Journal of Vegetation Science 3(3):375–382.

Loik, M. E., and Holl, K. D. 1999. Photosynthetic responses to light for rainforest seedlings planted in abandoned pasture, Costa Rica. Restoration Ecology 7:382–391.

Marod, D., Kutintara, U., Tanaka, H., and Nakashizuka, T. 2002. The effects of drought and fire on seed and seedling dynamics in a tropical seasonal forest in Thailand. Plant Ecology 161:41–57.

McLaren, K. P., and McDonald, M. A. 2003. Coppice regrowth in a disturbed tropical dry limestone forest in Jamaica. Forest Ecology and Management 180:99–111.

Meiners, S. J., and LoGiudice, K. 2003. Temporal consistency in the spatial pattern of seed predation across a forest — old field edge. Plant Ecology 168:45–55.

Melo, V. A., Griffith, J. J., Marco Júnior, P., Silva, E., Souza, A. L., Guedes, M. C., and Ozório, T. F. 2000. Efeito de poleiros artificiais na dispersão de sementes por aves. Revista Árvore 24(3):235–240.

Melo, V. A., Guedes, M. C., and Griffith, J. J. 1997. O Uso de Poleiros artificiais para atração de aves dispersoras de sementes em áreas degradadas.Ararajuba 5(2):229–232.

Murphy, P. G., and Lugo, A. E. 1995. Dry forests of central America and the Caribbean. In Seasonally Tropical Forests, ed. H. A. Mooney and S. H. Bullock, pp. 9–34. Cambridge: University of Cambridge Press.

Myster, R. W. 1997. Seed predation, disease and germination on landslides in neotropical lower montane wet forest. Journal of Vegetation Science 8(1):55–64.

Myster, R. W. 2002. Foliar pathogen and insect herbivory effects on two landslide trees in Puerto Rico. Forest Ecology and management 169:231–242.

Myster, R. W. 2003. Effects of species, density, patch-type, and season on post-dispersal seed predation in a Puerto Rican pasture. Biotropica 35(4):542–546.

Myster, R. W. 2004a. Regeneration filters in post-agricultural fields of Puerto Rico and Ecuador. Plant Ecology 172(2):199–209.

Myster, R. W. 2004b. Post-agricultural invasion, establishment, and growth of neotropical trees. Botanical Review 70:381–402.

Negreros-Castillo, P., Snook, L., and Mize, C. W. 2003. Regenerating mahogany (Swietenia macrophylla) from seed in Quintana Roo, Mexico: the effects of sowing method and clearing treatment. Forest Ecology and Management 183(1–3):351–362.

Nepstad, D. C., Uhl, C., and Serrão, E. A. S. 1990. Surmounting barriers to forest regeneration in abandoned, highly degraded pastures: a case study from Paragominas, Pará, Brazil.

In Alternatives to deforestation: steps toward sustainable use of the Amazon rain forest, ed. A. B. Anderson, pp. 215–229. New York, USA: Columbia University Press.

Nepstad, D. C., Uhl, C., Pereira, C. A., and da Silva, J. M. C. 1996. A comparative study of tree establishment in abandoned pasture and mature forest of eastern Amazonia. Oikos 76(1): 25–39.

Oliveira-Filho, A. T., and Ratter, J. 1995 A study of the origin of central brazilian forests by the analysis of plant species distribution patterns. Edinb. J. Bot. 52(2):141–194.

Osunkoya, O. O. 1994. Postdispersal survivorship of North Queensland rainforest seeds and fruits: effects of forest, habitat and species. Australian Journal of Ecology19:52–64.

Posada, J. M., Aide, T. M., and Cavelier, J. 2000. Cattle and weedy shrubs as restoration tools of tropical montane rainforest. Restoration Ecology 8:370–379.

Price, M. V., and Jenkins, S. H. 1986. Rodents as seed consumers and dispersers. In Seed Dispersal, ed. D. R. Murray, pp. 191–235. Sidney, Australia: Academic Press.

Putz, F. E., and Canham, C. D. 1992. Mechanisms of arrested succession in shrublands: root and shoot competition between shrubs and tree seedlings. Forest Ecology and Management 49:267–275.

Ray, G. J., and Brown, B. J. 1995. Restoring Caribbean dry forests: evaluation of tree propagation techniques. Restoration Ecology 3:86–94.

Saha, S., and Howe, H. F. 2003. Species composition and fire in a dry deciduous forest. Ecology 84:3118–3123.

Sampaio, E., Salcedo, I. H., and Kauffman, J. B. 1993. Effect of different fire severities on coppicing of Caatinga vegetation in Serra Talhada, PE, Brazil. Biotropica 25:452–460.

Sampaio, A. B., Holl, K. D., and Scariot, A. 2007. Does restoration enhance regeneration of seasonal deciduous forests in pastures in central Brazil? Restoration Ecology 15:462–471.

Sampaio, A. B., Holl, K. D., and Scariot, A. In press b. Regeneration of Seasonal Deciduous Forest Tree Species in Long-Used Pastures in Central Brazil. Biotropica.

Sanchez-Azofeifa, G. A., Daily, G. C., Pfaff, A. S. P., and Busch, C. 2003. Integrity and isolation of Costa Rica's national parks and biological reserves: examining the dynamics of land-cover change. Biol. Conserv. 109:123–135.

Scariot, A., and Sevilha, A. C. 2005. Biodiversidade, estrutura e conservação de florestas estacionais deciduais no Cerrado. In Cerrado: Ecologia, Biodiversidade e Conservação.Ministerio do Meio Ambiente, p. 439. Brasil: Brasília.

Schaller, M., Schroth, G., Beer, J., and Jimenez, F. 2003. Root interactions between young Eucalyptus deglupta trees and competitive grass species in contour strips. Forest Ecology and Management 179:429–440.

Schupp, E. W. 1995. Seed-seedling conflicts, habitat choice, and patterns of plant recruitment. American Journal of Botany 82:399–409.

Sevilha, A. C., Scariot, A., and Noronha, S. E. 2004. Estado atual da representatividade de unidades de conservação em Florestas Estacionais Deciduais no Brasil. In Silva. A. (Org.) Anais do 55 Congresso Nacional de Botânica (CD-ROM). Viçosa, Minas Gerais, Brasil.

Silva. L. A., and Scariot, A. 2003. Composição florística e estrutura da comunidade arbórea em uma floresta estacional decidual em afloramento calcário (Fazenda São José, São Domingos, GO, bacia do rio Paranã). Acta Bot. Bras. 17(2):305–313.

Sweeney, B. W., Czapka, S. J., and Yerkes, T. 2002. Riparian forest restoration: increasing success by reducing plant competition and herbivory. Restoration Ecology 10:392–400.

Uhl, C., Buschbacher, R., and Serrao, E. A. S. 1988. Abandoned pastures in eastern Amazonia. I. patterns of plant succession. Journal of Ecology 76:663–681.

Vesk, P. A., and Westoby, M. 2004. Sprouting ability across diverse disturbances and vegetation types worldwide. Journal of Ecology 92:310–320.

Vieira, D. L. M. 2006. Regeneração natural de florestas secas: Implicações para a restauração. Dr. Thesis. Departamento de Ecologia, Universidade de Brasília, Brasília, DF, Brazil.

Vieira, D. L. M., and Scariot, A. 2006a. Principles of natural regeneration of tropical dry forests for restoration. Restoration Ecology 14(1):11–20.

Vieira, D. L. M., and Scariot, A. 2006b. Effects of logging, liana tangles and pasture on seed fate of dry forest tree species in Central Brazil. Forest Ecology and Management 230 (1–3):197–205.

Vieira, D. L. M., Scariot, A., Sampaio, A. B., and Holl, K. D. 2006c. Tropical dry-forest regeneration from root suckers in Central Brazil. Journal of Tropical Ecology 22:353–357.
Wijdeven, S. M. J., and Kuzee, M. E. 2000. Seed availability as a limiting factor in forest recovery processes in Costa Rica. Restoration Ecology 8:414–424.
Woods, K., and Elliott, S. 2004. Direct seeding for forest restoration on abandoned agricultural land in northern Thailand. Journal of Tropical Forest Science 16(2):248–259.
Zahawi, R. A. 2005. Establishment and growth of living fence species: An overlooked tool for the restoration of degraded areas in the tropics. Restoration Ecology 13:92–102.
Zahawi, R. A., and Augspurger, C. K. 1999. Early plant succession in abandoned pastures in Ecuador. Biotropica 31:540–552.
Zimmerman, J. K., Pascarella, J. B., and Aide, T. M. 2000. Barriers to forest regeneration in an abandoned pasture in Puerto Rico. Restoration Ecology 8:350–360.

Part III
Restoration, Management, and the Future of Post-Agricultural Areas

Chapter 12
Microbial Biomass in Native Amazonian Ecosystems and its Adaptation to Deforestation and Pasture Introduction and Management

Brigitte J. Feigl, Carlos Eduardo P. Cerri, Carlos C. Cerri and Marisa C. Piccolo

12.1 Introduction

Changes in biogeochemical cycles after land conversion from native forest to pasture have been focused in several aspects, like soil organic matter (SOM) quantity and quality, emissions of greenhouse gases, changes in hydrological and limiting elements cycles (Feigl, Melillo and Cerri 1995, Matson et al. 1997, Neill et al. 1997a,b, 2001, Garcia-Montiel et al. 2000). The decomposition of SOM and the mineralization of nutrients are mediated by microorganisms, and the significant environmental changes associated with land conversion are quite likely to alter their three most critical functions, i.e. labile source and immediate sink of C, N, P and S, and substrate transformations (Dalal 1998). Microbial biomass has even been proposed as a sensitive indicator of soil quality and soil health (Doran and Zeiss 2000, Schloter et al. 2003). As with any reservoir, its size is important not only because it provides an indication of slower, less easily detectable SOM changes, but also because it represents an important labile pool of plant available nutrients. Temporal measures of these changes are currently scarce. Users and policymakers require measures of microbial biomass in order to feed models to develop sustainable cropping systems.

Assessment in soil microbial biomass (SMB) was facilitated in the last 20 years by physiological, biochemical and chemical techniques, including chloroform fumigation incubation (CFI) (Jenkinson and Powlson 1976) and chloroform fumigation extraction (CFE) (Brookes et al. 1985, Vance et al. 1987) among many others. As pointed out by Dalal (1998), it is difficult to compare soil microbial biomass values obtained by different methods in different laboratories due to variants of k_{EC} values (the fraction rendered extractable by fumigation, used to convert the measured data into biomass C), soil moisture contents, incubation temperatures, SOM content, and instrumentation and analytical techniques.

Most of the data presented in this chapter were obtained at Nova Vida Ranch (10°30‘S, 62°30‘W) in Rondônia state, Brazil (Fig. 12.1). The local climate is humid tropical, with an annual average temperature of 25.5°C and precipitation of 2,200 mm (Bastos and Diniz 1982) including a dry season from May to September. Mean daily temperature for the warmest and coolest months varies less than 5°C.

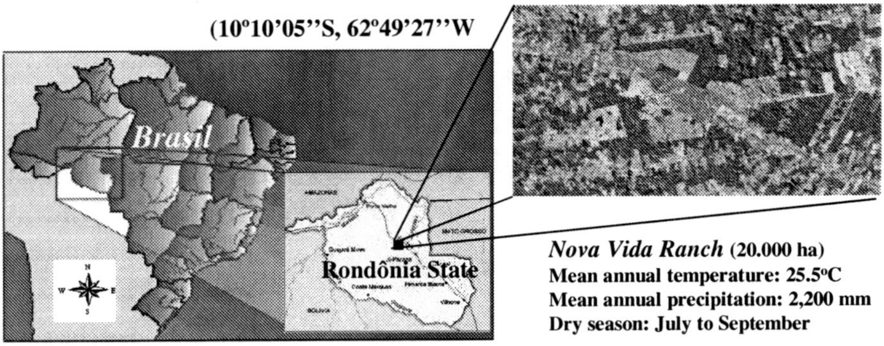

Fig. 12.1 Map of the southwestern Amazon region, indicating the localization of Nova Vida Ranch, origin of most of the microbial biomass data discussed in this chapter

Mean annual relative humidity is 89%. The predominant soil at the ranch is an Ultisol (Moraes et al. 1995), or Argissolo in the Brazilian soil classification (EMBRAPA 1999).

Soil samples under forest and different aged pastures were collected for microbial biomass quantification in four blocks. In most of the cases, five sub-samples were taken from the 0–10 and 10–20 cm soil layer of each block and bulked together to make four composite samples. Microbial biomass was estimated by CFI or CFE methods, processed by the same lab. When necessary, values were recalculated applying the k_{EC}=0.30 calculated by Feigl et al. (1995b) for Amazonian soils. Main soil characteristics are shown in Table 12.1.

Table 12.1 Soil characteristics under forest and pastures of different ages

Site	Layer	pH	Clay	d	C_{total}	N_{total}
	(cm)	(H_2O)	(%)	(g cm^{-1})	------g kg^{-1}	------
Forest	0–10	4.4	22	1.3	12.8	1.09
	10–20	4.1	30	1.5	7.6	0.78
2 month-old pasture	0–10	5.7	34	1.1	11.7	0.93
(4 month after burning)	10–20	4.6	38	1.3	8.6	0.70
3 year-old pasture	0–10	6.3	34	1.2	13.6	1.56
(created 1994)	10–20	6.0	38	1.2	7.8	0.89
10 year-old pasture	0–10	6.5	29	1.4	17.5	1.33
(created 1987)	10–20	6.1	35	1.4	9.3	0.88
15 year-old pasture	0–10	6.2	24	1.6	15.5	1.16
(created 1983)	10–20	6.1	32	1.6	11.8	1.10

12.2 The Role of Soil Microbial Biomass in Amazonian Native Ecosystems

Over three-fourths of the Amazonian rainforest grow on red and yellow clay-like Oxisols which are acidic and low in nutrients (Moraes et al. 1995, Bernoux et al. 2001). Most of the carbon and essential nutrients are fixed in the living vegetation.

As vegetation dies, dead wood, and decaying leaves are rapidly broken down by soil biota. Carbon availability thus often controls soil microbial growth, which positively correlated with aboveground forest litter input. In this way nutrients almost immediately returned to the system as they are taken up by roots that actually grow out of the ground to form a mat on the forest floor in order to more efficiently collect nutrients.

In spite of its importance, there have been few studies about litter production and decay in Amazonian rainforests. Total aboveground biomass (excluded dead wood and litter/rootmat) was estimated by Cummings et al (2002) as about 300 t dry matter ha^{-1}. This value corresponds to the mean of 20 sites located in the southwestern Brazilian Amazon. Annual litter production of this ecosystem was measured by Luizão and Schubart (1987), who reported a range between 6.5 and 7.4 t dry matter ha^{-1} y^{-1} for primary forest areas located near Manaus. Higher values, up to 8.8 or even 9.7 t dry matter ha^{-1} y^{-1} were found by Vasconcelos and Luizão (2004) and Smith et al. (1998a), respectively, in other sites of Amazonian undisturbed forest. The rates of decay and pathways of decomposition are determined by the quality of the litter material, the physical environment and the abundance and composition of decomposer organisms. Smith et al. (1998b) determined an exponential decay constant (k) of 0.605 for Amazonian forest litter. Potentially, 2–3% of total aboveground vegetation is recycled every year by microbial biomass.

A comparison of litter amount and microbial population was done at samples collected during the rainy season (December 1998) and the dry season (August 1999) from a forest site located at Nova Vida Ranch. The forest floor (i.e., L, F, and H, or O1 and O2 horizons) was sampled from five randomly placed 0.3 m^2 quadrates per site. The obtained data (Table 12.2) showed that relative participation of layers, as well as C:N ratio, remained almost the same over the year. Similar microbial biomass was supported by unit of plant dry matter over the year (5.04 × 4.68 g C_{micr} kg^{-1} soil, and 0.79 × 0.72 g N_{micr} kg^{-1} soil during the rainy and the dry season, respectively), although the difference in quantity of fallen leaves makes the dry season a larger reservoir of microbial C and N (Table 12.2). Apparently

Table 12.2 Seasonal changes in aboveground litter stand and supported microbial biomass C and N (kg ha^{-1}) under native forest conditions

Litter layer		Litter dry matter			Microbial biomass		C_{total}:C_{micr}
		Weight	Total C	Total N	C_{micr}	N_{micr}	
		kg ha^{-1}					%
			Rainy season				
(50%)*	L	2,400	1,100	38	8.3	1.1	0.75
(37%)	F	1,600	700	30	12.0	2.6	1.71
(18%)	H	1,300	400	18	6.4	0.5	1.60
Total		**5,300**	**2,200**	**86**	**26.7**	**4.2**	**1.21**
			Dry season				
(48%)	L	4,700	2,100	71	13.9	2.1	0.66
(35%)	F	3,400	1,400	65	23.9	4.4	1.71
(17%)	H	1,700	500	22	8.1	0.6	1.62
Total		**9,800**	**4,000**	**158**	**45.9**	**7.1**	**1.15**

* Relative distribution of litter layers

moisture was not a limiting factor for microbial population maintenance, suggesting that faster litter decay during the rainy period can be partially attributed to increase in the activity of decomposers. Luizão and Schubart (1987) pointed out that the macro-arthropods appear to be the most affected by adverse humidity conditions during the dry season. Ruan et al. (2004) also concluded that aboveground plant litter input had no effect on soil microbial biomass or on its pattern of fluctuation, since it is not directly regulated by soil temperature or moisture at local scales within a tropical wet forest. The authors found asynchronous fluctuation between soil microbial biomass and plant litter fall due to regulation by competition for soil nutrients between microorganisms and plants, and regulation by below-ground carbon inputs associated with the annual solar and drying-rewetting cycles in tropical wet forests.

At the forest site of Nova Vida ranch, microbial biomass in litter represented a pool of labile C and N of about 36 and 6 kg ha^{-1}, respectively, over the year. However, an unknown amount of the recycled material effectively reaches the first 5 cm of soil. Tree roots actively grow into the litter layer and, associated or not to mycorrhizae, take up the released nutrients. The poorly known root biomass itself also plays an integral role in forest floor decomposition process and N dynamics (Smith et al. 1998b).

Microbial biomass in soil is concentrated in the top 5 to 10 cm because substrates are more plentiful there. A survey of sites across Brazilian Amazon region (Fig. 12.2) showed that microbial biomass C (or C_{micr}) ranged from 200 to 2,000 kg ha^{-1} when the upper 10 cm soil layer is considered (Table 12.3).

The soil microbial biomass dynamics in tropical forests and savannas allows accumulation and conservation of nutrients in biologically active forms during the

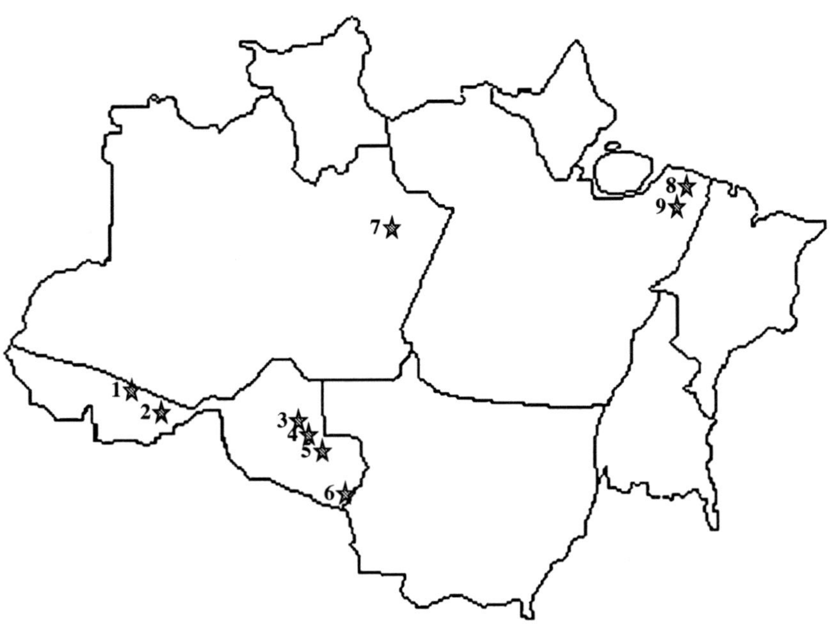

Fig. 12.2 Geographic location of the sites referred on Table 12.3. Samples represent major Amazonian soil types according to Moraes et al. 1995 and Koutika et al. 1999

Table 12.3 Survey of soil microbial biomass C and N under native forest. Data obtained by fumigation-extraction method were recalculated, using k_{EC}=0.30 and k_{EN}=0.54

Site	Season	Soil	Clay	Layer	Soil microbial	biomass	C_{micr}:C_{org}
			(%)	**(cm)**	**(mg C kg^{-1})**	**(mg N kg^{-1})**	**(%)**
1	Late rainy	Oxisol	53	0–10	292	41	0.7
2	Late rainy	Ultisol	67	0–10	579	125	2.3
3	Late rainy	Ultisol	35	0–10	426	86	2.2
4	Dry	Alfisol	20	0–5	293	46	1.7
	Dry	Ultisol	10	0–5	193	30	1.4
5	Dry	Ultisol	18	0–10	407	63	3.2
	Dry	Ultisol	25	0–10	337	50	3.3
	Dry	Alfisol	23	0–10	500	74	2.5
6	Rainy	Oxisol	65	0–5	1,388	36*	5.0
	Dry	Oxisol	65	0–5	784	47*	2.8
7	Rainy	Oxisol	70	0–5	1,035	–	1.4
	Annual	Oxisol	70	0–5	1,287**	–	2.5
	Dry	Oxisol	78	0–10	780	102	3.1
8	Rainy	Oxisol	35	0–5	586	–	4.5
9	Dry	Oxisol	80	0–10	524	35	2.1
	Dry	Oxisol	80	0–10	970	111	2.9

1 Unpublished data (10°05'S–67°15'W)
2 Unpublished data (03°16'S–48°26'W)
3 Unpublished data (10°14'S–62°10'W)
4 Unpublished data (10°44'S–62°13'W)
5 Feigl et al. 1995b
6 Carvalho et al. (in prep.)
7 Pfenning et al. 1992, Luizão et al. 1992, Feigl et al. 1995b
8 Pfenning et al. 1992
9 Geraldes et al. 1995, Feigl et al. 1995b
*Determined by ninhydrin reactive compounds
**Determined by fumigation-incubation method, using k_C= 0.45

dry period (high biomass, low turnover), when plant activities are low. At the beginning of the rainy period (low biomass, high turnover), nutrients are then released, being rapidly taken up by plants (Singh et al. 1989).

Factors influencing organic matter decomposition of topsoils from the Brazilian Amazon Basin were evaluated by several authors. Koutika et al. (1999) concluded that, independent of geographical location and vegetation cover, C decomposition rates are lower on topsoils of fine texture with a sand/clay ratio of less than 1, compared to topsoils of coarse texture, with a sand/clay ration ranging from 2 to 8. According to Telles et al. (2003) clay content exerts a major control on the amount of slowly cycling carbon and therefore influences the storage and dynamics of carbon in tropical forest soils.

12.3 Alteration of Soil Microbial Biomass During Deforestation Process

The cumulative deforested area in Brazilian Amazonia is estimated to have reached 55,000 km^2 by 2000 (INPE 2000). Cattle ranching is thus the leading cause of deforestation in the Brazilian Amazon, since more than 90% of this area has been used for this purpose. When forest is converted to pasture/grazing areas or agriculture, primary burning after deforestation cleans up the areas and promotes a transient increase in nutrient availability. This tillage practice prepares the field for sowing and fertilizes the soil by adding an important amount of nutrients in the form of ashes (Fearnside et al. 2001). In addition, burning organic matter releases the mineral nutrients as oxides or carbonates, which have alkaline reaction and reduce soil acidity, promoting higher phosphorus availability, depending on the type and amount of biomass. These significant environmental changes associated with land conversion are quite likely to alter the structure and activity of the soil microbial community, as indicated by Cleveland et al. (2003).

The effect of deforestation on several parameters, including microbial biomass, was measured during a field experiment realized at Nova Vida Ranch aiming a better understanding of the early biogeochemical alterations due to land-use changes (Neill et al. 1997b, Graça et al. 1999, Garcia-Montiel et al. 2001). The slash and burn procedure was carried out at one hectare area by traditional Amazonian methods, i.e. vegetation was cut down at the end of rainy season (May) and left on place till end of the dry season (September). Fire was lit at the edges of the clearing so the flames would meet in the middle and maximize the effect of burning (Graça et al. 1999).

Microbial biomass C and N, and other eco-physiological indicators were measured simultaneously in the upper soil layers (0–10, 10–20 and 20–30 cm) under intact forest, cut-down and dried plant material, and 2 days after burning (Table 12.4). Native forest soil data had slightly lower microbial biomass C and N than the perturbed sites, while basal respiration and metabolic quotient (or community respiration per biomass unit) decreased after disturbance. This is an indication that more carbon and nitrogen, originated mainly by leaves, were available for biomass production, although changes in percentage microbial carbon (C_{micr}) to total organic carbon (C_{org}) were not found.

Table 12.4 Changes in soil microbial biomass C and N ($n = 4$, sd in parenthesis) due to clear cutting, drying and burning native vegetation

Depth cm	Microbial C μg C g^{-1} soil		Microbial N μg N g^{-1} soil		C:N	*Respiration	**qCO^2	C_{micr}:C_{org}
Native forest site								
0–10	254.3	(7.0)	42.1	(2.8)	6.0	479.2	1.9	2.2
10–20	248.1	(7.8)	50.6	(1.9)	4.9	364.6	1.5	3.0
20–30	241.9	(3.5)	31.3	(2.2)	7.7	208.3	0.9	5.0
Dry cut-down vegetation site, before burning								
0–10	303.9	(11.1)	63.8	(2.2)	4.8	270.8	0.9	2.3
10–20	204.7	(3.7)	37.3	(1.9)	5.5	229.2	1.1	2.5
20–30	229.5	(5.5)	54.2	(2.4)	4.2	145.8	0.6	4.8
Carbonized vegetation site, 2 days after burning								
0–10	285.3	(8.9)	55.4	(1.6)	5.2	229.2	0.8	2.4
10–20	260.5	(6.7)	55.4	(1.6)	4.7	156.2	0.6	3.2
20–30	254.3	(6.2)	40.9	(1.2)	6.2	187.5	0.7	5.3

*(ng CO_2-C g^{-1} soil h^{-1})
**(ng CO2-C ug^{-1} C_{micr} h^{-1})

Different from other studies, fire had little immediate reduction effect on microbial biomass, even in the surface 0–10 cm soil layer. A review done by González-Pérez et al. (2004) reports drastic decrease mainly of fungal biomass after wildfire. Increase of available nutrients in soil, mainly in the form of water-soluble components of ash that became available to living organisms, also commented by the authors, was not observed, since evaluation was done before any rainfall occurred. In so far, the increase in soil pH associated to increase in exchangeable cations in soil which affects soil microbial populations could be observed only in later measurements (Table 12.5).

Table 12.5 Changes in soil microbial biomass C and N during early pasture cultivation

Depth (cm)	FE-C (μg C g^{-1} soil)		FE-N (μg N g$^{(-1)}$ soil)		C:N	*Respiration	**qCO^2	C_{micr}:C_{total}
3 months after burning, rainy season								
Native forest site								
0–10	274.7	(2.6)	41.1	(0.9)	6.7	257.3	0.9	2.0
10–20	285.5	(3.8)	60.6	(1.0)	4.7	157.9	0.6	3.5
20–30	209.7	(2.7)	51.9	(1.8)	4.0	297.0	1.4	4.4
Burnt site, before grass seeds were spread								
0–10	498.9	(2.7)	81.1	(1.0)	6.2	480.8	1.0	2.5
10–20	320.3	(6.1)	65.0	(1.0)	4.9	112.2	0.4	2.6
20–30	224.3	(7.7)	43.7	(1.0)	5.1	111.9	0.5	2.3
8 months after burning, late rainy season								
Native forest site								
0–10	252.1	(3.2)	61.1	(0.8)	4.1	324.0	1.3	1.8
10–20	246.5	(7.2)	53.0	(0.8)	4.7	136.5	0.6	3.0
20–30	179.3	(2.2)	39.2	(1.2)	4.6	230.2	1.3	3.7
5 months-old pasture site								
0–10	310.0	(63.2)	43.0	(17.4)	7.2	181.6	0.6	1.6
10–20	257.7	(21.5)	56.5	(6.1)	4.6	171.2	0.7	2.1
20–30	190.5	(60.3)	44.5	(5.9)	4.3	171.2	0.9	1.9
15 months after burning, rainy season								
Native forest site								
0–10	548.9	(5.7)	59.7	1.01	9.2	483.7	0.9	3.9
10–20	205.5	(2.9)	33.8	1.60	6.1	251.5	1.2	2.5
20–30	162.5	(2.8)	37.0	1.30	4.4	263.3	1.6	3.4
12 months-old pasture site								
0–10	276.6	(31.6)	45.6	10.59	6.1	188.6	0.7	1.4
10–20	219.9	(38.2)	32.7	4.14	6.7	259.2	1.2	1.8
20–30	179.3	(33.2)	27.7	4.23	6.5	253.8	1.4	1.8
27 month after burning, rainy season								
Native forest site								
0–10	328. 3	(2.5)	47.5	(2.6)	6.9	140.5	0.4	2.3
10–20	208.3	(5.6)	48.8	(1.8)	4.3	61.5	0.3	2.5
20–30	169.5	(10.8)	34.7	(0.9)	4.9	63.1	0.4	3.5
24-months old pasture site								
0–10	244.7	(28.1)	44.5	(4.3)	5.5	159.0	0.6	1.2
10–20	300.9	(27.1)	52.3	(7.4)	5.8	141.7	0.5	2.4
20–30	190.3	(46.9)	43.8	(4.5)	4.3	87.0	0.5	1.9

* (ng CO_2-C g^{-1} soil h^{-1})
**(ng CO_2-C ug^{-1} C_{micr} h^{-1})

Three months after burning, during rainy season, soils from burned sites had higher soil microbial biomass and activity than soils under forest. Microbial C and N in the 0–10 cm layer were about 90 and 100% greater, while respiration increased 93%. However, pools of readily available substrates were rapidly depleted. Eight months after burning, when grass plants were about 5 months old, microbial N already decreased to levels similar to forest. Measurements done 1 and 2 years after pasture plantation showed differences mainly in microbial C and C_{micr}:C_{total}, indicating that the young pasture site had less C available for population growth.

12.4 Evolution of Soil Microbial Biomass Content Under Aging Pastures

Differences in litter composition, besides environmental factors like temperature and humidity, determine changes in microbial biomass after forest conversion to pasture. In general, higher rates of cellulose and hemicellulose are associated with higher, whereas phenols are associated to lower microbial biomass.

An evaluation of C_{micr} and N_{micr} in the litter compartment of a 10 year-old pasture showed that forest (Table 12.2) and pasture litter (Table 12.6) maintained a similar level of C_{micr} over the year. However, pastures showed greater C_{total}:C_{micr} ratio, since total litter amount, as well as total C and N content, are lower. Pasture N_{mic} followed the widely observed tendency of reduced N dynamic under altered conditions (Neill et al. 1997b, Ellingson et al. 2000, Garcia-Montiel et al. 2001).

Soil microbial biomass content showed no significant changes which could be attributed to aging of pastures (Table 12.7), at least during wet season. However, C_{micr} was more affected by dryness in the older pasture, probably due to a less effective protection against water evaporation provided by plant cover. As can be verified in Table 12.8, during the dry season the soil of the 15 year-old pasture suffered a reduction in water content of more than 70% in relation to the wet season. In comparison, soil humidity under forest decreased 55% in the 0–10 cm soil layer. N_{micr} and P_{micr} were less affected by seasonality, although the amount of

Table 12.6 Changes in litter microbial biomass C and N in a 10 year-old pasture

Litter layer		Litter dry matter			Microbial biomass		C_{total}:C_{micr}
		Weight	Total C	Total N	C_{micr}	N_{micr}	
		-------------------------		kg ha^{-1} -----	-------------------------		%
				Rainy season			
(56%)	L	2,100	900	10	12.1	1.4	1.34
(44%)	F+H	1,900	700	13	13.4	1.8	1.91
Total		**4,000**	**1,600**	**23**	**25.5**	**3.2**	**3.25**
				Dry season			
(53%)	F+H	4,600	1,900	28	19.9	2.4	1.05
(47%)		4,100	1,600	29	27.8	3.4	1.74
Total		**8,700**	**3,500**	**57**	**47.7**	**5.8**	**2.79**

* Relative distribution of litter layers

Table 12.7 Changes in soil microbial biomass C, N and P under forest and pastures of different ages

Site	Soil layer	C_{micr}		N_{micr}		P_{micr}	
	(cm)	kg ha^{-1}					
		Early wet season					
Forest	0–10	633.7	(42.5)	133.0	(16.5)	22.4	(1.2)
	10–20	577.9	(10.2)	101.9	(7.7)	15.5	(0.0)
3 year-old pasture	0–10	609.7	(31.2)	58.5	(9.1)	19.5	(0.0)
	10–20	486.5	(54.2)	57.0	(12.5)	12.5	(1.4)
10 year-old pasture	0–10	628.3	(70.7)	137.2	(13.6)	19.0	(0.0)
	10–20	649.1	(12.4)	65.3	(12.5)	13.9	(1.4)
15 year-old pasture	0–10	523.0	(35.9)	74.5	(6.9)	15.2	(2.8)
	10–20	411.8	(35.5)	52.5	(5.7)	11.4	(1.4)
		Dry season					
Forest	0–10	520.4	(75.5)	107.4	(4.7)	20.1	(1.2)
	10–20	348.3	(43.9)	95.5	(11.6)	14.2	(2.6)
3 year-old pasture	0–10	357.5	(53.3)	52.0	(6.5)	18.2	(1.3)
	10–20	248.8	(26.4)	40.3	(11.1)	12.5	(0.0)
10 year-old pasture	0–10	364.5	(70.7)	99.3	(23.1)	16.3	(1.4)
	10–20	300.2	(16.7)	52.8	(8.3)	8.3	(1.4)
15 year-old pasture	0–10	265.0	(30.4)	74.5	(16.6)	11.0	(0.0)
	10–20	197.4	(25.5)	49.7	(12.8)	7.1	(1.4)

immobilized N followed the general tendency of the N cycle to be lowered after deforestation.

The replacement of the forest, which has mostly C3 photosynthetic pathway vegetation, by C4 pasture grasses results in an enrichment in ^{13}C of the soil organic matter over the years, observed by several authors (Feigl et al. 1995a, Moraes et al. 1996, Bernoux et al. 1998a). Isotopic composition of microbial biomass follows this pattern (Table 12.9), since it preferentially uses the most available organic matter and even concentrates C derived from pasture (Fernandes et al. 2007). Changes in $^{13}C_{micr}$ can already be seen 3 years after pasture plantation. At this time, about 40% of microbial C of the upper 10 cm soil layer was obtained from grass plants. Ten years after, this food source rose to 80% and apparently stabilized in this range over the next 5 years. $^{13}C_{micr}$ in the 10–20 cm soil layer changed slower but continuously during the considered period.

Table 12.8 Water content in soil layers under forest and pastures of different ages

Site	Layer	Wet season	Dry season	Difference
	(cm)	Humidity (g kg^{-1})		(%)
Forest	0–10	225.8	102.6	55
	10–20	237.7	120.1	49
3 year-old pasture	0–10	200.2	88.3	56
	10–20	211.3	111.2	47
10 year-old pasture	0–10	231.4	91.6	60
	10–20	221.1	111.1	50
15 year-old pasture	0–10	182.8	54.9	70
	10–20	172.7	72.9	58

Table 12.9 Changes in the $\delta^{13}C$ of soil microbial biomass and calculated contribution of C4 plants to its maintenance (adapted from Fernandes et al. 2007)

Site	Layer	$\delta^{13}C$ in MB	C_{micr} derived from C4 plants
	(cm)	(%)	(%)
Forest	0–10	–25.9	
	10–20	–25.5	
3 year-old pasture	0–10	–21.0	38
	10–20	–22.8	22
10 year-old pasture	0–10	–15.8	78
	10–20	–18.5	56
15 year-old pasture	0–10	–15.3	82
	10–20	–16.1	75

Adapted from Fernandes et al. (2007)

12.5 Pasture Soil Management Practices (Plowing, Crop Desiccation, Fertilization) Influencing Soil Microbial Biomass

Intensification of agriculture is emerging as a major issue in the Amazon Basin because it is likely to have broad-scale biological and biogeochemical effects. The focus is to make established pastures more productive through a variety of treatments including application of herbicides, disking and replanting of pasture grass.

An experiment to test the different strategies was conducted on an existing 63 ha area of pasture in process of degradation located at Nova Vida Ranch. Details of site characterization, experimental design and biological status were given by Cerri et al. (2004a, 2005) and Feigl et al. (2006). In short, here we considered three pasture reformations arranged in four blocks (replicates), besides the control plots. The four treatments were control (C), tillage (T), one cycle of no-till soybean (S) followed by grass renovation, and selective herbicide (H) application. Treatment C did not receive any management than mowing at 40 cm high. Treatment T was plowed twice, fertilized with N, P and K and had *Brachiaria brizantha* seeded in early rainy season, by November 2001. Treatment S involved complete desiccation of degraded pasture vegetation, fertilizer application like T except for N, no-tillage seeding of soybean followed by grass cultivation. Treatment H had the standing weeds eliminated by selective herbicide and application of N, P and K.

Microbial biomass C and N was measured during pasture restoration process by Augusti (2004). Both parameters were decreased by tillage treatment by about 25% in relation to the control during the first month, until the grass was seeded and fertilized (Fig. 12.3). After that, C_{micr} recovered faster than N_{micr}, although with no clear pattern in relation to the control. The grass dessecation (7.5 l ha^{-1} of glyphosate) process itself had no significant effect on C_{micr} or N_{micr} of treatment S. An increase after fertilization was observed mainly for N_{micr}, even no mineral N was added. Soybean plants maintained N_{micr} about 20% higher than control during cropping period (Fig. 12.3). Selective herbicide application (4.0 l ha^{-1} of 2,4-D Triet. + picloram) sharply decreased both, C_{micr} and N_{micr}, an expected effect, according

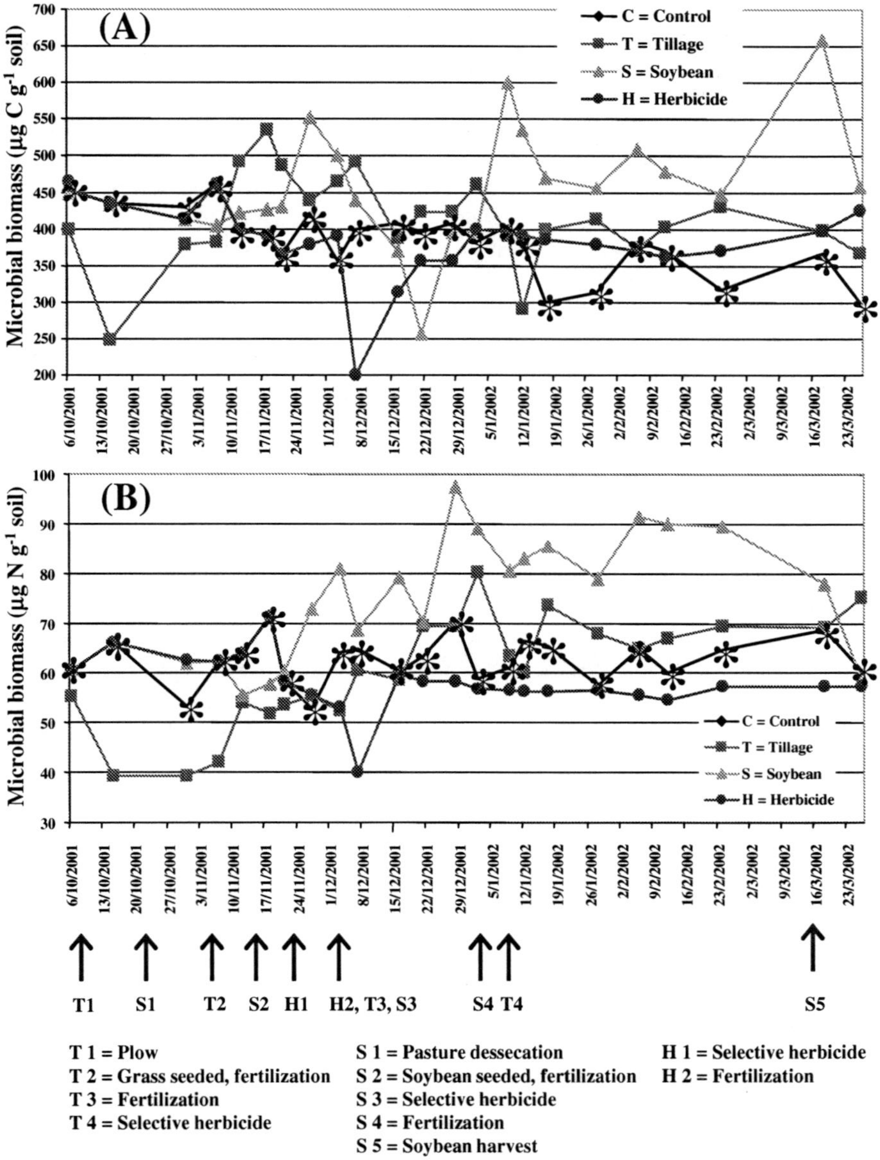

Fig. 12.3 Changes in C_{micr} (A) and N_{micr} (B) of the 0–5 cm soil layer during pasture restoration process

to Prado and Airoldi (2001). Subsequent fertilization, which included mineral N, returned C_{micr} to levels similar to control treatment, but N_{micr} remained lover than control over the next 3 months. It is likely that in this tropical Ultisol fertilizer and/or legume-derived N input stimulated C and N mineralization in the surface depth by increasing microbial activity. Similar results were obtained by Balota et al. (2004).

12.6 Simulating Total Soil C and Microbial Biomass C in a Forest to Pasture Chronosequence Using the Century Model

The effects of the conversion of tropical forest to pasture on total soil C was analyzed by Cerri et al. (2004b), using the Century ecosystem model and chronosequence data collected from the Nova Vida ranch. First, the model was applied to estimate equilibrium soil organic matter levels, plant productivity and residue carbon inputs under native forest conditions. Then the model was set to simulate the deforestation following slash and burn. Soil organic matter dynamics were simulated for pastures established in 1989, 1987, 1983, 1979, 1972, 1951 and 1911. The Century model predicted that forest clearance and conversion to pasture would cause an initial fall in the stock of soil C, followed by a slow rise to levels exceeding those under native forest (Fig. 12.4). The model predicted the longer-term changes in soil C under pasture close to those inferred from the pasture chronosequence. Mean differences between the simulated and observed values for the pasture chronosequence were about 17% for the total soil C (data not shown). Although variability in soil C observations for young pastures was high (Fig. 12.4), statistical evaluations suggested by Smith et al. (1996, 1997) show a satisfactory agreement between the simulated and measured data.

In addition, Century accurately simulated the level and long-term dynamics of microbial biomass C measurements for the pasture chronosequence, providing a further test of model performance. The active SOM fraction as conceptualized in Century includes microbes, microbial products and labile organic substrates (Parton

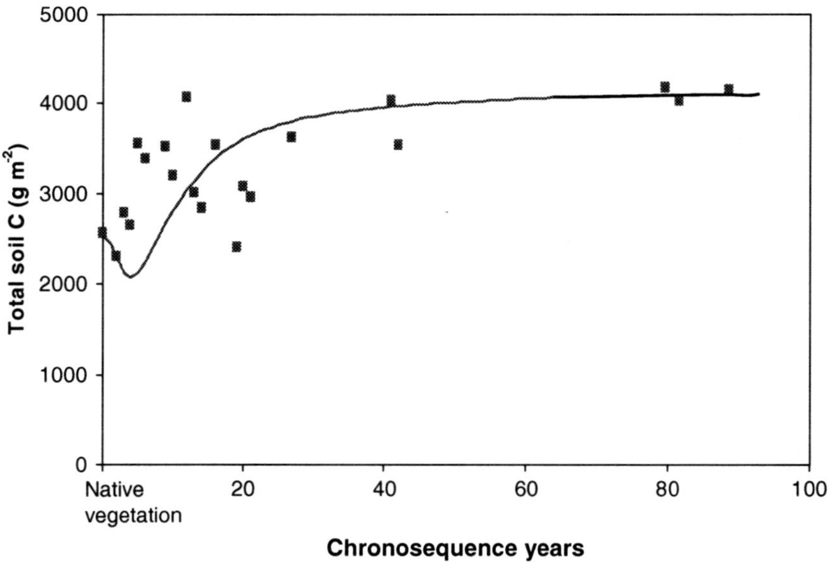

Fig. 12.4 Simulated (continuous line) and observed (square symbols) total soil C in the 0–20 cm layer from the 7 pasture sites at Nova Vida Ranch (modified from Cerri et al. (2004b)

et al. 1987) and can be roughly used to represent microbial biomass and associated metabolites.

Figure 12.5 shows simulated and measured values for soil microbial biomass C in the 0–10 cm layer of the Nova Vida chronosequence. Century simulates soil carbon dynamics for the 0–20 cm layer and our microbial biomass data were determined for the 0–10 cm layer (Feigl et al. 1995b, Fernandes 1999). Therefore, we multiplied the simulated results by 0.65, which is the percentage (65%) of the total soil C present in the 0–10 cm layer in relation to the total soil C contained in the 0–20 cm layer, to make the comparison.

The simulated quantities of soil microbial biomass C fit the measured data reasonably well ($r = 0.84$). The model under-estimated the concentration of microbial biomass C for the youngest pastures (created in 1989 and 1987) and overestimated values for the oldest pastures created in 1911 (Fig. 12.5), reflected by the mean difference between observations and simulation equals to 1.28 g C m^{-2}. However, Century simulated reasonably well (CD = 1.01, CRM = 0.021, and RMSE = 17.20) the soil microbial biomass C for the other pasture sites.

The general trend over time in microbial biomass is well represented by the model, but as with the C and N data from the chronosequences, there is considerable variability in the observed data which is likely associated with factors such as differences in soil temperature and moisture (Steudler et al. 1996), pasture grass status (Moraes et al. 1995) and pre-existing differences in the total amount of soil carbon between sites at the time of clearing (Neill et al. 1997a, Bernoux et al. 1998b).

In order to evaluate soil temperature and moisture effects we compared simulated and measured values for soil microbial biomass-C in the 0–10 cm layer of five soil

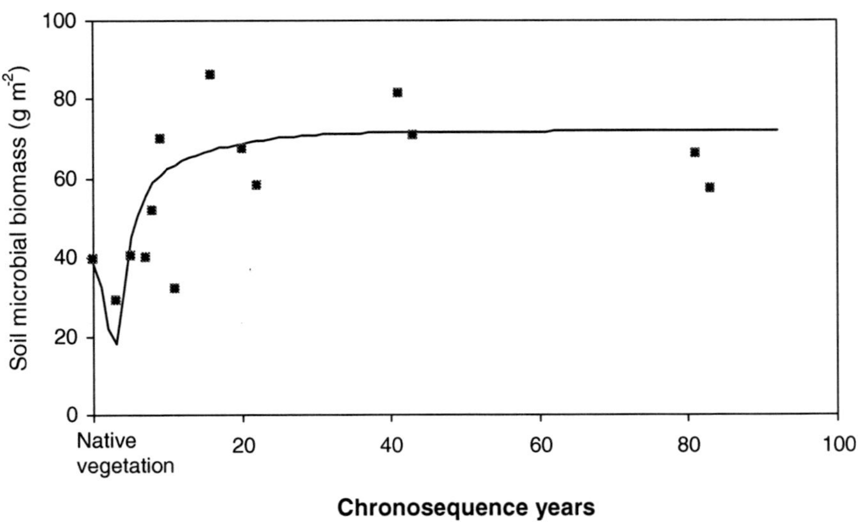

Fig. 12.5 Simulated (continuous line) and observed (square symbols) soil microbial biomass C in the 0–10 cm layer at Nova Vida Ranch treated as a chronosequence (compiled from Cerri et al. (2004b)

types under degraded pastures within the Nova Vida ranch area in both, wet and dry season . Measured soil microbial biomass-C contents varied greatly from 35.4 to 235 g C m^{-2} in the wet season. In the dry season, biomass C values showed a much smaller variation, ranging between 17.8 and 46.5 g C m^{-2} (Table 12.10). Our measured data are consistent with those reported from Amazonian soils by Feigl et al. (1995b) and Fernandes (1999). Results showed that soils have important differences in their soil microbial biomass contents within the same season and also between climate seasons.

The comparison of simulated (active SOM pool) and observed microbial biomass C showed that the model tended, in general, to underestimate the values for the wet season (excepted for the soils TP and XA) and overestimate for the dry season. Also, Century simulated very similar absolute biomass C values for both seasons. This is indicative of the fact that the active pool in Century is simulated as an organic matter fraction, governed by first-order kinetics. Thus, the fact that the simulated active pool was higher in the dry season than in the wet season is likely due to a reduced decomposition rate during the dry season. However, the observed microbial biomass shows a much more dynamic seasonality and likely a response to dry season moisture stress that is not captured in the model. Based on these comparisons, we conclude that the Century model is not totally suitable for studies of soil microbial biomass dynamics in different weather season conditions.

Finally, pasture development has become the largest anthropogenic disturbance of forests in the Amazon region. While our analysis represents a single 'case study' of pasture conversion, our results suggested that modeling approaches can help to evaluate the magnitude of impacts on C cycling, and determine the effect of management strategies on pasture sustainability.

Table 12.10 Observed and simulated soil microbial biomass values in the 0–10 cm layer for the five soil types within the 63 ha pasture area at Nova Vida Ranch

Soil type	Observed values		Simulated values	Difference	Mean difference
	($g\ m^{-2}$)				(%)
Wet season					
TP	35.4	(2.9) c	60.4	−25	−71
TA	124.2	(18.7) b	51.1	73	59
RA	215.3	(42.9) a	58.3	157	73
HX	235.0	(15.3) a	55.7	179	76
XA	64.5	(6.7) c	65.2	−1	−1
Dry season					
TP	17.8	(1.4) c	79.8	−62	−349
TA	12.5	(2.5) c*	73.7	−61	−488
RA	25.4	(2.3) b*	77.7	−52	−206
HX	23.7	(2.1) b*	76.5	−53	−222
XA	46.5	(4.6) a	82.2	−36	−77

Values in parenthesis are standard errors.
Means within the same season followed by different letters differ significantly at $P<0.05$
An asterisk indicates a statistically significant difference ($P<0.05$) between seasons.

12.7 Final Considerations

The present knowledge on microbial biomass in natural Amazonian ecosystems and its adaptation to deforestation and pasture introduction is still not very well understood. It would be necessary to better understand interactions between soil microbial abundance and activity, the quality and quantity of soil organic matter, and the effects of environment conditions. Study cases are disperses and do not cover the large combination of soils, vegetations and present climate conditions. In order to better explore this issue and scale it up, some aspects should be taken into consideration. For instance, tropical area could be split into homogeneous edapho-climatic zones. Differences in methodological aspects for microbial biomass measurements are another example that makes difficult to make extrapolations.

The situation gets even more complex once we include climate change scenarios. In a global warming environment it could be expected that an increase in microbial biomass activity would lead to accelerated soil organic matter decomposition and consequently, reducing soil carbon stocks and exacerbating greenhouse gas emissions. Nevertheless, other factors like increased net primary production, extreme events (rainfall, storms, temperature peeks), and adaptation capacity or resilience have to be taken into account to adequately predict soil microbial community dynamics under global change.

References

Augusti, K. C. 2004. Biomassa microbiana e suas relações com fontes de matéria orgânica do solo em sistemas de recuperação de pastagens em Rondônia. MSc Dissertation, Centro de Energia Nuclear na Agricultura, University of São Paulo, Brazil, p. 121.

Balota, E. L., Colozzi Filho, A., Andrade, D. S., and Dick, R. P. 2004. Long-term tillage and crop rotation effects on microbial biomass and C and N mineralization in a Brazilian Oxisol. Soil & Tillage Research 77:137–145.

Bastos, T. X., and Diniz, T. D. A. S. 1982. Avaliação do clima do Estado de Rondônia para desenvolvimento agrícola. (Boletim de Pesquisa no. 44). EMBRAPA-CPTU, Belém.

Bernoux, M., Cerri, C. C., Neill, C., and Moraes, J. F. L. 1998a The use of stable carbon isotopes for estimating soil organic matter turnover rates. Geoderma 82:43-58.

Bernoux, M., Arrouays, D., Cerri, C.C., Graça, P.M. de A., Volkoff, B., and Trichet, J. 1998b. Estimation des stocks de carbone des sols du Rondônia (Amazonie brésilienne). Études et gestion des sols 5:31–42.

Bernoux, M., Graça, P. M. A., Fearnside, P. M., Cerri, C. C., Feigl, B., and Piccolo, M. C. 2001. Carbon storage in biomass and soils. In The Biogeochemistry of the Amazon Basin and its Role in a Changing World, ed. M. McClain, p. 416. Oxford: Oxford University Press.

Brookes, P. C., Kragt, J. F., Powlson, D. S., and Jenkinson, D. S. 1985. Chloroform fumigation and the release of soil nitrogen: the effect of fumigation time and temperature. Soil Biol. Biochem. 17:831–835.

Cerri, C. E. P., Bernoux, M., Chaplot, V., Volkoff, B., Victoria, R. L., Mellilo, J. M., Paustian, K., and Cerri, C. C. 2004a. Assessment of soil property spatial variations in an Amazon pasture: basis for selecting an agronomic experimental area. Geoderma 123:51-68.

Cerri, C.E.P., Paustian, K., Bernoux, M., Victoria, R. L., Mellilo, J. M., and Cerri, C. C. 2004b. Modelling changes in soil organic matter in Amazon forest to pasture conversion, using the Century model. Global Change Biology 10:815–832.

Cerri, C. C., Melillo, J. M., Feigl, B. J., Piccolo, M. C., Neill, C., Steudler, P. A., Carvalho, M. C. S., Godinho, V. P., Cerri, C. E. P., and Bernoux, M. 2005. Recent history of the agriculture of the Brazilian Amazon Basin: prospects for sustainable development and a first look at the biogeochemical consequences of pasture reformation. Outlook on Agriculture 34(4): 215–223.

Cerri, C. E. P., Piccolo, M. C., Feigl, B. J., Paustian, K., Cerri, C. C., Victoria, R. L., and Melillo, J. M. 2006. Interrelationships among soil total C and N, microbial biomass, trace gas fluxes, and internal N-cycling in soils under pasture of the Amazon region. Journal of Sustainable Agriculture 27(4):45–69.

Cummings, D. L., Kauffman, J. B., Perry, D. A., and Hughes, R. F. 2002. Aboveground biomass and structure of rainforests in the southwestern Brazilian Amazon. Forest Ecology and Management 163:293–307.

Dalal, R. C. 1998. Soil microbial biomass – what do the numbers really mean? Australian Journal of Experimental Agriculture 38:649–665.

Doran, J. W., and Zeiss, M. R. 2000. Soil health and sustainability: managing the biotic component of soil quality. Applied Soil Ecology 15:3–11.

Ellingson, L. J., Kauffman, J. B., Cummings, D. L., Sanford Jr., R. L., and Jaramillo, V. J. 2000. Soil N dynamics associated with deforestation, biomass burning, and pasture conversion in a Mexican tropical dry forest. Forest Ecology and Management 137(1–3):41–51.

EMPRESA BRASILEIRA DE PESQUISA AGROPECUÁRIA (EMBRAPA). 1999. Sistema brasileiro de classificação de solos. Rio de Janeiro: Embrapa Solos, p. 412.

Fearnside, P. M., Graça, P. M. L. D., and Rodrigues, F. J. A. 2001. Burning of Amazonian rainforests: burning efficiency and charcoal formation in forest cleared for cattle pasture near Manaus, Brazil. Forest Ecology and Management 146:115–128.

Feigl, B. J., Melillo, J., and Cerri, C. C. 1995a. Changes in the origin and quality of soil organic matter after pasture introduction in Rondônia (Brazil). Plant and Soil 174:21–29.

Feigl, B. J., Sparling, G. P., Ross, D. J., and Cerri, C. C. 1995b. Soil microbial biomass in Amazonian soils: Evaluation of methods and estimates of pool sizes. Soil Biol. Biochem. 27: 1467–1472.

Feigl, B. J., Cerri, C. C., Piccolo, M., Noronha, N., Augusti, K., Melillo, J., Eschenbrenner, V., and Melo L. 2006. Biological survey of a low-productivity pasture in Rondônia state, Brazil. Outlook on Agriculture 35(3):199–208.

Fernandes, S. A. P. 1999. Propriedades do solo na conversão de floresta em pastagem fertilizada e não fertilizada com fósforo na Amazônia (Rondônia). Doctoral thesis, University of São Paulo, Centro de Energia Nuclear na Agricultura, Piracicaba, Brazil, 131pp.

Fernandes, S. A. P., Feigl, B. J., Bernoux, M., Cerri, C. C., Carvalho, M. C. S. 2007. Carbon and nitrogen of the soil microbial biomass, composition in ^{13}C, and fluxes of gases in Amazon (Rondônia). Submitted to Biology and Fertility of Soils.

Garcia-Montiel, D. C., Neill, C., Melillo, J., Thomas, S., Steudler, P. A., and Cerri, C. C. 2000. Soil phosphorus transformations following forest clearing for pasture in the Brazilian Amazon. Soil Science Society of America Journal 64:1792–1804.

Garcia-Montiel, D. C., Steudler, P. A., Piccolo, M. C., Melillo, J., Neill, C., and Cerri, C. C. 2001. Controls on soil nitrogen oxide emissions from forest and pastures in the Brazilian Amazon. Global Biogeochemical Cycles 15(4):1021–1030.

Geraldes, A. P. A., Cerri, C. C., and Feigl, B. J. 1995. Biomassa microbiana de solo sob pastagens na Amazônia. R.bras.Ci. Solo 19:55–60.

González-Pérez, J. A., González-Vila, F. J., Almendros, G., and Knicker, H. 2004. The effect of fire on soil organic matter – a review. Environment International 30:855–870.

Graça, P. M. L. A., Fearnside, P. M., and Cerri, C. C. 1999. Burning of Amazonian forest in Ariquemes, Rondônia, Brazil: biomass, charcoal formation and burning efficiency. Forest ecology and Management 120:179–191.

INPE, 2000. Deforestation estimates in the Brazilian Amazon. Instituto Nacional de Pesquisas Espaciais, São José dos Campos, Brazil.

Jenkinson, D. S., and Powlson, D. S. 1976. The effect of biocidal treatments on metabolism in soil. V. A method for measuring soil biomass. Soil Biol. Biochem. 8:209–213.

Koutika, L.-S., Choné, T., Andreux, F., Burtin, G., and Cerri, C. C. 1999. Factors influencing carbon decomposition of topsoils from the Brazilian Amazon Basin. Biol. Fertil. Soils 26: 436–438.

Luizão, F. J., and Schubart, H. O. R. 1987. Litter production and decomposition in a terra-firme forest of Central Amazonia. Experientia 43:259–265.

Luizão, R. C. C., Bonde, T. A., and Rosswall, T. 1992. Seasonal variation of soil microbial biomass – The effect of clearfelling a tropical rainforest and establishment of pasture in the central Amazon. Soil Biol. Biochem. 24:805–813.

Matson, P. A., Parton, W. J., Power, A. G., and Swift, M. J. 1997. Agricultural intensification and ecosystem properties. Science 277:504–509.

Moraes, J. F. L., Cerri, C. C., Melillo, J. M., Kicklighter, D., Neill, C., Skole, D. L., Steudler, P. A. 1995. Soil carbon stocks of the Brazilian Amazon basin. Soil Science Society of America Journal 59:244–247.

Moraes, J. F. L., Volkoff, B., Cerri, C. C., and Bernoux, M. 1996. Soil properties under Amazon forest and Changes due to pasture installation in Rondônia/Brasil. Geoderma 70: 63–81.

Neill, C., Cerri, C. C., Melillo, J., Feigl, B. J., Steudler, P. A., Moraes, J. F. L., and Piccolo, M. C. 1997a. Stocks and dynamics of soil carbon following deforestation for pasture in Rondonia. In Soil processes and the carbon cycle, ed. R. Lal, J. M. Kimble, R. F. Follett, and B. A. Stewart, pp. 9–28. Boca Raton: CRC Press.

Neill, C., Melillo, J. M., Steudler, P. A., Cerri, C. C., Moraes, J. F. L., Piccolo, M. C., and Brito, M. 1997b. Soil carbon and nitrogen stocks following forest clearing for pasture in the southwestern Brazilian Amazon. Ecological Applications 7:1216–1225.

Neill, C., Deegan, L. A., Thomas, S. M., and Cerri, C. C. 2001. Deforestation for pasture alters nitrogen and phosphorus in soil solution and streamwater of small Amazonian watersheds. Ecological Applications 11:1817–1828.

Parton, W.J., Schimel, D. S., Cole, C. V., and Ojima, D. S. 1987. Analysis of factors controlling soil organic matter levels in Great Plains grasslands. Soil Sci. Soc. of Am. J. 51:1173–1179

Pfenning, L., Eduardo, B. P., and Cerri, C. C. 1992. Os métodos da fumigação-incubação e fumigação-extração na estimativa da biomassa microbiana de solos da Amazônia. R.bras.Ci. Solo 16:31–37.

Prado, A. G. S., and Airoldi, C. 2001. Toxic effect caused on microflora of soil by pesticide picloram application. J. Environ. Monit. 3:394–397.

Ruan, H. H., Zou, X.,M., Scatena, F. N, Zimmerman, J. K. 2004. Asynchronous fluctuation of soil microbial biomass and plant litterfall in a tropical wet forest. Plant and Soil 260:147–154.

Schloter, M., Dilly, O., and Munch, J. C. 2003. Indicators for evaluating soil quality. Agriculture, Ecosystems and Environment 98:255–262.

Singh, J. S., Raghubanshi, A. S., Singh, R. S., and Srivastava, S. C. 1989. Microbial biomass acts as a source of plant nutrients in dry tropical forest and savanna. Nature 338:499–500.

Smith, J. U., Smith, P., and Addiscott, T. 1996. Quantitative methods to evaluate and compare soil organic matter (SOM) models. In Evaluation of Soil Organic Matter Models Using Existing, Long-Term Datasets, ed. D. S. Powlson, P. Smith, and J. U. Smith, pp. 181–200. NATO ASI Series I, Vol. 38, Berlin: Springer-Verlag.

Smith, P., Smith, J. U., Powlson, D. S., McGill, W. B., Arah, J. R. M., Chertov, O. G., Coleman, K., Franko, U., Frolking, S., Jenkinson, D. S., Jensen, L. S., Kelly, R. H., Klein-Gunnewiek, K., Komarov, S. A., Li, C., Molina, J. A. E., Mueller, T., Parton, W. J., Thornley, J. H. M, and Whitmore, A. P. 1997. A comparison of the performance of nine soil organic matter models using datasets from seven long-term experiments. Geoderma 8:153–225.

Smith, K., Gholz, H. L., and Oliveira, F. A. 1998a. Litterfall and nitrogen-use efficiency of plantations and primary forest in the eastern Brazilian Amazon. Forest, Ecology and Management 109:209–220.

Smith, K., Gholz, H. L., and Oliveira, F. A. 1998b. Fine litter chemistry, early-stage decay, and nitrogen dynamics under plantations and primary forest in lowland Amazonia. Soil Biology & Biochemistry 30:2159–2169.

Steudler, P. A., Melillo, J. M., Feigl, B. J., Neill, C., Piccolo, M. C., and Cerri, C. C. 1996. Consequence of forest-to-pasture conversion on CH_4 fluxes in the Brazilian Amazon Basin. Journal of Geophysical Research 101:18547–18554.

Telles, E. C. C., Camargo, P. B., Martinelli, L. A., Trumbore, S. E., Costa, E. S., Santos, J., Higuchi, N., Oliveira, R. C. 2003. Influence of soil texture on carbon dynamics and storage potential in tropical forest soils of Amazônia. Global Biogeochemical Cycles 17(2):9–11.

Vance, E. D., Brookes, P. C., and Jenkinson, D. S. 1987. An extraction method for measuring soil microbial biomass C. Soil Biol. Biochem. 19:703–707.

Vasconcelos, H. L., and Luizão, F. J. 2004. Litter production and litter nutrient concentrations in a fragmented Amazonian landscape. Ecological Applications 14:884–892.

Chapter 13
Management for Sustainability and Restoration of Degraded Pastures in the Neotropics

Florencia Montagnini

13.1 Introduction

According to recent statistics, today the total agricultural land area in the Neotropics amounts to about 500 million ha or nearly 30% of its territory, including crop and pastures (FAO 2000). Pastures and agrosilvopastoral land represent 77% of the total agricultural land, mostly on poor acid soils. Crops cover the remaining 23%, which are mostly located on better quality soils. The Neotropics have a pasture/crop land ratio of 3.4, higher than the world ratio of 2.3 (Amézquita et al. 2005).

Conversion of forests to crops and pastures has been the most important land–use change in the Neotropics during the second half of the twentieth century (Kaimowitz 1996). Cattle and pastures replaced large amounts of tropical forests during the decades of the 1960s, 1970s and 1980s, especially in Central America (Kaimowitz 1996, Harvey et al. 2005). A similar pattern has been reported over large areas of the Neotropics such as in the Brazilian Amazon (Hetch 1993, Dias-Filho et al. 2001).

Among soil nutrients, phosphorus is generally most critical to pasture productivity. Soil phosphorus availability increases with each successive burning, but at lower and lower levels. When pasture productivity declines significantly as a result of degradation, the cattle are taken elsewhere or the activity is discontinued altogether.

It is estimated that about 30% of these pastures are degraded as a result of stocking levels being higher than carrying capacity (Szott et al. 2000). Other sources give estimates of degradation on more than 60% of the pasture area of the Neotropics CIAT (1976–2000). Although pasture degradation in the Amazon basin has received the most attention, the surface area affected by pasture degradation may be much larger in the savannas of Bolivia, Brazil, Colombia, Guyana and Venezuela than in the Amazon rainforest (Vera et al. 1998). Locally, the extent of pasture degradation can be even higher: for example in the northern Caribbean zone of Costa Rica, more than 70% of the pastures were at advanced stages of degradation in the late 1990s (Jansen et al. 1997, CATIE 2002). Differences in perception of the degree and extent of degradation by farmers and technical personnel can account for significant differences in these estimates (Holmann et al. 2004).

Pasture development has been ranked one of the least desirable land uses of tropical forest regions because of land degradation caused by overgrazing. Formation of gullies can be a clear symptom of overgrazing. Gullies may appear in the landscape

R. W. Myster, *Post-Agricultural Succession in the Neotropics.*

in very humid regions following heavy rains due to lack of soil water infiltration caused by compaction. Furthermore, land affected by gullies is very difficult to restore.

In regions where natural grasses are used for cattle grazing, these grasses often have low nutrient content and are not preferred by cattle. In such situations, pastures are burnt periodically to encourage regrowth and also to increase nutrient levels in soils. The tender regrowth of pastures is eaten by cattle, and pasture productivity increases temporarily (Serrao and Toledo 1990, Nepstad et al. 1991a, Dias-Filho et al. 2001).

Abandoned pastures can remain in a state of arrested succession due to a number of factors that delay or impede their recovery (Uhl et al. 1988, Nepstad et al. 1991a, 1991b, Reiners et al. 1994, Aide et al. 1995, Dias-Filho et al. 2001). However, there are strategies that can be used to successfully manage tropical pastures to increase sustainability and avoid degradation. In cases where pastures have been abandoned due to low productivity, there are also a number of strategies that can be used to recover the land with the objectives of converting it back to forest and restoring plant and animal diversity. This chapter examines some of the factors involved in pasture degradation, and presents experiences used to increase pasture sustainability and restore degraded pastures in the Neotropics.

13.2 Pasture Degradation and Regeneration Following Abandonment

13.2.1 Factors Resulting in Pasture Degradation

The progressive displacement of cattle raising toward marginal areas of lower productive capacity is a generalized phenomenon in the Neotropics (Holmann et al. 2004). The low availability of adapted forage materials of high productivity together with the inadequate management of pastures has led to a rapid decrease of productivity and low revenues to livestock farmers, leading to pasture abandonment and/or conversion to other land uses.

Pasture degradation is defined as the temporary or permanent decrease of the soil productive capacity (Stocking and Murnaghan 2001). Pasture degradation can result from the interaction of the following factors: (1) establishment of pastures on fragile soils, e.g. on infertile soil or on steep slopes; (2) use of natural grasses of low productivity, or sowing of poorly adapted species; (3) soil compaction (use of soils that are shallow and prone to compaction, excess stocking, and improper management); (4) soil erosion (as result of compaction and establishment of pastures on steep slopes); (5) decrease in soil fertility (especially P, N), without addition of fertilizer; (6) uncontrolled and frequent burning; (7) overgrazing during the rainy season due to excess stocking, and use of inadequate animals or breeds; and (8) weed invasion (as result of the previously mentioned factors). The degradation of pastures has serious consequences for the producer: first, it reduces milk and/or beef yields and second, it increases production costs (Holmann et al. 2004).

Even in pastures that have been established on adequate soil and with proper management, changes in soil productivity may lead to pasture decline and eventual degradation unless other management techniques are introduced early enough to reverse the process. Nitrogen deficiency is the first factor that destabilizes the pasture, initiating the degradation process (Holmann et al. 2004). After nitrogen deficiency takes place, the quality and vigor of the pasture begin to decline and other nutrients, such as phosphorus and sulfur, can become deficient. When the pasture begins to lose vigor, invasive weeds appear, increasing the severity of the problem (Spain and Gualdrón 1991, Dias-Filho et al. 2001).

After a prolonged period of grazing, important changes in the physical structure of the soil may appear, such as compaction, which increases water runoff, diminishes root development, and decreases the extraction of nutrients located deeper in the soil. In addition, water runoff initiates the development of erosion and the pasture enters a process of severe degradation (Hoyos et al. 1995).

13.2.2 Obstacles to Natural Forest Regeneration in Abandoned Pastures

Following degradation and abandonment of pastures, the efforts needed to recover the site for continued use as pasture, or to restore the previous forest cover, depend on several factors. Some are related to specific site conditions prior to its use as pasture, and others depend on length of use as pasture and on type of pasture management. In general, the characteristics and rate of natural forest regeneration depend on the intensity, duration and specific nature of previous land use. The factors limiting tree regeneration can include scarcity of nutrients, soil compaction, lack of or excess of soil humidity, high solar radiation, and intra- and inter-specific competition (Uhl et al. 1988, Nepstad et al. 1991a, 1991b, Aide et al. 1995, Dias-Filho et al. 2001).

13.2.3 Seed Dispersal and Seedling Establishment

Availability or arrival of seed from trees can be a limiting factor to tree regeneration, especially in sites whose distance to seed sources may limit propagule dispersal (Holl 1999, 2002, Wijdeven and Kuzee 2000). In areas that are artificially seeded or that receive a natural seed input by seed dispersal agents or from nearby forests, still other factors can impede or delay seed germination and seedling establishment and growth following seed arrival. Seed and seedling mortality may occur as a result of fungal diseases or seed predation. In addition, the specific microsite where a seed falls may not fulfill the environmental requirements for seed germination, seedling growth, and seedling survival (Nepstad et al. 1991a, 1991b).

Competition for space, nutrient and light resources by aggressive vegetation may severely impede the regeneration of forest seedlings in abandoned pastures. For example, at La Selva Biological Station in the Caribbean lowlands of Costa Rica, very few woody species were found in abandoned pastures left to natural regeneration, in comparison to pasture areas planted with native trees (Figs 13.1 and 13.2). High light availability in the open pastures favored the growth of heliophillous species, such as grasses and ferns, impeding the growth of woody species (Cusack and Montagnini 2004). This effect was more pronounced 3–4 years following pasture abandonment (Guariguata et al. 1995), but dominance by grasses and ferns still persisted at 10–14 years after abandonment (Carnevale and Montagnini 2002, Cusack and Montagnini 2004, Orozco Zamora and Montagnini 2007a, 2007b). Factors that impeded the establishment of woody species in open pastures were a combination of lack of perches for seed dispersers and invasion by herbaceous vegetation which out-competed the tree seedlings in their growth (Montagnini 2001, Cusack and Montagnini 2004, Orozco Zamora and Montagnini 2006).

13.2.4 Intensity and Duration of Pasture Use

Intensity and duration of use are key factors influencing pasture recovery to forest. Recovery to forest is considerably slower following disturbances that heavily impact the soils as well as the aboveground vegetation. Activities such as bulldozing, heavy or long-term grazing, and severe fires, often have long-lasting effects on species

Fig. 13.1 At La Selva Biological Station in the Caribbean lowlands of Costa Rica, very few woody species were found in abandoned pastures left to natural regeneration, which were dominated by grasses and ferns even after 15 years following pasture abandonment. A 15-year-old plantation of *Jacaranda copaia* can be seen at the back. Photo: F. Montagnini

Fig. 13.2 At La Selva Biological Station, abandoned pasture areas planted with native trees developed a diverse understory including several woody species. This mixed-species plot is composed of *Jacaranda copaia*, *Calophyllum brasiliense* and *Vochysia guatemalensis*. The plantation was 15 years old. Photo: F. Montagnini

composition (Chazdon 2003). Several examples illustrate this point, often comparing forest recovery following pastures and other previous land uses. For example, Ferguson et al. (2003) compared forest recovery in El Petén, Guatemala, following abandonment from agroforestry, shifting agriculture, and pastures. They found that forest succession was much faster following use in agroforestry and shifting agriculture than following abandonment from pastures or from input-intensive monocultures. Their results suggest that initial ground cover by herbs inhibited recruitment by woody colonists, a result consistent with many other studies. In Puerto Rico, Aide et al. (1995) and Aide et al. (2000) studied changes in vegetation in chronosequences ranging from 0 to 60 years after pasture abandonment, and found that species richness, abundance and biomass were very low during the first 10 years. They concluded that recovery in pastures was greatly delayed in comparison with forest recovery following other types of human and natural disturbances. The successional trajectory was quite different, with an absence of the more typical pioneer species that are more commonly found in other areas. The presence of grasses and the rapid colonization and growth of ferns and herbaceous species in the abandoned pastures appeared to be a major factor inhibiting the establishment of secondary forest, and imparted a selective barrier on the colonizing woody species.

In a study conducted near Paragominas in the state of Para, eastern Amazon of Brazil, recovery to forest following pasture abandonment was examined in sites where pasture management intensity ranged from low to high, and duration of use was 4–10 years (Buschbacher et al. 1987, Uhl et al. 1988). In the low-intensity management (comprising about 20% of abandoned pastures in the region) following forest cutting and burning the area was seeded to pasture grasses, but grasses failed to establish due to environmental and management reasons, only few cattle were grazed, and abandonment usually occurred within 4 years of pasture formation. In the medium-intensity management (about 70% of pastures in the region), seeded pastures grasses established well, the pasture was weeded with machetes when needed, and was burnt every few years to eliminate woody seedlings and favor grass species. Burning increased soil pH and improved availability of nutrients. Grazing density was intermediate with about one animal per hectare, and abandonment occurred 6–8 years after pasture establishment. In the high intensity management (only about 10% of the cases), seeded pasture grasses also established well, and after several cycles of weeding and burning, the area was cleared with heavy machinery to remove woody species that had become established in spite of the weeding treatments. Thereafter, the sites were mowed mechanically each year and then burned. Grazing pressure was heavy with more than two heads of cattle per hectare, and abandonment happened after 10 or more years of use (Buschbacher et al. 1987).

Results of evaluations of plant biomass and diversity in the three categories of abandoned pastures showed that the abandoned low-intensity pasture sites were progressing toward forest, in contrast to the high-intensity sites. In the low-intensity management sites, recovery of one-fourth the original biomass in 8 years suggested that a rapid structural recovery was occurring, with most of the biomass being comprised of trees, as opposed to herbs, shrubs and grasses. In the sites of medium-intensity use, total biomass was less than in the low-intensity use site, and similarly most of the biomass was in trees, again suggesting recovery to forest should be fast, although not as rapid as in the low-intensity use site. In contrast, the high-intensity management sites were almost completely dominated by shrubs, herbs and grasses, with only 4% of total biomass in trees, therefore the potential for this site to recover to forest appeared to be much less than in the two other cases.

13.2.5 Influence of Soil Fertility

Sites with inherently poor, acid soil may be less conducive to woody regeneration of abandoned pastures than soils with more favorable conditions. For example, at our study site at La Selva, Costa Rica, abandoned pastures were dominated by ferns and grasses several years after abandonment, while in other abandoned pastures at La Selva and nearby in locations with better microsite conditions, natural regeneration was abundant and diverse (Chazdon et al. 2005). In other sites at La Selva of ages 12–25 years following pasture abandonment there were more than 1,000 trees larger than 5 cm diameter at breast height per hectare, and tree species density ranged from 64–121 species per hectare (Chazdon et al. 2005).

Other studies of vegetation and soil changes following conversion of forest to pasture show that recovery to forest may be relatively fast following abandonment, but soil properties may take a longer time to recover (Reiners et al. 1994). These authors examined soil and vegetation characteristics in primary forest, actively grazed pastures (10–36 years old), abandoned pastures (4–10 years following abandonment), and 10–20 year old secondary forest in the Atlantic zone of Costa Rica. They found decreases in soil acidity, some increases in bases, increases in soil compaction, decreases in porosity, and lower rates of nitrogen mineralization and nitrification in the actively grazed pastures. Their data suggest that with pasture abandonment and succession to secondary forests these changes can be reversible, although they may take a long time. They emphasized that some of the soil changes were positive, such as increases in soil pH, while some were negative such as lesser availability of soil nitrogen and lower porosity which will lead to higher erosion (Reiners et al. 1994).

13.2.6 Position of Pastures in the Landscape

In addition, the size and position of pastures in the landscape matrix will influence the speed and characteristics of natural regeneration following abandonment. Different types of matrices have a large influence on the ecology of recovering forests. Proximity of pastures to remnant forest patches promotes more rapid forest recovery, which depends heavily on seed dispersal (Chazdon 2003). In the Amazon, forest fragments surrounded by areas of natural regeneration 5–10 m in height had fewer micro-climate changes and lower tree mortality than did forest fragments surrounded by pastures (Didham and Lawton 1999). The fragments surrounded by this low-stature secondary forest also had greater bird abundance than did those surrounded by pastures. Certain types of matrices favor the fauna of tropical forests more than others. In the Amazon, natural regeneration areas dominated by *Cecropia* spp., which tend to grow tall and with a closed crown, are utilized more by birds, frogs, and ants than areas of more open *Vismia* spp.-dominated regeneration. In general, any type of natural regeneration that borders forest fragments favors fauna more than pastures do (Williamson et al. 1998). These patterns are found in lowland as well as in montane forest regions of the Neotropics (Sarmiento 1997).

A condition of interrupted or arrested succession may develop when several interacting factors leading to pasture degradation operate together, either simultaneously or in a sequence. For example, long and high intensity of use may lead to nutrient-depleted and/or compacted soil, over which and with high solar radiation, invasion by aggressive or weed vegetation will occur (Buschbacher et al. 1987, Nepstad et al. 1991b, Reiners et al. 1994, Dias-Filho et al. 2001). This may be compounded by large size and long distance from sources of seed that can impede or delay re-sprouting and seed arrival. In such situations, once the condition has been identified along with its interacting causes, active measures are needed to artificially encourage and favor forest regeneration. This may include seeding, soil treatments, control of invasive vegetation, and planting appropriate tree species to serve as perches for dispersal agents and to ameliorate soil and micro-environmental conditions, as discussed in following sections.

13.3 Pasture Restoration

The purpose of pasture restoration may involve recovery of the former vegetation (generally forest), which in turn has the compound objectives of increasing biodiversity, recovering soils, and attaining vegetation structure and functional characteristics as close as possible to the original forest. An alternative pathway, more properly named rehabilitation, involves managing to recover productivity of degraded pastures, which in turn involves delaying degradation, increasing and sustaining yields in the long term. The overall objective in this case is to recover the land for continued use in pasture.

13.4 Techniques for Restoration of Degraded Pastures: Soils, Biodiversity, Forest Structure and Function

The presence of trees, either in plantation, in groups, in lines or in isolated form may contribute to the recovery of environmental conditions favorable to tree regeneration processes (Guariguata et al. 1995, Parrotta 1995). The establishment of plantations may overcome some of the barriers to natural forest regeneration by attracting seed dispersal agents into the landscape and by ameliorating local microclimatic conditions within the area, and thereby accelerating the recovery of these lands (Parrotta 1995, Parrotta and Turnbull 1997, Powers et al. 1997, Lamb 1998). Strategies to encourage natural regeneration of degraded pastures and fulfilling the objectives named before (recovery of soils, biodiversity and forest structure and function) may involve (1) reforestation, (2) care and management of remnant trees in pastures; setting perches for birds; (3) planting individual trees/blocks/windbreaks.

13.4.1 Reforestation of Abandoned Pastures

13.4.1.1 Impacts on Soils and Nutrient Cycling

Some degraded pasture soils can be recovered through the use of fertilizers while others that are badly eroded, or that are fully covered by invasive species such as grasses or ferns, need more drastic rehabilitation techniques. There are also situations of extreme degradation where soils cannot be recovered at all. The recovery of the soil's productive capacity is frequently very expensive, thus the techniques involved must produce financial returns to ensure their adoptability by the local farmers (Montagnini 2002). Reforestation can be such alternative as the established plantations not only can contribute to recover soils and the land's productive capacity but they can also serve for cash income and long term savings to the owner. Plantations can also serve other environmental services such as carbon sequestration and recovery of biodiversity (Montagnini and Jordan 2005).

Abandoned pastures are often invaded by aggressive grass or ferns for the reasons discussed previously. The competitive advantage of grasses, combined with degraded soils and lack of nutrients, often prevents the germination and initiation of tree seedlings. These grassland areas are also often maintained by fires which inhibit natural colonization by tree species (Montagnini and Jordan 2005). In these cases, if the area is left for natural regeneration to take place, it may stay in an arrested stage of succession for many years. Often the best alternative to restore these areas and convert them back to forest is planting fast-growing trees, sometimes even using exotic species.

In many cases it is not feasible to plant tree seedlings without first removing the invasive vegetation. In such cases the use of herbicides, fire or mechanical weeding may be necessary to clear the area and allow tree seedlings to establish successfully. In some cases treatments may involve tillage, or root removal of the invasive plants (Montagnini and Jordan 2005).

In research to develop alternatives for the rehabilitation and use of abandoned pastures in the Neotropics, soil chemistry and nutrient cycling parameters were measured in sites planted with indigenous tree species and in abandoned pastures. At La Selva Biological Station in the Caribbean lowlands of Costa Rica, in just 3 years soil conditions improved in the young, native tree plantations compared to abandoned pasture. In the top 15 cm of soil, total nitrogen and organic matter were higher under the trees than in the degraded pastures, with values close to those found in adjacent 20-year-old forests. The highest values for soil organic matter, total N, Ca, and P were found under *Vochysia ferruginea*, a species common in mature and secondary forests in the region (Montagnini 2001, 2002). Subsequent measurements revealed similar trends in the soil parameters in the following years (Montagnini 2007). Based on the standards determined by the Ministry of Agriculture of Costa Rica for soil fertility assessments, the cation levels (Ca, Mg and K) under most of the trees were at or above the critical values needed for conventional agriculture in the region. In contrast, the cation levels in the adjacent abandoned pasture soils were too low for the growth of the subsistence crops preferred in the region (rice, beans). Soil organic matter also had positive influences on soil physical properties: soil bulk density was lower (i.e., lower compaction) while soil moisture was higher under the trees than in abandoned pastures.

Other research at La Selva Biological Station included mixed- and pure-species plantations of indigenous trees as strategies to recover abandoned pastures (Montagnini et al. 1995b). Early results of this research showed that 5 years after planting, there were decreases in the content of P, K, and Ca in the soil in pure plots of fast-growing tree species, such as *Jacaranda copaia* and *Vochysia guatemalensis*

(Montagnini 2007). However, in other cases there were beneficial effects upon the soil: for example, increases in Ca in the soil under *Terminalia amazonia* and *Virola koschnyi*, both species with a high content of Ca in their foliage and elevated rates of annual litterfall. In a similar fashion, soils under *Vochysia ferruginea* had greater concentrations of Ca, Mg, and organic matter in comparison to the other species. This result is consistent with other studies that include this species (Montagnini 2001, 2007).

Measurements over a long period are necessary to determine the effects of tree species on soils. When the plantations were 11–12 years old, results indicated that while many of these trends continued, some new ones were observed. For example, the soil under *Vochysia guatemalensis* had higher pH, less acidity, and greater Mg than other treatments. On the contrary, the soils under *Calophyllum brasiliense* had lower values for the same parameters. These results appear to be related to the high capacity of *Vochysia guatemalensis* to recycle cations, given its high quantity and rapid decomposition of leaf litter, while the opposite happens with *Calophyllum brasiliense*. The value of soil pH in *Vochysia guatemalensis* (5.03) was higher than in previous measurements; the values of soil Ca, Mg, K, organic matter, and N were similar and the values of P were less (Montagnini 2007).

The mixed plots in this research had average values for the nutrients examined, and even improved conditions for some soil nutrients, such as P (Montagnini 2007). In some cases there were lower values for nutrients in mixed plantations than in pure plantations, as was the case for Ca and Mg. This suggests that in mixed plots, soils have a more balanced nutrient status as a result of the complementary effect on nutrient cycling of the different species participating in the mixtures. In contrast, in pure-species plots, there were some decreases or increases of soil nutrients, depending on the species.

In similar studies conducted in Bahia, Brazil, values of at least five soil parameters under 15 out of the 20 species of the plantations were similar or higher to those found under primary or secondary forest. Several species contributed to increased C and N, including: *Inga affinis*, *Parapiptadenia pterosperma*, *Plathymenia foliolosa* (leguminous, N-fixing species), *Caesalpinia echinata*, *Copaifera luscens* (leguminous, non-N-fixing), *Eschweilera ovata*, *Pradosia lactescens* (of other families). Others increased soil pH and/or some cations, such as *Copaifera luscens*, *Eschweilera ovata*, *Lecythis pisonis* and *Licania hypoleuca* (Montagnini et al. 1994, 1995a). Most of the species identified in this research for their positive influence on soil properties are currently being used as components in productive land use systems such as commercial plantations and agroforestry in each region (Montagnini 2001).

Exotic tree species can also play an important role in the restoration of degraded pastures, especially when conditions of extreme degradation persist, such as extremely eroded, denuded soil. If there is not enough time or resources to test the performance of local species, a well-known exotic species that can thrive in highly adverse conditions can be planted initially and later serve as nurse or facilitator for the introduction of native species. For example, in southern Costa Rica, Carpenter et al. (2004) recommended planting *Pinus tecunumanii*, an exotic tree, on the most deeply eroded, bare areas of pasture to ameliorate hostile micro-sites quickly. Two or more years after planting the pines, seedlings of shade-loving native species could be intermixed. In areas that have suffered less extreme erosion, native species such as *Vochysia guatemalensis*, *Terminalia amazonia*, *Calophyllum brasiliense* are preferred, according to results of long-term studies by the authors in the region. The native trees were interplanted with cover crops or with edible beans in controlled experiments, where these systems rapidly restored cover to eroded pastures without fertilization (Nichols et al. 2001). These authors suggest a system of rotations, planting pine as a rapidly growing, frequently harvested species, and inter-planting more slow-growing but more valuable species (Carpenter et al. 2004). However, when using exotic species for restoration one has to be aware that the qualities that make these species adaptable to harsh environmental conditions can also allow them to become invasive. For example in Misiones, NE Argentina, *Leucaena leucocephala*, a well-known N-fixing species broadly used in agroforestry systems worldwide, has been introduced to improve degraded soils, and today this species has expanded beyond the planted areas and in some case has become a weed.

Reforestation can serve diverse economic, social, political and ecological functions in the Neotropics. With considerably higher yields than managed native forests, tropical and subtropical plantations can make substantial contributions to world timber and pulp production (Montagnini and Jordan 2005). Tree plantations can also be a source of cash, savings and insurance for individual farmers. In combination with subsistence and commercial crops (agroforestry) or cattle (agrosilvopastoral systems), plantations have been used as tools in rural development projects worldwide. Plantations are often seen as alternatives to deforestation as they can provide products that otherwise would be taken from natural forests. If plantation species and management strategies are chosen with knowledge of their nutrient-use efficiencies and recycling capacities, they can be highly productive and also serve a function in ecological restoration projects.

13.4.1.2 Restoration of Biodiversity

Several authors have reported on the role of tree plantations as accelerators or "catalysts" of natural succession in tropical and subtropical sites (Parrotta 1995, Jussi et al. 1995, Parrotta and Turnbull 1997, Keenan et al. 1999, Parrotta 1999, Otsamo 2000). For example, in Southeast Asia, Jussi et al. (1995) reported on the spontaneous and fast growth of indigenous tree species under plantations of exotic trees. On the other hand, in north Queensland, Australia, a greater diversity of species was found in the understory of plantations of native species than in plantations of exotic species (Keenan et al.1999). In Puerto Rico, in the understory of plantations of the exotic *Albizia lebbek*, 22 species of trees and shrubs were found, in comparison with just one species in control plots without trees (Parrotta 1992). At La Selva Biological Station, Costa Rica, results of some studies also suggest that tree plantations have a good potential for accelerating the processes leading to recovery of biodiversity in degraded soils (Guariguata et al. 1995, Powers et al. 1997, Carnevale and Montagnini 2002, Cusack and Montagninin 2004, Montagnini et al. 2005). On the other hand, high establishment and maintenance costs are potential disadvantages of the use of plantations for accelerating natural regeneration, given the intensive management that is needed especially during the first 2–3 years (Montagnini et al. 1995b).

Mixed plantations may offer a more favorable environment for natural regeneration than pure plantations due to their multi-strata architecture. Mixed plantations may have a higher variety of microhabitats for seed dispersers and potentially create a greater variety of ecological niches allowing for the establishment of diverse regeneration. At La Selva Biological Station, mixed plantations with native tree species had relatively high abundance and high numbers of regenerating species in their understory in comparison with pure plantations (Carnevale and Montagnini 2002, Cusack and Montagnini 2004) (Table 13.1). Higher plant species richness accumulated under *Vochysia guatemalensis*, *Virola koschnyi*, *Terminalia amazonia*, *Hyeronima alchorneoides* and *Vochysia ferruginea*, all species commonly planted by farmers in the region. Natural regeneration was higher in understories with low or intermediate light availability.

Results of seed rain and seed dispersal studies showed that most of the seeds entering the open pastures were wind-dispersed, while most seeds entering the plantations were bird or bat-dispersed (Orozco Zamora and Montagnini 2007a) (Table 13.2). In terms of seed rain abundance, plantation plots with either pure or mixed species had greater abundance than the abandoned pasture plots left for natural regeneration. The pure species plantations that had greater seed rain abundance

Table 13.1 Abundance of woody seedlings and saplings in three plantations with native tree species in pure plots, mixtures of thee species, and open pastures left for natural regeneration at La Selva Biological Station. Plantations were 7 years old at time of sampling

Treatment	No. Individuals/32 m^2	No. Individuals/ha
Plantation 1		
Vochysia guatemalensis	90	28,125
Calophyllum brasiliensis	78	24,375
Jacaranda copaia	57	17,812
Mixture of 3 species	88	27,500
Natural Regeneration	29	8,937
Plantation 2		
Virola koschnyii	38	11,875
Dypterix panamensis	18	5,625
Terminalia amazonia	88	27,500
Mixture of 3 species	34	10,625
Natural Regeneration	14	4,375
Plantation 3		
Vochysia ferruginea	18	5,703
Hyeronima alchorneoides	25	7,891
Balizia elegans	13	4,219
Mixture of 3 species	32	10,156
Natural Regeneration	2	703

Sources: Carnevale and Montagnini (2002); Cusack and Montagnini (2004).

Table 13.2 Distribution of dispersal vectors by treatment in pure plantations, mixed plantations, and open pastures left for natural regeneration at La Selva Biological Station. Plantations were 12–13 years old at the time of the sampling

	Categories of dispersal vectors						
Treatment	Birds (%)	Mammals (%)	Wind (%)	Mammals and birds (%)	Mammals, birds, and wind (%)	Others (%)	Total (%)
J. copaia	0	2	8	70	14	6	100
C. brasiliense	12	–	–	64	10	14	100
V. guatemalensis	94	–	1	–	3	2	100
Mixed 1	13	–	–	80	–	7	100
Natural Regen. 1	79	1	–	–	10	10	100
T. amazonia	–	–	–	99	–	1	100
V. koschnyi	–	–	–	98	1	1	100
D. panamensis	2	–	–	98	–	–	100
Mixed 2	5	–	3	88	2	2	100
Natural Regen. 2	–	–	97.7	0.3	2	–	100
B. elegans	–	–	–	100	–	–	100
H. alchorneoides	44	–	–	50	2	4	100
V. ferruginea	98	–	–	1	–	1	100
Mixed 3	18	0	0	79	1	2	100
Natural Regen. 3	–	–	100	–	–	–	100

(Source: Orozco Zamora and Montagnini 2007a,b)

were *Balizia elegans*, *Dipteryx panamensis*, and *Jacaranda copaia.* Apparently, plantation trees in either mixed or pure plots are more effective in attracting birds and other dispersal agents than the grass and other low-stature vegetation that predominates in the abandoned pastures left for natural regeneration. In terms of seed species richness, the abandoned pastures ranked lower than both the pure and mixed plantation plots, again suggesting that plantations provide a more attractive habitat for seed dispersal agents than natural regeneration plots (Orozco Zamora and Montagnini 2007a, 2007b). The lower species richness of seeds found in the abandoned pasture treatments, especially in those populated by grasses and ferns, suggests that natural regeneration can be delayed by the absence of seeds and seed dispersal agents. This suggests that plantations facilitate tree regeneration by attracting seed-dispersing birds and bats into the area.

The different species of the plantations created different conditions of shade and litter accumulation, which in turn affected forest regeneration (Carnevale and Montagnini 2002). Competition from grasses is a major factor influencing woody invasion under these plantations. High accumulation of litter on the plantation floor may help diminish grass growth and thus encourage the establishment of other woody species under tree canopies; however, canopy shading apparently had a greater influence on tree regeneration than litter depth (Carnevale and Montagnini 2002). In addition, results of further studies in this site suggest that there was no significant correlation between leaf litter in the plantation floor and abundance of regeneration (Cusack and Montagnini 2004). Powers et al. (1997) and Parrotta (1995) found litter depth to be negatively correlated with abundance and richness of understory regeneration. Parrotta (1995) suggested that high litter biomass suppresses regeneration of small-seeded species.

Although the results of several studies suggest that plantations can accelerate or "catalyze" secondary forest succession, it is still unknown as to what extent they can promote the establishment of species of more advanced stages of succession. In our research at La Selva, the majority of seeds collected under the mixed and pure plantations were from heliophillous species, which are associated with early secondary forests. (Orozco Zamora and Montagnini 2007a, 2007b). A total of 35% of the species of the seeds collected in this research belonged to secondary forests (*Renealmia alpinia*, *Psychotria bracheata*, *Palicourea guianensis*, among others) (Fig. 13.3). A total of 20% of the collected species of seeds belonged to a highly varied functional stage or location in the landscape as many species can be found in abandoned fields, advanced secondary forests, riparian zones, open areas, minimally altered forests, and other functional stages (e.g. *Ficus* spp., and species of the Araceae and Cucurbitaceae families). Approximately 15% of the species of the collected species were associated only with abandoned fields and pastures (e.g. grass species, Poaceae family). The rest of the species (5–10%) are found in one to three functional stages, all of which had secondary forests in common (Fig. 13.3).

Recent studies suggest that there may be a limit on the use of plantations to foster biodiversity at a particular site, although such plantations often catalyze successional development in the plantation understory (Lamb et al. 2005). Subsequent studies can verify whether the forest species composition favored by plantations approximates more mature stages of natural forest succession in the region.

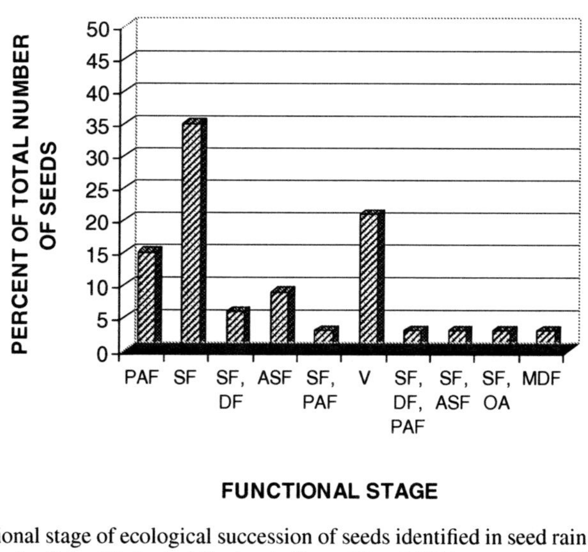

Fig. 13.3 Functional stage of ecological succession of seeds identified in seed rain studies in native tree plantations at La Selva Biological Station in Costa Rica. PAF (pastures and abandoned fields), SF (secondary forest), DF (disturbed forests), ASF (advanced secondary forest), V (various), OA (open areas) and MDF (minimally disturbed forests) (Source: Orozco Zamora and Montagnini 2007b)

In addition to considering the ecological factors influencing tree regeneration, the socioeconomic feasibility of these plantations as practical systems for the recovery of biodiversity in deforested landscapes in each region should be assessed.

13.4.2 The Value of Isolated Trees in Pastures for Restoration

13.4.2.1 Role in Favoring Tree Regeneration

In many tropical pastures there are isolated trees that can be remnants of the original forest. They are conserved by farmers for their functions as shade for cows and humans, and as sources of firewood, timber and fruit. Often the trees are left uncut because they are too large or the farmer may not have the tools or labor available to remove them. For example, isolated trees of *Dipteryx panamensis*, which has relatively dense wood, are often present in many pastures in Central America, because they are harder to cut than other forest trees.

Often the microenvironment below tree crowns is favorable for the establishment of species that cannot grow in pastures. For example, in the lowlands of the Pacific coast of Chiapas (Mexico), researchers found a total of 65 individual trees in surveys done in a total of 15 pastures of 1 ha in size each, grazed with a cattle density of two animals per ha (Otero Arnaiz et al. 1999). The trees belonged to 20 different species in 11 families. The timber species were typically from the primary forests common in the region. In addition, under the crown of the trees they found 134 species and 45 families of plants, predominantly herbs (62 species) followed by shrubs (32), trees (31) and lianas (3). The authors recommended that future efforts to restore the forest from isolated trees already present in the pastures should put emphasis on maintaining a great diversity of tree species in the pasture fields.

In a similar study conducted on 24 dairy farms in Monteverde, Costa Rica, Harvey and Haber (1999) examined a total of 237 ha of pastures. They found a total of 5,583 trees (25 trees/ ha on average). The total number of tree species was 190, of which 57% belonged to primary forest of the region. They also examined the farmers' uses of the trees: 37% were used for timber, 36% for firewood, and 20% for posts. About 90% of the trees also provided food for birds. The authors concluded that although the tree densities were apparently high in the studied pasture fields, the species diversity would probably decrease with time, since the farmers were harvesting trees for the uses already mentioned. There should be incentives for planting native trees in the pastures to replace those that were harvested or died during previous years (Harvey and Haber 1999).

13.4.2.2 Influence on Seed Dispersal

Several species of native trees growing isolated in pastures can be effective as perches to attract birds and aid in the establishment of other species of native forest.

In a study conducted in "Las Alturas" biological station in southern Costa Rica, Holl (1998) compared the effectiveness of *Inga* trees growing naturally in pastures, with artificial posts set as perches. The artificial perches were cross-posts 6 m in height with two attached 2 m length crossbars installed in abandoned pastures at different distances from the border of the forest. In addition, dead branches of *Inga* were used in the same manner, separated by 10 m between them. The quantity of seeds dispersed by birds was greater beneath the branches than beneath the crosses or in the pasture without perches. Nevertheless the number of plants of these species was similar below perches and in the pasture. Despite the fact that the perches increased the dispersion of seeds, they could not compensate for other barriers to the recuperation of the tropical forest, such as the depredation of seeds, low germination rate of seeds, and low survival of plants.

The type of seeds dispersed by the birds should also be considered. Were they from species that could promote successional processes or that could be used by local farmers? Plantations with tree species are the most expensive, but they can be the most effective strategy for recuperating the forest (Holl 1998).

Crown size and shape, tree height and other characteristics may influence the ability of trees to attract birds and facilitate tree regeneration in abandoned pastures. In a seed rain study at finca La Suerte in the Caribbean region region of Costa Rica, Slocum and Horvitz (2000) found 25 species of seeds from trees and shrubs under trees that were producing fruit, with larger numbers than those found in open pastures. These authors found larger abundance and species richness of seeds under *Ficus* trees than under *Cordia* trees. They found a positive correlation between height of trees and seed rain abundance and diversity, with larger seed input under the taller trees. Apparently birds prefer taller trees as perches to have a better view of the surroundings and to avert potential predators. In our study at La Selva, trees of *Jacaranda copia* and *Vochysia guatemalensis*, which were the tallest, had approximately twice as much seed rain abundance and seed species richness as the shortest trees, *Callophylum brasiliense* and *Dipteryx panamensis* (Orozco Zamora and Montagnini 2007a).

However, other characteristics of trees apart from height may be important in their role as recruitment foci. For example, Slocum (2001) found that the amount of shade cast was more important than the type of fruit produced (dry or fleshy), with lower recruitment under the shadier canopies. In the same study site, Slocum and Horvitz (2001) found that canopy architecture did not appear to be a significant factor affecting seed arrival with no difference in seed input beneath trees of dense canopy (*Ficus* and *Pentaclethra*) and those of sparse canopy (*Cordia* and *Cecropia*). On the other hand, results of a recent study in the Pacific region of Costa Rica suggested that crown diameter was more important than height of trees in attracting birds (Francesconi 2006). Thus, a combination of several characteristics may determine the role of each tree species as recruitment foci. For example, at La Selva, Guariguata et al. (1995) suggested that although *Vochysia guatemalensis* trees do not fruit early, this species looks promising as an effective restoration tool because it appears to provide perching and roosting sites for frugivorous birds and bats while rapidly suppressing grass cover.

13.4.2.3 Influence on Soil Nutrients

In research on the role of isolated trees on soil nutrients in pastures in lower montane ecosystems in Ecuador, the density of trees was about 30 individuals/ha, of which about 75% were species of the nitrogen-fixing *Inga* spp., and the rest were guajava trees, *Psidium guajaba* (Rhoades et al. 1998). Below the trees there were moderate air temperatures and intensity of light similar to that of the forest. The concentration of nitrates in soils beneath *Inga* spp. trees was four times higher than in the open pastures. There was also increased organic matter in the soil under the *Inga* trees. In other work done in the same area it had been shown that the activity of birds was 20 times higher in pastures containing trees than in pastures without trees. As birds can be a source of seed, their activity can favor the establishment and growth of timber species in the pastures (Rhoades et al. 1998). The number and diversity of species of trees could be increased in order to accelerate regeneration of forest in abandoned pastures. These results and recommendations are relevant since the species found in these pastures are common to many tropical regions of the Neotropics.

13.4.2.4 Use of Nurse-Trees and Artificial Protection

Addition of artificial structures can sometimes increase survival of regenerating seedlings when used for protection against excess insolation, drought or frost. For example, in abandoned pastures at high elevations in Hawaii, the aim of restoration projects is to return the original mixed forest in order to serve as habitat for threatened birds. In a study conducted in the Wildlife Refuge of Hakalon, on the island Mauna Kea at 1,840 m elevation, with 2,000 mm of rain per year, the meadows that remain after more than 100 years of human intervention (forest clearcut, pastures) and conversion to grassland are generally dominated by exotic grasses (Scowcroft et al. 2000). At high elevations, planted seedlings of native species can suffer frost damage and desiccation when planted in the open. In their study site, Scowcroft et al. (2000) transplanted seedlings of *Metrosideros polymorpha* (Myrtaceae), a native tree of Hawaii from nearby woodlands. They planted the seedlings in the open and within a corridor of 8-year-old, 4 m tall trees of the native species *Acacia koa* that were sown earlier in order to serve as protection against frost for the *Metrosiderus* trees. The use of *Acacia koa* trees or artificial mechanisms, such as protection with cloth on the sides, increased the successful establishment of *Metrosiderus polymorpha* seedlings in these degraded, abandoned rangelands. The authors recommend *Acacia koa* as a nurse tree because it is endemic to Hawaii, fixes nitrogen and is capable of rapid growth, which allow it to quickly dominate a site when planted in groves or corridors and enhance soil N availability. Within 2–3 years the trees should be tall enough to be out of danger from frost, and if planted at 4–5 m intervals, crowns should close after 10 years resulting in reduced herbaceous competition in the understory. The improved temperature, nutrient content, and herbaceous competition regimes under canopy of *Acacia koa* may favor establishment of other, more sensitive native species.

13.4.2.5 Role of Windbreaks to Accelerate Succession in Pastures

In regions with large agricultural fields that are far from sources of propagules, windbreaks in pastures and agricultural fields may be important reservoirs of native tree species (Holl 1998, Harvey and Haber 1999, Harvey 2000). For example, the effects of planted windbreaks on seed deposition patterns were examined in dairy farms in Monteverde, Costa Rica by Harvey (2000). The windbreaks were planted strips of trees about 5 m wide and 9 m tall and were 7–8 years old at the time of the study. Trees and rows within the windbreaks were spaced at 1.5 m. The most common species were *Cupressus lusitanica*, *Croton niveus*, *Casuarina equisetifolia* (all exotic species) and *Montanoa guatemalensis* (native). Windbreaks were found to receive significantly greater densities and species richness of seeds from tree and shrub species than open pastures. Windbreaks received an average of 39 times as many tree seeds, and 67 times as many shrub seeds as pastures. The differences in the seeds entering windbreaks versus pastures appeared to be due almost entirely to the enhanced activity of birds in windbreaks. Bird-dispersed seeds occurred in greater densities (about 100 times greater) and number of bird-dispersed species was three times greater in windbreaks than in pastures. The high densities of bird-dispersed seeds within windbreaks suggest that windbreaks increase forest seed recruitment by serving as habitat and/or movement corridors for seed-dispersing birds (Harvey 2000).

Windbreaks may serve as sources of woody colonists if the agricultural lands are later abandoned. Positioning of windbreaks within the landscape may affect seed deposition patterns by influencing the movements of seed-dispersing birds. Tree recruitment may be higher in windbreaks that are connected to forests. Windbreaks could be made more attractive to birds by including native, fruit-producing trees, by increasing their species and structural complexity, and by positioning them between forest patches to facilitate bird movement (Harvey 2000).

13.5 Alternatives to Delay Pasture Degradation and Increase Pasture Productivity

Pasture productivity and longevity in the Neotropics seem to be closely related to soil fertility and efficient nutrient recycling. CIAT (International Center for Tropical Agriculture) in Cali, Colombia; EMBRAPA-CPATU (Agency for Agricultural Research of Brazil, Center for Studies of the Humid Tropics); CATIE (Tropical Agriculture Research and Higher Education Center) in Turrialba, Costa Rica; INTA (National Institute for Agricultural Technology) in Misiones, Argentina and other international and local agencies in countries of the Neotropics have developed recommendations based on long-term research carried at their stations and in local farms. These strategies are intended to increase productivity and avoid or delay degradation, and include the following:

1) Use of adequate stocking (animal type/breeds, weight, numbers per hectare)
2) Planting good pastures (palatable, nutritious), often including exotic pasture species
3) Use of cover crops to protect against erosion especially on slopes
4) Use of mixed pasture species, intercropping with legumes
5) Rotational pastures (including use of electric fencing)
6) Fertilization as needed (especially N, P, K)
7) Use of agrosilvopastoral systems

13.5.1 Adequate Management Techniques to Ensure Pasture Sustainability

13.5.1.1 Method of Land Clearing

The method of land clearing can have important and long-lasting consequences for pasture productivity and longevity. Land clearing for pasture establishment on forested areas of the Amazon basin generally involves cutting the forest with chain saws, generally after extraction of the commercially valuable woody species (Serrao and Toledo 1990, Dias-Filho et al. 2001). Mechanical clearing with bulldozers is seldom practiced because of the high costs and the detrimental effects on soil properties, as was shown by studies done in the 1970s when clearing forest for pasture in the Amazon first began.

For example, in a study in the Amazon forest of Peru, biomass production of the pasture grass *Panicum maximum* was 68% higher in a pasture on land that had been cleared manually (slash and burn method) in comparison with a pasture that had been established on an area cleared with bulldozer (Seubert et al. 1977 cited by Dias-Filho et al. 2001). Further studies have confirmed this assertion. The fire employed in the manual clearing not only contributes to remove weeds but also improves nutrient content of the pasture soils with inputs of nutrients in the ash. However it should be noted that the combustion of vegetation also results in large losses of nutrients such as nitrogen, sulfur and carbon (Dias-Filho et al. 2001), with negative consequences to long-term pasture productivity, and air pollution as well.

In addition, the initial input of nutrients to the soil following burning is generally short-lived. The early claims from the 1970s that conversion to cattle pastures results in improved soil fertility of once forested soils following clearing and burning were refuted by Fearnside (1980). Fearnside reviewed the available evidence and concluded that long-term sustainability of the pastures was not possible under those conditions, evidence that was supported by this same author's subsequent articles as well as by many others. Improvement of soil pasture nutrient status must be obtained through the use of N-fixing species and/or the addition of fertilizer (Serrao and Toledo 1990, Dias-Filho et al. 2001).

13.5.1.2 Use of Adequate Pasture Species Mixtures

Alternatives to delay pasture degradation and increase productivity developed by EMBRAPA-CPATU in Brazil, by CIAT in Colombia and by CATIE in Central America recommend intercropping leguminous herbs such as *Centrosema pubescens*, *C. macocarpus*, *C. acutifolium*, *Stylosantes guianensis*, *Arachis pintoi* and *Desmodium ovalifolium* with productive grasses such as *Brachiaria brizantha*, *B. dictyoneura*, *Andropugon gayanus*, *Panicum maximum* (Serrao and Toledo 1990, Vera et al. 1998, Bouman et al. 1999, CIAT 1997–2000, Amézquita et al. 2005). In unfertile soils, N-fixing species are better competitors than grasses and they tend to dominate the pasture. Thus, in these mixtures the appropriate proportion of legumes/grasses has to be sown, and the timing of planting the legumes/grasses has to be set to fulfill the objectives of producing a nutritious pasture and maintaining or improving soil nutrients. To stimulate the early growth of the leguminous herbs, they are inoculated with *Rhyzobium* bacteria to ensure the formation of root nodules and increase rates of N fixation.

The species of pasture and leguminous herbs will vary according to the specific region and the varieties and cultivars that have been developed by the corresponding institutions in each region. For example, in the Andean hillsides of Colombia improved pastures of *Brachiaria decumbens* are used, while in the Atlantic humid region of Costa Rica mixtures of *Brachiaria brizantha* + *Arachis pintoi* are more commonly used. In drier regions of the Neotropics, such as in the Pacific region of Costa Rica *Brachiaria decumbens* is more common, among other alternatives (Bouman et al. 1999, Amézquita et al. 2005). *Brachiaria decumbens* is also commonly used in the savanna regions of eastern Colombia (Vera et al. 1998).

13.5.1.3 Management of Soil Nutrients

Fertilization with P, Ca, Mg and K is often recommended for sown pastures, however depending on the region and cost of fertilizer, these recommendations are seldom followed, except for the application of P (Vera et al. 1998). Inputs of N to the pasture are expected to be sufficient when using mixtures with leguminous herbs. However the leguminous herbs also may need a "starter" fertilizer application with N, P and K to stimulate their growth and to improve the efficiency of N fixation.

Soil phosphorus availability has been shown to be of particular importance in maintaining pasture productivity, as this element is most limiting (Garcia-Montiel et al. 2000, Dias-Filho et al. 2001). P can be in different forms in the soil and some may be more or less available to plants. In a study describing changes in total P and different soil P fractions that occurred after the conversion of forest to pasture in the Brazilian Amazon state of Rondonia, Garcia-Montiel et al. (2000) used chronosequences of forest and pasture of different ages to document patterns of labile, occluded, and organic P pools using a sequential P-fractionation technique. These authors found that P released from the forest biomass left on the ground substantially increased soil available P during the first 3 to 5 years following clearing and burning. Since large portions of P are released from the burned and decomposing biomass, management could be directed to improve the distribution of decomposing biomass on the site, and to adding other sources of organic P once the residual biomass is gone.

Of the approximately 200 kg of P per hectare measured in soils of one pasture in the eastern Amazon, less than 1% was thought to be available to plants (Dias-Filho et al. 2001). Since fertilization is expensive, management techniques should be geared towards effective P recycling and conservation. Much of the P taken up by plants is returned to soils via animal manure, however cattle tend to deposit dung in concentrated areas rather than uniformly across the pasture. Pasture creation also impacts carbon and nitrogen cycles by reducing standing stocks and cycling of these elements (Dias-Filho et al. 2001). C and N stocks may increase in pasture soils, but such increases are unpredictable and dependent on native soil fertility, fertilization, climate, fire frequency, and grazing intensity. Even when standing stocks increase, research from the eastern and western Amazon shows that cycling rates of these elements consistently decrease (Dias-Filho et al. 2001).

P supply to plants in pastures is strongly influenced by type of management. In a study done in soils of pastures in Colombia, after little P fertilization and 15 years of continuous growth of a grass alone (*Brachiaria decumbens*) and a grass/legume (*B. decumbers* + *Pueraria phaseoloides*) pasture, it was found that reserves of labile organic P fractions associated with humic and fulvic acids was higher in the grass/legume pasture than in the grass alone (Guggenberger et al. 2000). Earthworm casts were highly abundant in the grass/legume pasture, and the authors concluded that earthworm activity improves the supply of P in the soils under tropical pastures by creating an easily available organic P pool (Guggenberger et al. 2000).

13.5.1.4 Pasture Rotations

Leaving grazed fields or paddocks to fallow is key to slowing pasture degradation. The length of the fallow period differs according to the region and the practices used. For example, in surveys conducted in pastures located in the eastern savannas of Colombia, a 3–4 week fallow period followed P fertilization, while a fallow period following mechanical reclamation was 6–9 weeks depending on the specific site conditions (Vera et al. 1998). The fallow period has to be balanced by the opportunity cost represented by the gain in weight or other productivity parameter of the cattle that has to be taken to another grazing field. Use of electric fencing often improves the feasibility of moving the cattle to nearby fields, especially when relatively small grazing paddocks are used.

13.5.1.5 Animal Stocking

The choice of the right type of animal and its stocking is necessary to attain the best productivity and sustainability of any cattle system. Much of the earlier discussion has referred to beef or dairy cattle of either European or Asian origin (*Bos taurus* and *Bos indicus* respectively). CATIE and many of the other institutions mentioned previously have promoted the use of mixed breeds that can combine greater rusticity and adaptability to tropical conditions with higher productivity (beef and dairy products).

In semiarid and sub-humid regions of the Neotropics, other animals are used successfully for meat and dairy production, and wool in the case of sheep. In many regions, goats and sheep are also used with success. In many cases these animals are part of small-scale, self-subsistence systems. However, they are also used at larger scales for commercial purposes. For example the pelibuey, a type of sheep commonly raised in the Caribbean and Mexico, successfully replaces sheep and goats of other breeds due to higher reproduction rates, better production of wool, and greater adaptability and rusticity (Fig. 13.4).

13.5.2 Agrosilvopastoral Approaches to Sustain Pasture Productivity and Avoid Degradation

Agrosilvopastoral systems—the combination of timber, fuelwood or fruit trees with animals, and with or without crops—are practiced at many scales. A large scale system may include timber plantations with grazing to control weeds and to obtain a more immediate return from the sale of animal products. Cattle-raising can also

Fig. 13.4 In Atotonilco, in the state of Hidalgo, Mexico, mixed breeds of pelibuey are grown using electric fencing for pasture rotations, on fields sown with African star grass (*Cynodon nlemfluensis* and *C. plectostachyus*). The owner and other farmers in the region apparently consider this a successful type of operation due to high productivity of the pelibuey and soil conservation. Photo: F. Montagnini

complement subsistence agriculture, with animals often integrated in home gardens or in systems of fodder production to feed animals in stables (Montagnini 1992). In some regions, the incorporation of trees—especially multiple-purpose trees (MPTS)—can change cattle raising from an inefficient use of land to a more ecologically and economically feasible activity. The incorporation of trees can improve system productivity either by increasing pasture yields through more efficient nutrient recycling, or through the production of fodder from leaves and fruits. The introduction of cattle in plantations, generally when the trees are big enough so that cattle cannot damage them, can be a more efficient use of land than plantations alone. Cattle can also help to control weeds and early economic returns from cattle products help to pay for the costs of reforestation (Montagnini et al. 2003) (Fig. 13.5). In some cases young calves are preferred over adult cattle because they cannot damage the trees as easily. The smaller animals also decrease soil compaction problems. Cattle prefer some tree species over others for browsing and can also damage young trees when they scratch against them or strip their bark.

Fig. 13.5 Agrosilvopastoral system established to offset costs of reforestation: beef cattle (bulls) grazing eucalypt plantations at CATIE in Turrialba, Costa Rica. Photo: F. Montagnini

Generally, in agrosilvopastoral systems there are short and long-term benefits from cattle and tree products. Economically, the system is favored through product diversification. In semiarid ecosystems, fodder trees have the additional advantage over grasses, of being able to reach deeper layers of soil thus remaining green for longer periods and providing fodder at critical times.

Depending on the relative proportion of trees to cattle and the emphasis given to the principal productive objective of the systems, three main types of agrosilvopastoral systems can be found: 1-Trees in pastures, 2-Cattle grazing plantations or forest, and 3-Integrated agrosilvopastoral systems.

13.5.2.1 Trees in Pastures

In this type of agrosilvopastoral system, the principal objective is cattle production (e.g., beef, milk) with timber, fruit or fuelwood as secondary products. Often the trees in the pasture are remnants from the previous forest, or have regenerated in the pasture and the farmers keep them to provide shade for the cattle, improve the pasture grasses, and obtain tree products. For example guacimo (*Guazuma ulmifolia*) trees grow naturally in pastures in Hojancha, in the Pacific region of Costa Rica. In pasture recovery projects in Hojancha, dense plantations of guacimo are established to replace the old trees (personal observation). The examples cited in previous sections (4.2.3) on isolated trees of *Psidium guajaba* and *Inga* spp. in pastures in Ecuador, and on other species in Chiapas, Mexico, and in Monteverde, Costa Rica (4.2.1), are also agrosilvopastoral systems of this category. In all these cases the trees serve multiple functions in terms of soil improvement, shelter for cattle, and also contribute to biodiversity as they are used as perches by birds.

Nitrogen-fixing trees are often planted in degraded pastures for the main purpose of soil restoration, and also for their fodder. In degraded agricultural fields near Guapiles, Costa Rica, researchers from CATIE use a restoration system they call "Poró acostado" ("laid poró"). They plant stakes of an N-fixing tree, ("poró"), *Erythrina poeppigiana*, laid horizontally in a trench to obtain rapid growth and nitrogen inputs to the soil (personal observations). This tree can sprout quickly from a horizontal stake giving multiple stems that will form clumps of young trees in the pasture, where they can shelter the cattle, improve growth of native or planted grasses and also their fodder can be harvested and fed to dairy or beef cattle.

13.5.2.2 Cattle Grazing in Managed Forests and Tree Plantations

In these types of systems, the principal objective is to obtain tree products; cattle serve to control weeds and to obtain an earlier financial return from the system. Cattle are often used to complement reforestation projects in the Caribbean lowlands of Costa Rica, such as calves used to graze plantations of native timber species (*Hyeronima alchorneoides*) in Horquetas, Costa Rica; and beef cattle (bulls) grazing eucalypt plantations at CATIE in Turrialba, Costa Rica (Fig. 13.5).

Examples under forest include beef cattle grazing under natural forest that has been exploited and later thinned to allow for pasture growth in Santiago de Liniers, in the subtropical forest of Misiones, Argentina. *Axonopus* spp. ("pasto Jesuita") is a native grass that is shade-tolerant and thus can grow well and stand the grazing when planted under the forest canopy. Cattle density is high with about 200 calves (250–400 kg each) per hectare. At these high densities calves are kept grazing in the forest for just half a day or a day at most, and electric fencing is used to allow for pasture rotations (personal observations). In another location near the town of Puerto Rico in the same region of subtropical forest, fully grown (about 20 years old) *Araucaria angustifolia* plantations are also underplanted with *Axonopus* spp. and grazed by beef calves (120–150 kg each). These calves are also fed a corn supplement (personal observations). In this region, the araucaria trees, which are native, take about 30–35 years to reach full size for timber harvest, and income from the cattle contributes to make the plantations more profitable. In these forest or plantation grazing systems, the forest or plantations are the principal management objective of the farmer, and the cattle are only introduced after the plantations are already established, or the forest has already been managed for timber production.

13.5.2.3 Integrated Agrosilvopastoral Systems

These systems are designed with the purpose of having an integrated system from the start, where cattle production is enhanced by planting pastures in association with trees that can benefit the pastures and cattle and at the same time provide important tree products. Tree planting distances, pasture species and cattle management are all carefully selected with these objectives in mind. Examples from the subtropical region of Misiones, Argentina include plantations of *Toona australis* (Australian cedar) planted at 9 m×3 m, a planting distance that gives enough room between the lines of trees to allow plantations of a shade-intolerant grass for the cattle. In this case the combination was with *Brachiaria* spp. grass to feed beef cattle (personal observation). In this same region, close to the locality of Puerto Rico in Misiones, *Paulownia* spp. trees ("kiri"), 9 years old, planted at 9 m ×6 m were seen in combination with *Brachiaria brizantha* grass to feed beef cattle. Both the Australian cedar and kiri trees (exotic) are highly profitable, and although neither of the two species tends to increase yields of grass, they do provide shelter for the cattle especially during the hot summer months. The trees, if grown in a pure plantation, could be planted at higher densities and thus the owner would obtain higher profits from selling a higher volume of timber at the end of the rotation. However, the farmers prefer to sacrifice some of the timber profits that they would obtain only at the end of a rotation in order to include the grasses and cattle and obtain earlier returns on the investment.

In other examples of this same type of system, trees are planted to improve the growth of grasses and not necessarily for timber or other products. For example, near Guapiles in the Caribbean lowlands of Costa Rica, projects by CATIE researchers for restoration of degraded pastures include plantations of *Acacia mangium*, an

N-fixing tree planted for soil erosion control. *A. mangium* is associated with native or planted pasture species such as *Ischaemum ciliare*, *Brachiaria brizantha*, and the herbaceous legume *Arachis pintoi* (Amézquita et al. 2005). *Arachis pintoi*, a type of peanut, is often planted to control soil erosion because it forms a dense carpet covering the soil, helps to recover soil nutrients, and also improves pasture and cattle productivity (Jansen et al. 1997, Bouman et al. 1999) (Fig. 13.6).

In the same region, *Gliricidia sepium*, an N-fixing tree that is an important firewood species and a good soil improver, has been found to have excellent qualities for fodder production and thus is often planted as a living fence. This is another type of integrated agrosilvopastoral system where the trees are purposefully chosen to serve their function as a fence and at the same time to improve soils, control erosion and serve as fodder (often in a cut-and-carry system).

Still another type of integrated agrosilvopastoral system is the "forage bank" or "fodder bank" generally composed of tree species with highly nutritious foliage for cut-and-carry systems. The trees are planted very densely to encourage foliage production. They are pruned periodically (often weekly), ground, occasionally mixed with cut grasses and then fed to cattle. Examples of tree species that can sustain high fodder productivity in this condition include *Trichanthera gigantea*, *Morus spp.*, *Erythrina edulis*, *E. berteroana*, *E. fusca*, *Boehmeria nivea*, which are used with success in Colombia and in Costa Rica (CIAT 1976–2000, Amézquita et al. 2005).

Fig. 13.6 *Arachis pintoi*, a type of peanut, is often planted in the humid regions of Costa Rica to control soil erosion because it forms a dense carpet covering the soil and contributes to soil nitrogen and organic matter inputs. In other regions of Costa Rica and Colombia it is also intercropped with trees or grasses to improve pasture and cattle productivity. Photo: F. Montagnini

Agrosilvopastoral systems can also be conceived at a landscape scale, where the trees are kept in a managed forest and cattle raising makes use of trees in a complementary manner. For example, models of low-external input agrosilvopastoral systems that incorporate farmer preferences, and use well-adapted grass-legume pastures, rotational grazing and the management of natural forest regeneration have been proposed for regions of the Amazon basin (Loker 1994). The systems provide farmers with the benefits of both annual crops and cattle raising, avoid land degradation and effectively incorporate trees in the production system.

Agrosilvopastoral systems are important components of the economies of Tropical America's countries (Amézquita et al. 2005). When improved and well managed, they can become key land-use systems for the provision of environmental services, particularly the recovery of degraded areas and carbon sequestration. In addition, they have the capacity to provide viable economic alternatives to farmers.

13.6 Conclusions

In the Neotropics, pastures represent a high proportion of the total agricultural land, and they are generally located on poor, acid soils. Conversion of forests to crops and pastures has been a predominant land–use change in the Neotropics during the second half of the twentieth century. Pasture development has been ranked one of the least desirable land uses of tropical forest regions because of land degradation caused by overgrazing and mismanagement. However, cattle ranching is important for local and regional economies, therefore it is crucial to understand the factors limiting pasture productivity, and to design ways of improving productivity and sustainability.

Abandoned pastures can remain in a state of arrested succession due to a number of factors that delay or impede their recovery. The factors limiting tree regeneration in abandoned pastures can include scarcity of nutrients, soil compaction, lack of or excess soil humidity, high solar radiation, and intra- and inter-specific competition. To restore abandoned pastures, strategies have to consider these factors limiting tree regeneration. Restoring abandoned pastures can involve planting trees in lines, groups or blocks. Tree plantations can improve soil conditions, ameliorate microclimatic conditions and serve as perches for birds, bats and other seed dispersers. Native tree species are preferred over exotics, except in cases where the performance of native species is still not well known.

A number of strategies can be used to manage tropical pastures to increase sustainability and avoid degradation. Several local and international research institutions, such as EMBRAPA (Brazil), CIAT (Colombia), CATIE (Costa Rica) and others, have designed pasture management strategies, mostly focusing on the use of improved pasture grasses, grass and leguminous herb mixtures, and proper management techniques. The use of the right animal breed and size is also key to avoid overgrazing and to sustain productivity. Agrosilvopastoral systems, or the combination of animals with trees in the landscape, offer many of the advantages of sound management and provide environmental services, such as carbon sequestration and conservation of biodiversity.

References

Aide, M. T., Zimmerman, L., Herrera Rosario, M., and Serrano, M. 1995. Forest recovery in abandoned tropical pastures in Puerto Rico. Forest Ecology and Management 77:77–86.

Aide, M. T., Zimmerman, L., Pascarella, J. B., Rivera, L., and Marcano-Vega, H. 2000. Forest regeneration in a chronosequence of tropical abandoned pastures: Implications for restoration ecology. Restoration Ecology 8(4):328–338.

Amézquita, M. C., Ibrahim, M., Llanderal, T., Buurman, P., and Amézquita, E. 2005. Carbon Sequestration in Pasture and Silvo-pastoral Systems in Four Different Ecosystems of the Latin American Tropics. Journal of Sustainable Forestry 21 (1):31–49.

Bouman, B. A. M., Nieuwenhuyse, A., and Ibrahim, M. 1999. Pasture degradation and restoration by legumes in humid tropical Costa Rica. Tropical Grasslands 33:98–110.

Buschbacher, R. J., Uhl, C., and Serrao, E. A. S. 1987. Large-scale development in Eastern Amazonia. Case study No. 10: Pasture management and environmental effects near Paragominas, Para. In Amazonian Rain Forests. Ecosystem Disturbance and Recovery. Case Studies of Ecosystem Dynamics Under a Spectrum of Land Use- Intensities, ed. C. F. Jordan, pp. 90–99. Springer, New York.

Carnevale, N. J., and Montagnini, F. 2002. Facilitating regeneration of secondary forests with the use of mixed and pure plantations of indigenous tree species. Forest Ecology and Management 163:217–227.

Carpenter, F. L., Nichols, J. D., and Sandi, E. 2004. Early growth of native and exotic trees planted on degraded pasture. Forest Ecology and Management 196:367–378.

CATIE. 2002. Multi-stakeholder participatory development of sustainable land use alternatives for degraded pasture lands in Central America. Tropical Agriculture Research and Higher Education Center (CATIE). Turrialba, Costa Rica.

Chazdon, R. L. 2003. Tropical forest recovery: legacies of human impact and natural disturbances. Perspectives in Plant Ecology, Evolution and Systematics 6(1,2):51–71.

Chazdon, R. L., Redondo Brenes, A., and Vilchez Alvarado, B. 2005. Effects of climate and stand age on annual tree dynamics in tropical second-growth rain forests. Ecology 86(7):1808–1815.

CIAT (Centro Internacional de Agricultura Tropical). 1976–2000. Tropical Pastures Program and Tropical Forages Project Annual Reports. Cali, Colombia.

Cusack, D., and Montagnini, F. 2004. The role of native species plantations in recovery of understory diversity in degraded pasturelands of Costa Rica. Forest Ecology and Management 188: 1–15.

Dias-Filho, M. B., Davidson, E. A., and Reis de Carvalho, C. J. 2001. Linking biogeochemical cycles to cattle pasture management and sustainability in the Amazon basin. In The Biogeochemistry of the Amazon Basin, ed. M. E. McClain, R. L. Victorio, and J. E. Richey, pp. 106–121. New York: Oxford University Press.

Didham, R. K., and Lawton. J. H. 1999. Edge structure determines the magnitude of changes in microclimate and vegetation structure in tropical forest fragments. Biotropica 31:17–30.

FAO. 2000. Food Balance Sheets. Rome, Italy: Food and Agriculture Organization of the United Nations (FAO).

Fearnside, P. M. 1980. The effects of cattle pasture on soil fertility in the Brazilian Amazon: consequences for beef production sustainability. Tropical Ecology 21 (1):125–137.

Ferguson, B. G., Vandermeer, J., Morales, H., and Griffith, D. M. 2003. Post-agricultural succession in El Petén, Guatemala. Conservation Biology 17(3):818–828.

Francesconi, W. 2006. The potential of living fences to connect the fragmented agricultural landscape in Esparza, Central Pacific Region of Costa Rica. MSci Thesis, School of Forestry and Environmental Studies, Yale University, p. 19.

Garcia-Montiel, D. C., Neill, C., Melillo, J., Thomas, S., Steudler, P. A., and Cerri, C. 2000. Soil phosphorus transformations following forest clearing for pasture in the Brazilian Amazon. Soil Science Society of America Journal 64:1792–1804.

Guariguata, M., Rheingans, R., and Montagnini, F. 1995. Early invasion under tree plantations in Costa Rica: implications for forest restoration. Restor. Ecol. 3:252–260.

Guggenberger, L. Haumaier, L, Thomas, R. J., and Zech, W. 2000. Assessing the organic phosphorus status of an oxisol under tropical pastures following native savanna, using ^{31}P NMR spectroscopy. Biology and Fertility of Soils 23:332–339.

Harvey, C. A. 2000. Windbreaks enhance seed disperal into agricultural landscapes in Monteverde, Costa Rica. Ecological Applications 10:155–173.

Harvey, C. A., and Haber, W. A. 1999. Remnant trees and the conservation of biodiversity in Costa Rican pastures. Agroforestry Systems 44:37–68.

Harvey, C., Alpizar, F., Chac6n, M., and Madrigal, R. 2005. Assessing linkages between agriculture and biodiversity in Central America. Historical overview and future perspectives. Mesoamerican and Caribbean region, Conservation Science Program. The Nature Conservancy (TNC). San José, Costa Rica, p. 138.

Hetch, S. B. 1993. The logic of livestock and deforestation in Amazonia. BioScience 43(10): 687–695.

Holl, K. D. 1998. Do bird perching structures elevate seed rain and seedling establishment in abandoned tropical pasture? Restoration Ecology 6:253–261.

Holl, K. D. 1999. Factors limiting tropical rain forest regeneration in abandoned pasture: seed rain, seed germination, microclimate, and soil. Biotropica 31:229–241.

Holl, K. D. 2002. Tropical moist forest restoration. In: Handbook of Ecological Restoration, Vol. II, ed. A. J. Davy and M. Perrow, pp. 539–558. Cambridge: Cambridge University Press.

Holmann, F., Argel, P., Rivas, L., White, D., Estrada, R. D., Burgos, C., Perez, E., Ramirez, G., and Medina, A. 2004. Is it worth to recuperate degraded pasturelands? An evaluation of profits and costs from the perspective of livestock producers and extension agents in Honduras. Livestock Research for Rural Development. Vol. 16, Art. #90. Retrieved August 28, 2006, from http://www.cipav.org.co/lrrd/lrrd16/11/holm16090.htm

Hoyos, P., García, O., and Torres, M. I. 19956. Manejo y utilizaci6n de pasturas en suelos ácidos de Colombia. Fascículo 4, Serie: Capacitaci6n en Tecnología de Producción de Pasturas. Centro Internacional de Agricultura Tropical (CIAT). Cali. p. 120.

Jansen, H. G. P., Ibrahim, M. A., Niewenhuyse, A., 't Manetje, L., Joenje, M., and Abarca, S. 1997. The economics of improved pasture and silvipastoral technologies in the Atlantic zone of Costa Rica. Tropical Grasslands 32:34–44.

Jussi, K., Goran, A., Yasuf, J., Antti, O., Kari, T., and Risto, V. 1995. Restoration of natural vegetation in degraded *Imperata cylindrica* grassland: understorey development in forest plantations. J. Veg. Sci. 6:205–210.

Kaimowitz, D. 1996. Livestock and Deforestation. Central America in the 1980s and 1990s: A policy perspective. CIFOR Special Publication. Center for International Forestry Research, Bogor, Indonesia, p. 88.

Keenan, R. J., Lamb, D., Parrotta, J., and Kikkawa, J. 1999. Ecosystem Management in Tropical Timber Plantations: Satisfying Economic, Conservation, and Social Objectives. J. Sustainable For. 9:117–134.

Lamb, D. 1998. Large scale ecological restoration of degraded tropical forest lands: the potential role of timber plantations. Restoration Ecology 6:271–279.

Lamb, D., Erskine, P. D., and Parrotta, J.A. 2005. Restoration of degraded forest landscapes. Science 310: 1626–1632.

Loker, W. M. 1994. Where is the beef? Incorporating cattle into sustainable agroforestry systems in the Amazon Basin. Agroforestry Systems 25:227–241.

Montagnini, F. 1992. Sistemas Agroforestales. Principios y Aplicaciones en los Tr6picos. 2da. ed. OTS/CATIE. San José, Costa Rica, p. 622.

Montagnini, F. 2001. Strategies for the recovery of degraded ecosystems: experiences from Latin America. Interciencia 26(10):498–503.

Montagnini, F. 2002. Tropical plantations with native trees: their function in ecosystem restoration. In Management of Tropical Plantation-Forests and their Soil Litter System. Litter, Biota and Soil-Nutrient Dynamics, ed. M. V. Reddy, pp. 73–94. Plymouth, UK: Science Publishers, Inc, Enfield (NH) USA.

Montagnini, F. 2007. Soil sustainability in agroforestry systems: experiences on impacts of trees on soil fertility from a humid tropical site. In Ecological Basis of Agroforestry, ed. D. R. Batish, R. K. Kohli, S. Jose, and H. P. Singh pp. 239–251. CRC Press Boca Raton.

Montagnini, F., Cusack, D., Petit, B., and Kanninen, M. 2005. Environmental services of native tree plantations and agroforestry systems in Central America. Journal of Sustainable Forestry 21(1):51–67.

Montagnini, F., Fanzeres, A., and da Vinha, S. G. 1994. Studies on restoration ecology in the Atlantic Forest region of Bahia, Brazil. Interciencia 19(6):323–330.

Montagnini, F, Fanzeres, A., and da Vinha, S. G. 1995a. The potential of twenty indigenous tree species for reforestation and soil restoration in the Atlantic Forest region of Bahia. J Appl Ecol 32:841–856.

Montagnini, F., González, E., Rheingans, R., and Porras, C. 1995b. Mixed and pure forest plantations in the humid neotropics: a comparison of early growth, pest damage and establishment costs. Commonwealth For Rev 74(4):306–314.

Montagnini, F., Ugalde, L., and Navarro, C. 2003. Growth characteristics of some native tree species used in silvopastoral systems in the humid lowlands of Costa Rica. Agroforestry Systems 59:163–170.

Montagnini, F., and Jordan, C. F. 2005. Tropical Forest Ecology. The Basis for Conservation and Management. Springer-Verlag, Berlin-New York. 295 pp.

Nepstad, D., Uhl, C., and Serrao, E. A. S. 1991a. Recuperation of a degraded Amazonian landscape: forest recovery and agricultural restoration. Ambio 20(6):248–255.

Nepstad, D., Uhl, C., and Serrao, E. 1991b. Surmounting barriers to forest regeneration in abandoned, highly degraded pastures: a case study from Paragominas, Pará, Brasil. In Alternatives to Deforestation: Steps Toward Sustainable Use of the Amazon Rain Forest, ed. A. B. Anderson, pp. 215–229. New York: Columbia University Press.

Nichols, J. D., Rosemeyer, M. E., Carpenter, F. L., and Kettler, J. 2001. Intercropping legume trees with native trees rapidly restores cover to eroded tropical pasture without fertilization. Forest Ecology and Management 152:195–209.

Orozco Zamora C., and Montagnini, F. 2007a. Seed rain and seed dispersal agents in pure and mixed plantations of native trees and abandoned pastures at La Selva Biological Station, Costa Rica. Restoration Ecology. 15(3):453–461

Orozco Zamora, C., and Montagnini, F. 2007b. Lluvia de semillas y sus agentes dispersores en plantaciones forestales de nueve especies nativas en parcelas puras y mixtas en la Estación Biológica La Selva, Costa Rica. Recursos Naturales y Ambiente (Costa Rica) 49: 131–140.

Otero Arnaiz, A., Castillo, S., Meave, J., and Ibarra-Martinez, G. 1999. Isolated pasture trees and the vegetation under their canopies in the Chiapas coastal plain, Mexico. Biotropica 31: 243–254.

Otsamo, R. 2000. Secondary forest regeneration under fast-growing forest plantations on degraded *Imperata cylindrica* grasslands. New Forests 19:69–93.

Parrotta, J. A. 1992. The role of plantation forests in rehabilitating degraded tropical ecosystems. Agric. Ecos. Environ. 41:115–133.

Parrotta, J. A. 1995. Influence of understory composition on understory colonization by native species in plantations on a degraded tropical site. J. Veg. Sci. 6:627–636.

Parrotta, J. A. 1999. Productivity, nutrient cycling, and succession in single- and mixed-species plantations of *Casuarina equisetifolia*, *Eucalyptus robusta*, and *Leucaena leucocephala* in Puerto Rico. For. Ecol. Manage. 124:45–77.

Parrotta, J. A., and Turnbull, J. 1997. Catalizing native forest regeneration on degraded tropical lands. For. Ecol. Manage. 99:1–290.

Powers, J. S., Haggar, J. P. and Fisher, R. F., 1997. The effect of overstory composition on understory woody regeneration and species richness in seven year old plantation in Costa Rica. For. Ecol. Manage. 99:43–54.

Reiners, W. A., Bouwman, A. F., Parsons, W. F. J., and Keller, M. 1994. Tropical rain forest conversion to pasture: changes in vegetation and soil properties. Ecological Applications 4(2): 363–377.

Rhoades, C. C., Eckert, G. E., and Coleman, D. A. 1998. Effects of pasture trees on soil nitrogen and organic matter: implications for tropical montane forest restoration. Restoration Ecology 6(3):262–270.

Sarmiento, F. O. 1997. Arrested succession in pastures hinders regeneration in Tropeandean forests and shreds mountain landscapes. Environmental Conservation 24(1):14–23.

Scowcroft, P. G., Meinzer, F. C., Goldstein, G., Melcher, P. J., and Jeffrey, J. 2000. Moderating radiative cooling reduces frost damage to *Metrosiderus polymorpha* seedlings used for forest restoration in Hawaii. Restoration Ecology 8:161–169.

Serrao, E. A., and Toledo, J. M. 1990. The search for sustainability in Amazonian pastures. In Alternatives to Deforestation: Steps Toward Sustainable Use of the Amazon Rain Forest, ed. A. B. Anderson, pp. 195–214. New York: Columbia University Press.

Slocum, M. G. 2001. How tree species differ as recruitment foci in a tropical pasture. Ecology 82:2547–2559.

Slocum, M. G., and C. C. Horvitz. 2000. Seed arrival under different genera of trees in a neotropical pasture. Plant Ecology 149:51–62.

Spain, J. M., and Gualdrón, R. 1991 Degradación y rehabilitación de pasturas. In Establecimiento y renovación de pasturas. VI Reunión del Comité Asesor de la Red International de Evaluación de Pasturas Tropicales (RIEPT), ed. C. E. Lascano and J. Spain, p.283. Cali, Colombia: Centro Internacional de Agricultura Tropical (CIAT).

Stocking, M., and Murnaghan, N. 2001. Handbook for the Field Assessment of Land Degradation. London: Earthscan Publications Ltd.

Szott, L., Ibrahim, M., and Beer, J. 2000. The hamburger connection hangover: cattle pasture land degradation and alternative land use in Central America. Serie Técnica. Informe Técnico No. 313. CATIE, Turrialba, Costa Rica

Uhl, C., Buschbacher, R., and Serrao, E. A. S. 1988. Abandoned pastures in eastern Amazonia. I. Patterns of plant succession. Journal of Ecology 76:663–681.

Vera, R. R., Hoyos, P., and Moya, M. C. 1998. Pasture renovation practices of farmers in the neotropical savannahs. Land Degradation and Development 9:47–56.

Wijdeven, S. M. J., and M. E. Kuzee. 2000. Seed availability as a limiting factor in forest recovery in Costa Rica. Restoration Ecology 8(4):414–424.

Williamson, G. B., Mesquita, R., Ickes, K., and Ganade, G. 1998. Estratégias de árvores pioneiras nos Neotrópicos. In Floresta Amazônica: dinâmica, regeneraçâo e manejo, ed. C. Gascon and P. Moutinho, pp. 131–144. Brazil: Instituto Nacional de Pesquisas da Amazônia. Manaus.

Chapter 14
Conclusions, Synthesis, and Future Directions

Randall W. Myster

14.1 Conclusions

In Chapter 1, we saw how individual plants—critical to the functioning of an ecosystem—respond to various abiotic and biotic environment cues that are often "bundled together" on gradients and after disturbances (Myster 2001b). These responses combine to define a plant species' niche, and these probabilistic niches express themselves fundamentally by determining plant-plant replacements over time and space. Replacement of individual plants creates changes in plant abundances, which may also lead to changes in plant species composition. Consequently, over larger spatial and temporal scales plant-plant replacements produce such plant patterns as species distributions, coexistence of species, associations among species, and succession.

Also in Chapter 1, the analysis of data from Neotropical permanent plots after pasture and after banana and sugarcane cultivation suggested that facilitation is more common than competition in early old field succession, that grasses are strong competitors after abandonment especially if they persist as the past crop, and that old field plant communities are best thought of as loosely-structured assemblages of species rather than as a higher-order "unit" of interacting species.

Many of these themes—plot data and successional patterns, change in plant resources, plant tolerances and mechanisms—were discussed and developed in the ensuing chapters. The remainder of the first section of the book underscored how influential the past crop of a field can be in affecting the field's environmental conditions after abandonment. We saw that after pasture above-ground carbon (C) accumulates quickly, leading to secondary forests with more above-ground C than primary forests, and that this accumulation may be even faster if exotics are common. Below ground, however, soil C has much more modest gains, which may be correlated not with age after abandonment, but with years the field was used as pasture. As another example, after sugarcane cultivation in Puerto Rico an entire new forest type dominated by *Spathodea campanulata* has arisen. Such forests have high basal area, medium tree density, but low species richness. Finally, conversion of forests to either pastures or agricultural use reduces the soil biota of fungi, microbes, and earthworms, both in raw numbers and in species diversity. In particular, while

R. W. Myster, *Post-Agricultural Succession in the Neotropics.*

earthworms prefer moist fields/pastures, termites prefer drier sites and exotic earthworms are more common after tree plantations than after pastures and crops.

The second section of the book involved reference to the multiple mechanisms and tolerances that determine the process of plant-plant replacement in old fields. The authors discussed the importance of dispersal in Neotropic old fields, with major emphasis on the behavior of dispersing animals and the role of remnant "nurse" trees. Cattle and horses are an understudied dispersal vector in pastures and areas recovering from agriculture in the Neotropics, especially in tropical dry forest. They mainly disperse large, dry indehiscent fruits of the Legume family. Remnant trees greatly increase density and diversity of dispersed seeds in pastures by forming vegetation "islands." Seed predation, competition, and facilitation were also discussed as first-tier mechanisms that are critical to old fields derived both from wet and dry Neotropical forests. In Panama pastures, for example, the exotic grass *Saccharum spontaneum* invades, competes against, and inhibits the germination and growth of old field plants. Experiments demonstrate that facilitation is common at the early stage of early old field succession when neighbor vegetation frequently improves the performance of tree seedlings. For dry forest, in particular, resprouting may be an important regeneration option because trees in those areas are specifically adapted for above-ground mortality and drought. Further species-specific investigation into the sources of variation in the workings of the discussed mechanisms is needed.

The last section of the book opened with the analysis of soil microbes (controlled by soil C which is more easily accessible on coarser soils), their availability after agriculture (soil microbes stabilize 10 years after abandonment), and their possible role in restoration. Then authors offered management options for manipulating succession so as to speed up restoration to natural forest. One of the options is to keep isolated trees in pastures and plant trees in blocks or in plantations in order to reforest abandoned pasture land, to ameliorate microclimatic conditions and to attract seed dispersal agents. In addition, measures can be taken to either delay pasture degradation or convert unproductive pasture into more sustainable land uses. For example, several types of agrosilvopastoral systems can contribute to increased productivity and sustainability while also helping to conserve biodiversity and sequester carbon in vegetation and soils.

14.2 Synthesis

At least two things are true of Neotropic old fields. First, after abandonment they have a low level of advanced regeneration from either seedling banks, sapling banks, or from resprouting, which suggests that the analysis of Neotropic old field succession should focus on the dispersal and fate of seeds and seedlings (Grubb 1977). Second, the dynamics of the structurally and compositionally dominant woody plants, especially the functionally dominant trees, needs to be brought to the fore in old field investigations in areas that were once forest, which is the most common scenario.

This focus on tree regeneration highlights the limiting mechanisms, or "barriers," to tree invasion and establishment in old fields (as in the American temperate zone: Myster 1993). It has been suggested that other plants found during old field succession may remove some of these barriers and help facilitate tree invasion. For example, some shrub species (such as *Miconia* spp.) for the most part recruit asexually, thereby passing over any dispersal limitations that are common to trees, especially those tree species that require animals for dispersal. Indeed, any species that can sprout from rootstocks or from small trees/shrubs never cleared from the field at the beginning of agriculture would have a similar advantage. As reviewed, these shrubs, sprouts, remnant "nurse" trees, and downed trees and logs (Peterson and Haines 2000) have a variety of functions (for instance, acting as bird perches or seed dispersal foci, shading out grasses and other competitive old field species, and increasing soil nitrogen levels) that can expedite tree invasion.

Decades of old field research in temperate America allow further synthesis by providing solid ground for comparison between old field successions in the temperate and tropical zones. Following the outline of the present volume, such a comparison can be made in three ways: plant species patterns, plant mechanisms and tolerances, and plant restoration. In particular for plant patterns, permanent plots in temperate and tropical locations (Buell et al. 1971, Uhl 1987, Myster and Pickett 1990a,b, Myster 1993, 2003, 2007) provide the best data.

The temporal patterns in Neotropic old fields of Puerto Rico and Ecuador and in the temperate fields show similar life-form changes: (1) grass and past crops dominate, peaking early in succession and then declining; (2) fern and herbaceous cover peaks early in succession, but at lower levels than grass; (3) woody vines continue to increase in cover over the first 5 years; and (4) trees and shrubs enter, peak, and decline in an individualistic pattern.

Both temperate and tropical fields also demonstrate a similar spatial pattern: an exponential decline of tree density with distance to forest that is followed by a patchy pattern of tree establishment, especially for bird-dispersed species. Tropical old fields have a low total number of associations which are mainly positive, with negative associations involving mostly pasture grasses. There is further correspondence between tropical and temperate fields: the proportion, number, and level of significant associations decline with time; plants of short life-cycles are involved in more associations, and exotic and rare species are more likely to be involved in associations.

Key plant-plant replacement mechanisms and tolerances in the Neotropics include competition with trees, shrubs, and large herbs for light and with grasses for water and soil nutrients due in part to the high grass root mass per volume of soil, drought stress, seed predation, facilitation, and seed dispersal. However, seed and seedling pathogens may be important as well (Myster 2004). This compares well with studies in the American temperate zone (see Myster 1993) where dispersal is also key, along with seed predation, herbivory on seedlings (especially when it interacts with competition), and competition. Whereas light is critical both in the temperate zone and in the Neotropics, water may be more limiting in the Neotropics.

In terms of restoration, comparisons between neotropic and temperate old fields are difficult to make because in the temperate zone restoration is generally not needed due to the deep fertile soils of the area that are not degraded by long-term use without fertilizer as is common in the Neotropics. In the Neotropics, conversely, restoration is necessary to return land to high productivity in as short a period of time as possible and then maintain its sustainability. One way to manage succession towards an end that better satisfies human needs is to retard succession. Another way is to quicken the rate of recovery, perhaps by increasing facilitating mechanisms, such as dispersal, and decreasing inhibitory mechanisms, such as predation and competition (as discussed above), in order to generate rainforest timber to be used as a sustainable resource or to create rainforest habitat for rare species. One more option is to hold succession in a state when agriculture is less costly to maintain. Yet another approach is to alter the natural successional trajectory itself to match soil conditions for maximum crop production. Mechanisms of tree replacement, as discussed in the present book, can be manipulated either as inhibitors or facilitators towards any of these ends.

Whatever restoration approach is used, studies of post-agricultural succession are indispensable to the success of restoration attempts because they provide information about how Neotropic areas regenerate and function. Management of old field succession has to have scientific underpinnings in order to be effective. Hence, the successional processes and mechanisms that have been discussed in this book should be the bedrock of any restoration effort.

Studies of old field succession in the Neotropics, when combined with decades-long similar studies in the temperate zone, become instrumental in our understanding of basic ecological paradigms. Despite the fact that the definition of ecosystem function is generally accepted among ecologists, the definition of ecosystem structure is still hotly debated (Myster 2001a). Plants are the cornerstone of the ecosystem functions of productivity and decomposition, and an important agent in the functions of energy/carbon dynamics (e.g., tropic exchanges) and nutrient cycling in the soil (Wedin and Tilman 1993) which must play a critical role in determining ecosystem structure as well.

Questions of structure and process naturally fall into two ecological paradigms (Table 14.1). The subject matter for the first paradigm is variation of individuals within a species, the species itself, and the species' population. The paradigm's common structures include assimulation rates and population growth rates. Among its mechanisms, which are individualistic in nature, the ecophysiology of individuals, and the interaction of species populations with the environment are essential. The paradigm's cardinal process—the combination of mechanisms that produce the basic nature of plant communities under this perspective—is individual plant-plant replacements that create the terrestrial vegetation patterns of plant distribution, abundance, and association. Scientists who follow this paradigm combine individual responses of plants to their environment or focus on dominant species in a plant community in order to define ecosystem function.

The subject matter of the second paradigm revolves around communities and ecosystems (Table 14.1). The paradigm's common structures encompass species richness, interaction coefficients, and biogeochemical cycles. Its mechanisms are interactive, and interactions among species groups prevail. The paradigm's key processes are changes in plant resources and biogeochemical cycling, both of which

Table 14.1 The two paradigms of ecology and their associated characteristics. Plant examples are often used for illustrative purposes.

Characteristic	Paradigm I	Paradigm II
unit of study	individual variation within species, species populations	groups of species energy, nutrient dynamics
prime mechanism	species tolerances	species interactions
some structures	leaf area, biomass, growth rate, demographic parameters	species strategies, richness interaction coefficients biogeochemical sources/sinks
key process(es)	plant-plant replacements	higher-order interactions plant resource availability biogeochemical cycles
functions	productivity, decomposition	productivity, decomposition other energy/nutrient aspects

are thought to drive vegetation patterns. Followers of the second paradigm look for relationships between "emergent" structures of a higher order and ecosystem functions, for example, between diversity and productivity.

The two paradigms have recently been relabeled as a "Schimper" world view (Schimper 1998) and a "Hubbell" world view (Hubbell 2001), for Schimper laid emphasis on measurable traits of a species' individual plants (e.g., leaf mass per area, leaf lifespan, seed size, height at maturity) and plant evolution while Hubbell emphasizes diversity, ignoring species differences (Westoby and Wright 2005).

Early plant ecology was dominated by the second paradigm. F. E. Clements (1911) believed that interactions among species produced community and ecosystem structures with the ultimate result of a "super-organism." Later, H. A. Gleason (1926) focused on responses and structures of individual plants and on individual species' tolerances to environmental conditions. Gleason, having in mind successional plant communities, maintained that the biology of the species and the variation among individuals of that species were most important in determining how plants grew and reproduced. R. H. Whittaker (1977) was the first to apply this "individualistic" view to all plant communities, creating a "continuum" view. Some ecologists today proceed from the more holistic Clementian view and affirm that species interactions and resource cycling dominate structure and define function, thus giving priority to physics and chemistry over biology in ecosystems.

Are ecologists doomed to swing back and forth between the two paradigms forever, making the science of ecology circular rather than directional? To answer this question, we need to measure delineated plant communities over time, and the analysis of permanent plot data allows us to do just that. In Chapter 1, permanent plot data in post-agricultural areas in Puerto Rico and Ecuador were used to examine species associations over time and after different types of crops. The results supported the first paradigm overwhelmingly. This tentative conclusion, of course, needs further verification with other plant communities and in other locations. Wherever such testing takes place, however, permanent vegetation plots—diligently monitored over long periods of time—will be critical in ascertaining the nature of plant communities and, ultimately, the utility of this or any other ecological paradigm.

In addition to viewing a plant community as an "entity" with numerous interactions among its members, studies that are conducted within the framework of Paradigm II often contain highly synthetic parameters that are not measured directly (also a concern of Westoby and Wright 2006), such as "stability," "resilience," and "invasibility." These parameters can only have validity, however, if they are shown to be the product of plant-plant replacements. For example, one old field may be more resilient than another, returning to previous functional levels faster than the other field after agriculture, but mechanisms behind such dissimilarity must stem from differences in the plant-plant replacements within the fields. These dissimilarities are determined either by a difference in the pool of species (i.e., their niches) or by a difference in the local environmental conditions that activate different slices of plant niches. Changes in the niche pool and/or in the environment cause plant-plant replacements to "line up" in a particular way, creating plant patterns of abundance and composition that translate into such higher-level structures as resiliency. Another old field example is when one field has faster tree invasion than another field—greater invasibility—which must be caused also by either niche and/or environmental differences between the fields. These two examples illustrate how old field studies can act as a template in our general understanding of the ways plant niches are activated to produce plant-plant replacements and, ultimately, higher-level structures.

During these investigations it would be wrong to frame the important issue of whether it is the plants that produce environmental changes or visa versa as nothing more than a difference in philosophical opinion (a "dialectic": Bazzaz 1996). Priority should always be given to plants, their responses, and their niches because an ecosystem is first and foremost biological in nature and plants are the vast majority of that biology. Terrestrial plant communities are dynamic, and plant-plant replacements are the main agents—the fundamental process—of that continuous change.

Field experiments that are carried out at community or ecosystem levels provide ground for further concern about Paradigm II because plots to which treatments are applied change constantly and significantly throughout an experiment. As stated in Chapter 1, plant-plant replacements occur in any plot over time, so by the end of a field experiment many of the plants in the treatment plots have, in fact, either not received any treatments or have received a reduced application of the treatments, which in either case makes it impossible to test the plants statistically for any treatment affects. A review of the literature reveals that few researchers have checked if individual plants in their study plots are old or new inhabitants. Since replacement levels are often greater than p-values, many results that have been published from these kind of experiments are invalid. Moreover, as the length of the experiment increases, the likelihood of its results being invalid also increases, no matter whether the experiment deals with natural perturbation or involves applied treatments.

Another field practice for these kinds of experiments is to measure one individual plant of a study species in each plot at each sampling period. Since plants are not individually marked, however, odds are great that a different plant is being measured each time! As before, that plant may have been introduced late in the experiment, receiving reduced treatments or no treatments at all. The usual statistical remedies of a control plot, more replication, or a "blocked" design will not solve this problem because replacements and the plant patterns that result from them are probabilistic, not random.

Finally, Paradigm II allows the questionable approach of grouping species by function or strategy (e.g., Grime 1977) that treats all plant species within a group as the same, to the point of looking at parts of the world as green packages where a patch of moss would be equivalent to a tree (the "big leaf" perspective: Bazzaz 1996). This approach accommodates the risk of ignoring evolution because it downgrades plants to nothing more than temporary way-stations for energy and ions and thus passes over species adaptations and differences that are important in any ecosystem. It needs to be remembered that evolution is a process of creating individual species that have to be functionally different and important within an ecosystem in order to survive.

It appears logical then that for the purpose of avoiding significant limitations in research practices of Paradigm II (e.g., unmeasured parameters, replacements in treatment plots, species groupings) and of conducting accurate and useful experiments in general, researchers should either concentrate their attention on individual marked plants or direct their efforts to short-term projects. The outlined basic methodological issues, on the one hand, and the results of the analysis of permanent plot data in Chapter 1, on the other, cast considerable doubt on the validity of Paradigm II. It seems that this paradigm wants to present ecosystems as a "jumble" of interactions where everything is connected to everything else and all is equally important (or unimportant), that is to say there is no over-arching structure, set of priorities, or (ultimately) meaning.

14.3 Future Directions

The conceptual framework and the studies of post-agricultural succession in the Neotropics that are presented in this book hold the following recommendations for future research of Neotropic old fields:

(1) set up permanent vegetation plots in more Neotropic ecosystems and monitor them for decades with the aim of investigating the pace and nature of land recovery after common past crops, including other aspects of the agricultural disturbance regime (e.g., the year of abandonment, the season of abandonment, and whether the field was plowed or not when abandoned);
(2) explore the ecophysiology of Neotropic old field plants (see Burton and Bazzaz 1991, 1995 for similar studies in temperate old fields), the tolerances of their niches (see Myster and Everham 1996 for the germination niche of tropical trees), and other aspects of the biology and ecology of tropical plants (see Brokaw 1998 for an examination of *Cecropia schreberiana*); combine these studies with further mechanistic investigation of old field plants in the Neotropics for a complete definition of plant niches, using data from permanent plots to identify important plants for further investigation;

(3) sample environmental factors in study old fields, with a focus on the abiotic environment (e.g., energy pathways and biogeochemical cycles that lead to changes in plant resources) and on the abundance, composition, and behavior of the biotic environment (e.g., dispersers, predators, herbivores, pathogens, other plants) in the neighborhood;
(4) monitor those components of the environment (e.g., carbon) that indirectly affect the immediate abiotic and biotic conditions of a plant's life;
(5) perform restoration experiments in the field.

In this book, two approaches in investigation of plant communities have been contrasted: a "static" investigation that engages in snapshot growth and allocation field experiments, focusing on the largest or most common plants present at that point in time and space, and a "dynamic" view focusing on how plant niches and the local environment determine plant-plant replacements over time and space. For all plant communities, a major future challenge for scientists will be to find out how, under the influence of the local environment, plant niches generate the plant-plant replacements which produce the earth's plant patterns. The mere demonstration of these patterns will not suffice (Myster 2001a). There can be no shortcuts such as the use of "surrogate" parameters instead of direct measurements of plant function and structure.

Results from decades-long post-agricultural studies and the conceptual framework from Chapter 1 suggest how to proceed: key species and replacement patterns can be ascertained by using permanent plots, plant niches can be determined through various kinds of field and greenhouse experiments, and plant-plant replacements can be investigated by sampling local environmental factors and combining them with plant niches. Followers of Paradigm II may continue to look only to the abiotic environment for answers, yet the primary emphasis should be on the ways in which plants respond to and interact with both the living and non-living constituents of the environment.

As the analysis of data from old field permanent plots and the results of old field experiments and studies in various places throughout North, South, and Central America have shown, plant communities are much more individualistic (Pickett 1982, Myster and Pickett 1988) and much more dynamic than is often assumed.

References

Bazzaz, F. A. 1996. Plants in Changing Environments: Linking Physiological, Population and Community Ecology. Cambridge University Press. Cambridge, UK.

Brokaw, N. V. L. 1998. *Cecropia schreberiana* in the Luquillo Mountains of Puerto Rico. The Botanical Review 64:91–120.

Buell, M. F., Buell, H. F., Small, J. A., and Siccama, T. G. 1971. Invasion of trees and secondary succession on the New Jersey Piedmont. Bulletin of the Torrey Botanical Club 98:67–74.

Burton, P. J., and Bazzaz, F. A. 1991. Tree seedling emergence on interactive temperature and moisture gradients and in patches of old-field vegetation. American Journal of Botany 78: 131–149.

Burton, P. J., and Bazzaz, F. A. 1995. Ecophysiological responses of tree seedlings invading different patches of old-field vegetation. Journal of Ecology 83:99–112.

Grime, J. P. 1977. Evidence for the existence of three primary strategies in plants and its relevance to ecological and evolutionary theory. American Naturalist 111:1169–1194.
Grubb, P. J. 1977. The maintenance of species-richness in plant communities: the importance of the regeneration niche. Biological Reviews of the Cambridge Philosophical Society 52:107–145.
Hubbell, S. P. 2001. The Unified Neutral theory of Biodiversity and Biogeography. Princeton: Princeton University Press.
Myster, R. W. 1993. Tree invasion and establishment in old fields at Hutcheson Memorial Forest. The Botanical Review 59:251–272.
Myster, R. W. 2001a. What is Ecosystem Structure? Carribbean Journal of Science 37:132–134.
Myster, R. W. 2001b. Mechanisms of plant response to gradients and after disturbances. The Botanical Review 67(4):441–452.
Myster, R. W. 2003. Vegetation dynamics of a permanent pasture plot in Puerto Rico. Biotropica 35(3):422–428.
Myster, R. W. 2004. Post-agricultural invasion, establishment, and growth of Neotropical trees. The Botanical Review 70:381–402.
Myster, R. W. 2007. Early successional pattern and process after Banana, Sugarcane and Pasture in Ecuador. The New Zealand Journal of Botany 45:35–44.
Myster, R. W., and Everham, E. M. III. 1999. Termination cues across the distrubance regime in the Puerto Rican rain forest. Tropical ecology 110:89–98.
Myster, R. W., and Pickett, S. T. A. 1988. Individualistic patterns of annuals and biennials in early successional old fields. Vegetatio 78(1–2):53–60.
Myster, R. W., and Pickett, S. T. A. 1990a. The Buell successional study at the Hutcheson Memorial Forest Center. Permanent Plotter 3:2–3.
Myster, R. W., and Pickett, S. T. A. 1990b. Initial conditions, history and successional pathways in ten contrasting old fields. American Midland Naturalist 124(2):213–238.
Myster, R. W., and Pickett, S. T. A. 1992a. Effects of palatability and dispersal mode on spatial patterns of trees in old fields. Bulletin of the Torrey Botanical Club 119(2):145–151.
Myster, R. W., and Pickett, S. T. A. 1992b. Dynamics of associations between plants in ten old fields during 31 years of succession. Journal of Ecology 80(2):291–302.
Peterson, C. J., and Haines, B. L. 2000. Early successional patterns and potential facilitation of woody plant colonization by rotting logs in premontane Costa Rican pastures. Restoration Ecology 8:361–369.
Pickett, S.T.A. 1982. Population patterns through twenty years of old field succession. Vegetatio 49:45–59.
Schimper, A. F. W. 1998. Pflanzengeographic auf Physiologischer Grundlage. G. Fisher.Uhl, C. 1987. Factors controlling succession following slash-and-burn agriculture in Amazonia. Journal of Ecology 75(2):377–407.
Wedin, D. A., and Tilman, D. 1990. Species effects on nitrogen cycling: a test with perennial grasses. Oecologia 84:433–441.
Westoby, M., and Wright, I. J. 2005. Land-plant ecology on the basis of functional traits. Trends in Ecology and Evolution 21:261–268.

Index

Printed in the United Kingdom
by Lightning Source UK Ltd.
132469UK00001B/237/A

9 780387 336411